Metal Matrix Composites: A Modern Approach to Manufacturing

Edited by

Virat Khanna

Department of Mechanical Engineering, MAIT
Maharaja Agrasen University
Himachal Pradesh
India

Prianka Sharma

Department of Physics, School of Basic & Applied Sciences
Maharaja Agrasen University
Solan, Himachal Pradesh
India

&

Santosh Kumar

Department of Mechanical Engineering
Chandigarh Group of Colleges, Landran, Mohali
Punjab
India

Metal Matrix Composites: A Modern Approach to Manufacturing

Editors: Virat Khanna, Prianka Sharma and Santosh Kumar

ISBN (Online): 978-981-5223-43-9

ISBN (Print): 978-981-5223-44-6

ISBN (Paperback): 978-981-5223-45-3

need for a court order if at any point you breach any terms of this License Agreement. In no event will any delay or failure by Bentham Science Publishers in enforcing your compliance with this License Agreement constitute a waiver of any of its rights.

3. You acknowledge that you have read this License Agreement, and agree to be bound by its terms and conditions. To the extent that any other terms and conditions presented on any website of Bentham Science Publishers conflict with, or are inconsistent with, the terms and conditions set out in this License Agreement, you acknowledge that the terms and conditions set out in this License Agreement shall prevail.

Bentham Science Publishers Pte. Ltd.
80 Robinson Road #02-00
Singapore 068898
Singapore
Email: subscriptions@benthamscience.net

BENTHAM SCIENCE

CONTENTS

PREFACE

Metal matrix composites (MMCs) are a class of advanced materials that have gained significant attention in recent years due to their unique properties and superior performance characteristics. They are composed of a metal matrix, typically aluminium, titanium, or magnesium, reinforced with a high-strength ceramic or metallic material such as silicon carbide, carbon fibers, or aluminium oxide. This combination results in a material that exhibits exceptional mechanical properties, including high strength, stiffness, and wear resistance, making them suitable for a wide range of applications in modern industry.

The relevance of metal matrix composites to modern industry can be traced back to their ability to provide high-performance solutions to some of the most challenging problems faced by engineers and designers. In aerospace, MMCs are used for structural components in aircraft engines, landing gear, and airframe structures, as well as space applications such as rocket nozzles and thermal protection systems. These components need to withstand extreme conditions, such as high temperatures, pressure, and high-impact loads, making MMCs an ideal choice. In the automotive industry, MMCs are used in brake rotors, engine components, and suspension systems, where high wear resistance, low friction, and improved fuel efficiency are key requirements. MMCs have been shown to offer significant weight savings and improved performance over traditional materials, such as cast iron or steel, which can improve fuel efficiency and reduce greenhouse gas emissions. MMCs' unique properties have also found applications in the electronics and microelectronics industry, where they are used in heat sinks, packaging materials, and electronic substrates. These components need to dissipate heat efficiently and reliably, and MMCs have been shown to exhibit superior thermal conductivity and excellent dimensional stability under high temperatures, making them ideal for these applications. In addition, MMCs have shown potential in the defence and military industry for their superior properties. They are used in ballistic armour and vehicle protection systems, where they provide excellent protection against high-velocity projectiles, mines, and improvised explosive devices. MMCs have also been used in cutting tools and moulds, where they provide high-wear resistance and dimensional stability. The development of MMCs has been facilitated by advancements in materials science and manufacturing technologies. Advanced fabrication techniques, such as powder metallurgy, casting, and additive manufacturing, have enabled the production of complex shapes and sizes, as well as the incorporation of multiple reinforcement materials, allowing for tailored properties and performance characteristics.

In short, the use of metal matrix composites in modern industry has been instrumental in the development of high-performance materials and products that meet the demanding requirements of various applications. The unique properties of MMCs, such as high strength, stiffness, and wear resistance, combined with advancements in manufacturing technologies, have enabled their use in critical applications, such as aerospace, automotive, electronics, and defense. As technology continues to evolve, it is expected that MMCs will play an increasingly important role in the development of innovative products and solutions in various industries.

The motivation to prepare an edited book on this topic was to provide a comprehensive resource for students and professionals in the field of materials science and engineering. The book covers a broad range of topics related to metal matrix composite manufacturing, including the various fabrication methods, characterization techniques, and applications. The intended audience for this book includes students and professionals in the field of materials science and engineering, as well as researchers and engineers working in the field of metal matrix composite manufacturing. The book provides a comprehensive resource for those seeking to gain an in-depth understanding of metal matrix composite manufacturing, including the fundamental principles, latest developments, and future trends in the field.

In conclusion, this book provides a comprehensive overview of metal matrix composite manufacturing, covering the fundamental principles, latest developments, and future trends in the field. It is designed to be an essential resource for students and professionals in the field of materials science and engineering, as well as researchers and engineers working in the field of metal matrix composite manufacturing. We hope that this book will be a valuable resource for those seeking to gain an in-depth understanding of metal matrix composites with its relevance to the modern industry

Virat Khanna
Department of Mechanical Engineering, MAIT
Maharaja Agrasen University
Himachal Pradesh
India

Prianka Sharma
Department of Physics, School of Basic & Applied Sciences
Maharaja Agrasen University
Solan, Himachal Pradesh
India

&

Santosh Kumar
Department of Mechanical Engineering
Chandigarh Group of Colleges, Landran, Mohali
Punjab
India

List of Contributors

Anupam Thakur Department of Mechanical Engineering, MAIT, Maharaja Agrasen University, H.P., India

Abhishek Kandwal School of Physics and Materials Science, Shoolini University, Bajhol, Solan, H.P., India

Ankit Verma Faculty of Science and Technology, ICFAI University, Baddi, Himachal Pradesh, India

Abhimanyu Singh Rana Centre for Advanced Materials and Devices, School of Engineering & Technology, BML Munjal University, Sidhrawali, Gurgaon, Haryana, India

Dilbag Singh Department of Mechanical Engineering, Sardar Beant Singh State University, Gurdaspur, India

Gaurav Luthra Department of Regulatory Affair and Quality Assurance, Auxein Medical Pvt. Ltd. Sonipat, Haryana, India

Gurpreet Singh Department of Mechanical Engineering, St. Soldier Institute of Engineering and Technology, Jalandhar, Punjab, India

Harsh Kumar Department of Mechanical & Project Management, Kalyan Project Construction Company, Mohali, Punjab, India

Harpreet Singh School of Physics and Materials Science, Shoolini University, Bajhol, Solan, H.P., India

Himanshi School of Physics and Materials Science, Shoolini University, Bajhol, Solan, India

Jatinder Kumar Department of Mechanical Engineering, Modern Group of Colleges, Mukerian, punjab, India

Jahangeer Ahmed Department of Chemistry, College of Science, King Saud University, Riyadh, Saudi Arabia

Kirandeep Kaur PSE Sales & SVCS/Distribution Associate, New Jersey, USA

Kamaljit Singh Department of Mechanical Engineering, MAIT, Maharaja Agrasen University, H.P., India
Nurture education solutions pvt ltd, Bengaluru, India

Mohit Kumar Department of Regulatory Affair and Quality Assurance, Auxein Medical Pvt. Ltd. Sonipat, Haryana, India

Manish Taunk Department of Physics, Chandigarh University, Mohali, India

Neeraj Kumar Sharma Centre for Advanced Materials and Devices, School of Engineering & Technology, BML Munjal University, Sidhrawali, Gurgaon, Haryana, India

Nirmal S. Kalsi Department of Mechanical Engineering, Shri Vishwakarma Skill University, Palwal, Haryana, India

Prianka Sharma Department of Physics, School of Basic & Applied Sciences, Maharaja Agrasen University, Solan, H.P., India

Pawan Kumar School of Physics and Materials Science, Shoolini University, Bajhol, Solan, H.P., India

Qasim Murtaza   Mechanical Department, Delhi Technological University, Delhi, India

Rakesh Kumar   Department of Regulatory Affair and Quality Assurance, Auxein Medical Pvt. Ltd. Sonipat, Haryana, India

Rohit Jasrotia   School of Physics and Materials Science, Shoolini University, Bajhol, Solan, H.P., India

Rahul Mehra   Department of Mechanical Engineering, Chandigarh Group of Colleges, Landran, Mohali, Punjab, India

Satish Kumar   Department of Mechanical Engineering, Chandigarh Group of Colleges, Landran, Mohali, Punjab, India

Santosh Kumar   Department of Mechanical Engineering, Chandigarh Group of Colleges, Landran, Mohali, Punjab, India

Suman   Department of Mathematics, School of Basic and Applied Sciences, Maharaja Agrasen University, Baddi, H.P., India

Sachin Kumar Godara   Department of Chemistry, Guru Nanak Dev University, Punjab, Amritsar, India

Susheel Kalia   Department of Chemistry, ACC Wing (Academic Block), Indian Military Academy, Dehradun (Uttarakhand), India

Sarabjeet Kaur   Department of Physics, School of Basic & Applied Sciences, Maharaja Agrasen University, Solan, H.P., India
Department of Applied Science, Chandigarh Engineering College, Landran, Mohali, Panjab, India

Vidushi Karol   Department of Applied Science, Chandigarh Engineering College, Landran, Mohali, Panjab, India

Virat Khanna   Department of Mechanical Engineering, MAIT, Maharaja Agrasen University, H.P., India

Metal Matrix Composites: An Introduction and Relevance to Modern Sustainable Industry

Virat Khanna[1,*], **Rakesh Kumar**[2] and **Kamaljit Singh**[1,3]

[1] Department of Mechanical Engineering, MAIT, Maharaja Agrasen University, H.P., India

[2] Department of Regulatory Affair and Quality Assurance, Auxien Medical Pvt, Ltd. Sonipat, Haryana, India

[3] Nurture education solutions pvt ltd, Bengaluru, India

Abstract: Metal matrix composites (MMCs) are a family of strong yet lightweight materials that have many industrial uses, particularly in the automotive, aerospace, and thermal management industries. By choosing the best combinations of matrix, reinforcement, and manufacturing techniques, the structural and functional features of MMCs may be adjusted to meet the requirements of diverse industrial applications. The matrix, the interaction between them, and the reinforcement all affect how MMCs behave. Yet, there is still a significant problem in developing a large-scale, cost-effective MMC production method with the necessary geometrical and operational flexibility. This chapter provides an overview of Metal Matrix Composites (MMCs), their historical development, properties of MMCs, classification of MMCs, diverse applications, and the relevance of MMCs to sustainable industries.

Keywords: Artificial intelligence, Composite materials, Industry 4.0, MMC, Machine learning, Sustainability.

COMPOSITE MATERIALS

Composites, also known as composite materials, are materials made up of two or more distinct materials that are combined to form a new material with enhanced characteristics [1]. The use of composite materials can be traced back to ancient times, with examples including mud bricks reinforced with straw, and boats made from reeds and papyrus [2, 3]. In the 20th century, composites began to be used more widely in various industries owing to their desirable characteristics such as high resistance against corrosion, high strength-to-weight ratio, and stability.

---

* **Corresponding author Virat Khanna:** Department of Mechanical Engineering, MAIT, Maharaja Agrasen University, H.P., India; E-mail: khanna.virat@gmail.com

During World War II, composites were used in the construction of aircraft, such as the De Havilland Mosquito, which was made with a plywood composite [4].

After the war, the aerospace industry continued to be a major user of composites, with materials such as fiberglass and carbon fiber being used in the construction of aircraft and spacecraft. In the 1960s and 1970s, composites began to be used in the construction of sports equipment, such as tennis rackets and golf clubs [5, 6]. This trend continued into the 1980s, with composites being used in the construction of high-performance racing yachts and Formula One cars. Composites are employed in a variety of sectors today, including sports equipment, construction, automotive, marine, and aerospace. New materials and manufacturing processes continue to be developed, expanding the range of applications for composites and making them increasingly important in modern industry.

Composite materials are constructed of two or more different types of constituent materials that are combined in a way that produces a new material with superior properties compared to individual materials [7, 8]. The constituent materials can be organic or inorganic and can include fibers, resins, metals, ceramics, and polymers. There are several types of composite materials, each with unique properties and applications [9 - 11].

Polymer Matrix Composites (PMCs)

Fiber-reinforced polymers, sometimes referred to as polymer matrix composites (FRPs), are made up of a polymer matrix and a reinforcing fiber, such as carbon or glass fibers. The fibers are embedded in the polymer matrix to create a material with high strength and stiffness, making PMCs ideal for use in aerospace, automotive, and sports equipment applications.

Metal Matrix Composites (MMCs)

A metal matrix and a reinforcing substance make up MMCs, such as ceramic or carbon fibers. MMCs are known for their high strength and stiffness, as well as their resistance to high temperatures and wear. These properties make them ideal for use in aerospace, automotive, and military applications.

Ceramic Matrix Composites (CMCs)

A ceramic matrix is made up of ceramic matrix composites and a reinforcing material, such as carbon or silicon carbide fibers. CMCs are known for their high strength and stiffness at high temperatures, making them ideal for use in high-temperature applications, such as in gas turbines and heat exchangers.

Carbon Fiber Reinforced Polymer (CFRP)

One example of a carbon fibre reinforced polymer is carbon fibre reinforced plastic. PMC uses carbon fibers as the reinforcing material. CFRP is known for its stiffness, making it ideal for use in aerospace, automotive, and sports equipment applications.

Glass Fiber Reinforced Polymer (GFRP)

It is a type of PMC that uses glass fibers as the reinforcing material. GFRP is known for its high strength and stiffness, as well as its resistance to corrosion, making it ideal for use in marine and construction applications.

Natural Fiber Composites (NFCs)

Natural fiber composites are made up of a natural fiber, such as bamboo or wood, and a matrix material, such as a polymer or resin. NFCs are known for their low cost, biodegradability, and renewable nature, making them ideal for use in sustainable applications.

Hybrid Composites

Hybrid composites are made up of two or more different types of reinforcing materials, such as fibers or particles, in a single matrix material. Hybrid composites can have a range of properties, depending on the combination of materials used, and are often used in aerospace, automotive, and military applications.

Fig. (**1**) shows various types of composites along with various types of MMCs based on their matrix material. In conclusion, composite materials have revolutionized the world of engineering and technology by providing materials with superior properties than traditional materials. The different types of composite materials allow engineers and designers to choose the appropriate material for a given application based on the required properties, cost, and environmental impact. As technology advances, new composite materials and manufacturing techniques will continue to be developed, expanding the range of applications for composites and making them increasingly important in modern industry.

METAL MATRIX COMPOSITES

One kind of composite material is metal matrix composites (MMCs), consisting of a metal matrix, usually a light metal such as aluminium, magnesium, or titanium, reinforced with a secondary phase, which can be a ceramic, metal, or organic material [12, 13]. The reinforcing stage is typically in the form of fibers, whiskers, or particles, which are dispersed throughout the metal matrix to enhance its mechanical, thermal, or electrical properties [10]. The resulting material has improved strength, stiffness, wear resistance, and thermal stability compared to the base metal while retaining some of its ductility and toughness. MMCs are utilized in an extensive range of applications, including automotive, aerospace, electronic packaging, and sporting goods, among others.

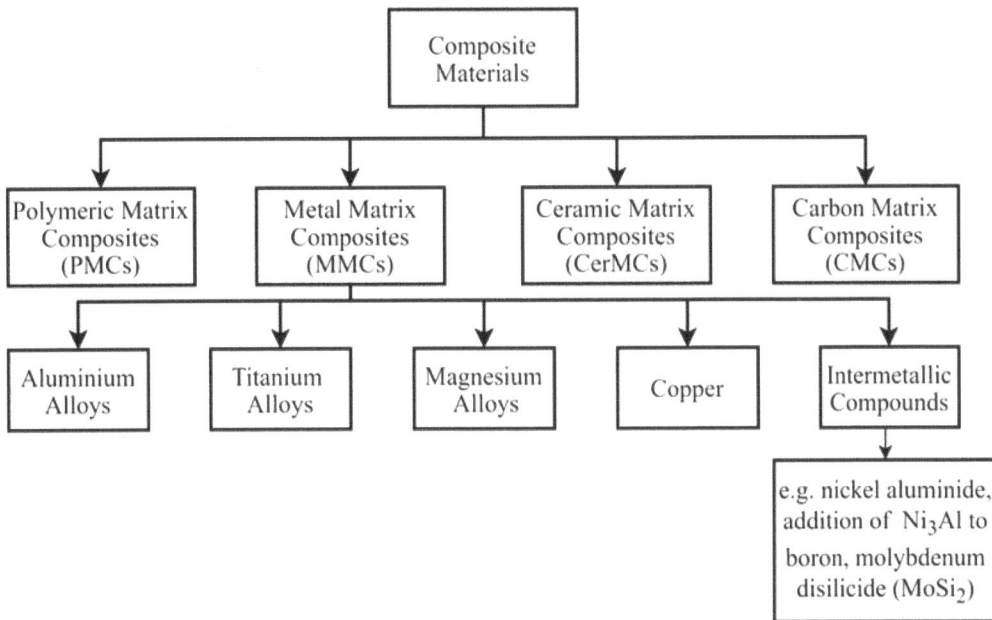

Fig. (1). Various types of Composite Materials [10].

History of Metal Matrix Composites

The history of MMCs dates back to the early 1900s when metal-polymer composites were first developed. These early composites were made by embedding fibers or particles of one material in a matrix of another material to create a new material with enhanced properties [14]. In the 1940s and 1950s, researchers began exploring the use of MMCs in various uses, particularly in the aerospace and defense industries [15, 16]. One of the first successful applications of MMCs was the development of the beryllium-aluminium composite material used in the construction of the X-15 hypersonic aircraft in the 1950s. In the 1960s, aluminium-based MMCs were developed for use in the aerospace industry, with the first aluminium-silicon carbide MMC being developed in 1967. These materials were found to have improved mechanical and thermal properties compared to traditional aluminium alloys, making them ideal for use in high-temperature applications. In the 1970s, MMCs began to be used in the automotive industry, particularly in racing and high-performance vehicles. The use of MMCs in the automotive industry was initially limited due to high costs and complex manufacturing processes, but advances in manufacturing technology and material development have made MMCs more affordable and accessible. In the 1980s and 1990s, the use of MMCs continued to expand into new applications, such as electronic packaging and sporting goods. Advances in material development and manufacturing techniques led to the development of MMCs with a wide range of properties consisting of high stiffness, strength, and wear resistance. Today, MMCs are employed in an extensive range of applications, including automotive, electronic packaging, and sporting goods [15, 17]. New materials and manufacturing processes continue to be developed, expanding the range of applications for MMCs and making them increasingly important in modern industry. Some of the commonly used MMCs include Al, Mg, Cu, Ti, and Ni matrix composites. The reinforcing materials used in MMCs can include ceramic fibers, metal fibers, or particulates such as silicon carbide, alumina, or graphite.

In conclusion, MMCs development has revolutionized the materials industry by providing materials with superior properties than traditional metals. MMCs have a long history of use in the aerospace and defense industries, and their applications have expanded to many other industries [15, 18]. As technology advances, new MMCs and manufacturing techniques will continue to be developed, expanding the range of applications for MMCs and making them increasingly important in modern industry.

PROPERTIES OF MMCs

The properties of MMCs depend on several factors, including the kind of reinforcement material, the composition of the matrix, the volume proportion of reinforcement, and the production method [19 - 21]. Nonetheless, the following is a discussion of some general characteristics of MMCs:

Mechanical Properties

MMCs exhibit stiffness, high strength, and wear resistance than conventional metals. This is due to the presence of the reinforcement material that strengthens the metal matrix. The strength of the composite material depends on the volume fraction, aspect ratio, orientation, and size of the reinforcement particles. The stiffness of the composite material is also influenced by these factors, as well as the modulus of elasticity of the matrix material.

Thermal Properties

The thermal characteristics of MMCs are determined by the matrix material and the reinforcement material. The COTE of the composite material can be controlled by the volume fraction and type of reinforcement material. The thermal conductivity of the composite material is enhanced due to the high thermal conductivity of the reinforcement material.

Electrical Properties

The electrical conductivity of MMCs is determined by the matrix material and the reinforcement material. The electrical conductivity of the composite material can be improved by increasing the volume fraction of the reinforcement material. The composite material can also exhibit improved electrical resistivity due to the presence of insulating reinforcement materials.

Corrosion Resistance

The corrosion resistance of MMCs is determined by the matrix material and the reinforcement material. The composite material can exhibit improved corrosion resistance due to the presence of ceramic reinforcement materials that are resistant to corrosion.

Fatigue Properties

The fatigue properties of MMCs based on the type of matrix material, reinforcement material, and the manufacturing process. The composite material can exhibit improved fatigue properties due to the reinforcement material that strengthens the metal matrix and improves its crack resistance.

TYPES OF METAL MATRIX MATERIAL COMPOSITES

Based on Matrix Material

Aluminium Metal Matrix Composites (AMCs)

AMCs are a type of metal matrix composite that uses aluminium alloys as the matrix material. The reinforcing fibers can be made of various materials, such as silicon carbide, boron, alumina, or graphite. AMCs are lightweight, have high strength-to-weight ratios, and exhibit good wear resistance, making them useful in applications such as aerospace, automotive, and sporting goods. AMCs are a popular type of metal matrix composite due to their low density, high strength, and good wear resistance [22 - 24].

Magnesium Metal Matrix Composites (MMCs)

MMCs use magnesium alloys as the matrix material. The reinforcing fibers can be made of silicon carbide, alumina, or carbon. MMCs have low density, good stiffness and strength, and good heat resistance, making them useful in applications such as aerospace, automotive, and electronic packaging [22 - 24].

Titanium Metal Matrix Composites (TMCs)

TMCs use titanium alloys as the matrix material. The reinforcing fibers can be made of silicon carbide, alumina, or graphite. TMCs have high strength, stiffness, and corrosion resistance, making them useful in aerospace, biomedical, and sporting goods applications.

Copper Metal Matrix Composites (CMCs)

CMCs use copper alloys as the matrix material. The reinforcing fibers can be made of tungsten, graphite, or silicon carbide. CMCs have high thermal and electrical conductivity, good wear resistance, and good machinability, making them useful in electronic, automotive, and aerospace applications [22 - 24].

Nickel Metal Matrix Composites (NMCs)

NMCs use nickel alloys as the matrix material. The reinforcing fibers can be made of alumina, silicon carbide, or carbon. NMCs have good corrosion resistance, high strength, and good high-temperature properties, making them useful in aerospace and chemical processing applications.

Iron Metal Matrix Composites (IMCs)

IMCs use iron or steel alloys as the matrix material. The reinforcing fibers can be made of silicon carbide, alumina, or carbon. IMCs have high strength, good toughness, and good wear resistance, making them useful in automotive, aerospace, and machinery applications [22 - 24].

Zinc Metal Matrix Composites (ZMCs)

ZMCs use zinc alloys as the matrix material. The reinforcing fibers can be made of alumina, silicon carbide, or carbon. ZMCs have good machinability, good damping properties, and good wear resistance, making them useful in automotive and machinery applications.

Tin Metal Matrix Composites (TMCs)

TMCs use tin alloys as the matrix material. The reinforcing fibers can be made of carbon or silicon carbide. TMCs have good stiffness, good wear resistance, and good corrosion resistance, making them useful in electronics and machinery applications.

Lead Metal Matrix Composites (LMCs)

LMCs use lead alloys as the matrix material. The reinforcing fibers can be made of tungsten, graphite, or silicon carbide. LMCs have good radiation shielding properties, good damping properties, and good machinability, making them useful in nuclear and medical applications [22 - 24].

Tungsten Metal Matrix Composites (TWCs)

TWCs use tungsten alloys as the matrix material. The reinforcing fibers can be made of carbon or silicon carbide. TWCs have high strength, good radiation shielding properties, and good high-temperature properties, making them useful in nuclear and aerospace applications [22 - 24].

Based on Reinforcement Material

Carbon Fiber Reinforced Metal Matrix Composites (CFR-MMCs)

These composites offer high stiffness, strength, and low density. They are typically used in aerospace, automotive, and sporting goods applications due to their favorable mechanical characteristics [22 - 24].

Silicon Carbide Reinforced Metal Matrix Composites (SiC-MMCs)

These composites exhibit high strength and good wear resistance. They are employed in electronic packaging, automotive, and aerospace applications owing to their favorable mechanical and thermal properties [22 - 24].

Aluminium Oxide Reinforced Metal Matrix Composites (Al$_2$O$_3$-MMCs)

These composites have high strength, stiffness, and good wear resistance. They are mainly utilized in aerospace, automotive, and machinery applications owing to their favorable mechanical characteristics.

Boron Reinforced Metal Matrix Composites (B-MMCs)

These composites have high stiffness, high strength, and good thermal properties. They are typically used in aerospace and automotive applications owing to their favorable mechanical and thermal characteristics [22 - 24].

Titanium Carbide Reinforced Metal Matrix Composites (TiC-MMCs)

These composites exhibit high hardness, high wear resistance, and good thermal properties. They are typically employed in aerospace, automotive, and cutting tool applications owing to their favorable mechanical and thermal characteristics.

Tungsten Reinforced Metal Matrix Composites (W-MMCs)

These composites offer high strength, stiffness, and good radiation shielding properties. They are typically used in nuclear and aerospace applications owing to their favorable mechanical and radiation shielding characteristics.

Graphite Reinforced Metal Matrix Composites (Gr-MMCs)

These composites have strong lubricating properties, a low COTE, and high heat conductivity. Due to their advantageous thermal and tribological properties, they are frequently utilised in mechanical, automotive, and electronic packaging [22 - 24].

Molybdenum Reinforced Metal Matrix Composites (Mo-MMCs)

These composites are strong, rigid, and have good thermal characteristics. Because of their advantageous mechanical and thermal characteristics, they are frequently employed in nuclear and aerospace applications [22 - 24].

Nickel Reinforced Metal Matrix Composites (Ni-MMCs)

These composites have high strength, good ductility, and good corrosion resistance. They are mainly employed in chemical processing and aerospace applications owing to their favourable mechanical and chemical characteristics.

Ceramic Reinforced Metal Matrix Composites (C-MMCs)

These composites have high strength/stiffness, and good wear resistance. They are usually employed in cutting tools, armour, and aerospace applications owing to their favourable mechanical and thermal characteristics.

APPLICATIONS OF MMCS

MMCs are highly developed materials that combine the mechanical properties of a metal matrix with the enhanced properties of one or more reinforcement materials. These composites have excellent mechanical and physical characteristics, which make them attractive for a wide range of industrial applications. Here are some of the most common applications of MMCs [25 - 27].

Aerospace

MMCs are widely employed in the aerospace industry owing to their excellent mechanical characteristics, such as high strength, stiffness, and resistance to high temperatures. They are commonly used in aircraft engine components, such as fan blades, compressor blades, and turbine blades, as well as in spacecraft components, such as rocket nozzles and heat shields.

Automotive

MMCs are used in the automotive industry to improve the performance and efficiency of vehicles. They are used in engine components, such as pistons, connecting rods, and cylinder liners, as well as in suspension components, such as control arms and brake rotors. MMCs can reduce the weight of these components while maintaining their strength and stiffness, which can improve fuel efficiency and handling.

Electronics

MMCs are used in the electronics industry to improve the performance of electronic packaging. They are employed in printed circuit boards and semiconductor packaging to improve thermal management and reduce the risk of thermal damage. MMCs can also improve mechanical stability and reduce the warping of electronic components.

Défense

MMCs are used in the defence industry to improve the performance and durability of weapons and vehicles. They are utilized in armour components to improve the resistance to ballistic and blast damage. MMCs can also be used in military vehicles, such as tanks and armoured personnel carriers, to improve the strength and stiffness of the vehicle.

Medical

MMCs are used in the medical industry to improve the performance and durability of medical implants. They are used in orthopaedic implants, such as hip and knee replacements, to improve the strength and stiffness of the implant. MMCs can also be used in dental implants to improve the resistance to wear and corrosion.

Sporting Goods

MMCs are used in the sporting goods industry to improve the performance and durability of sports equipment. They are used in golf club heads, tennis racket frames, and bicycle frames to improve the strength, stiffness, and impact resistance of the equipment. MMCs can also be used in athletic shoes to improve the shock absorption and durability of the sole.

Energy

MMCs are used in the energy industry to improve the performance and durability of energy generation and storage systems. They are used in wind turbine blades to improve the strength and stiffness of the blade. MMCs can also be used in batteries to improve the thermal management and durability of the battery.

Machinery

MMCs are used in the machinery industry to improve the performance and durability of machine components. They are used in gears, bearings, and bushings to enhance the wear resistance and durability of the components. MMCs can also be used in cutting tools to improve the hardness and wear resistance of the tool.

Construction

MMCs are used in the construction industry to improve the performance and durability of building components. They are used in structural components, such as beams and columns, to improve the strength and stiffness of the component. MMCs can also be used in roofing and siding to improve durability and resistance to weathering.

Marine

MMCs are used in the marine industry to improve the performance and durability of boat components. They are used in hulls, propellers, and shafts to improve the strength, stiffness, and resistance to corrosion of the component. MMCs can also be used in marine electronics to improve the thermal management and durability of electronic components.

Overall, the unique properties of MMCs make them suitable for a wide range of applications in various industries, and their application is expected to increase in the future as more research is done to develop new and innovative MMC materials. Fig. (2) shows various applications of MMCs.

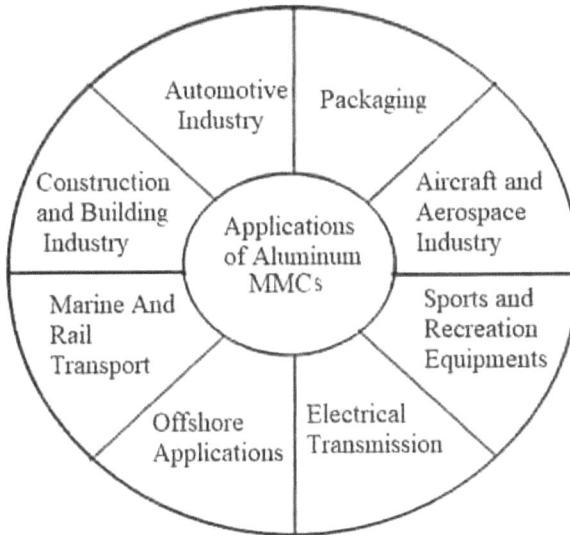

Fig. (2). Applications of MMCs.

RELEVANCE OF MMCs FOR SUSTAINABLE INDUSTRY

MMCs are advanced materials that have unique properties that make them ideal for sustainable industry practices. They are made by combining a metal matrix, typically aluminium, magnesium, or titanium, with a reinforcing material such as ceramic, metallic, or organic fibers [28, 29]. In comparison to typical metals, the resultant composites are lighter, more rigid, and stronger, have superior wear resistance, and have better thermal and electrical conductivity. The relevance of MMCs to sustainable industry can be seen in a variety of applications, including transportation, electronics, renewable energy, construction, and manufacturing [30]. In each of these industries, MMCs offer potential benefits to improve efficiency, reduce waste, and minimize environmental impacts.

Light Weighting

One of the key benefits of MMCs is their lightweight properties, which can decrease energy use, greenhouse gas emissions, and transportation costs significantly. The transportation industry, for example, is a major contributor to global greenhouse gas emissions, with automobiles, trucks, and aircraft being significant sources. The use of MMCs in these industries can help to reduce weight and improve fuel efficiency, which can lead to lower emissions and operating costs. In the aerospace industry, MMCs are already being used in the production of aircraft parts such as engine components, landing gear, and wing structures. By reducing weight and improving performance, MMCs can help reduce the environmental impact of air travel, which is expected to continue to grow in the coming years.

Longer Lifespan

The use of MMCs can also help to extend the lifespan of products and structures, reducing the need for replacements and minimizing waste. The improved mechanical and wear properties of MMCs can lead to longer lifespans of products and structures, which can be particularly important in industries such as construction and manufacturing. In the construction industry, MMCs can be used to reinforce concrete and other building materials, improving their durability and lifespan. This can lead to reduced maintenance costs and lower environmental impacts associated with the replacement of building materials.

Recycling

MMC materials can also be recycled, reducing the amount of waste sent to landfills and minimizing the need for virgin materials. The recycling of MMCs can help to reduce the environmental impact of manufacturing and construction industries, which are significant contributors to global waste generation.

Reduced Energy Consumption

The use of MMCs can also reduce energy consumption in manufacturing processes due to their lighter weight and improved performance. This can lead to lower greenhouse gas emissions and reduced energy costs, which can be particularly important for industries such as renewable energy and electronics. In the renewable energy industry, MMCs can be used to produce more efficient wind turbine blades, which can help to reduce the cost of wind energy production. In the electronics industry, MMCs can be used to manufacture heat sinks, electronic packaging, and interconnects, which can help to dissipate heat and reduce energy consumption associated with electronic devices.

Corrosion Resistance

Many MMCs have good corrosion resistance, which can lead to longer lifespans of products and structures in harsh environments, reducing the need for replacements and minimizing waste. The use of MMCs in marine environments, for example, can help to reduce the environmental impact of marine infrastructure and reduce the need for costly maintenance and replacements.

ROLE OF MACHINE LEARNING (ML) AND ARTIFICIAL INTELLIGENCE (AI) IN THE DEVELOPMENT OF MMCs

The goal of sustainable development in the modern industry has raised the emphasis on cutting-edge materials that offer superior mechanical qualities, lower environmental impact, and more energy efficiency. In this context, machine learning (ML) and artificial intelligence (AI) have become essential tools for creating new materials, particularly in the field of MMCs. MMCs offer a special chance to create lightweight, highly-stable components with specialised qualities for a variety of applications spanning from the aerospace to the automotive sectors. By speeding up the processes of material discovery and development, AI and ML approaches have completely changed the way MMCs are designed, synthesised, and optimised. These technologies provide better-informed choice-making throughout the material selection and design phases by enabling researchers to mimic the behaviour of MMCs under various settings. AI and ML systems find patterns, correlations, and ideal combinations that human intuition alone would miss by analysing enormous datasets comprising material qualities, processing parameters, and performance attributes [31]. Due to the quicker identification of viable MMC compositions, experimentation time and expenses are decreased.

Additionally, AI-driven automation streamlines manufacturing procedures for MMCs, guaranteeing consistency and repeatability in material production. Manufacturers may precisely adjust production parameters and produce desired material attributes with more precision because of ML algorithms' ability to forecast the effects of processing variables on material microstructure and qualities. By reducing waste and energy use, this level of control improves the overall quality of MMCs and is consistent with sustainable production practises. Furthermore, during the whole lifecycle of MMCs, AI and ML are crucial for monitoring and quality control. Analysing real-time data during manufacture and performance testing enables quick modifications and minimises faults by identifying departures from required standards. This proactive method reduces resource waste and encourages the adoption of MMCs with the best mechanical qualities, increasing the materials' general effectiveness and service life. In conclusion, the incorporation of AI and ML into the creation of metal matrix composites represents a fundamental transition in the modern industry towards environmentally friendly methods. These technologies speed up the search for new materials, streamline production methods, improve quality assurance, and make it easier to produce MMCs with superior mechanical qualities and minimal environmental effects. Industries can support sustainable development by promoting resource efficiency, eliminating waste, and accelerating the manufacture of cutting-edge products that solve the challenges of a quickly changing world by utilising the potential of AI and ML [32].

CONCLUSION

In conclusion, the relevance of MMCs to sustainable industry is significant due to their potential to reduce material consumption, improve energy efficiency, and minimize waste. Their lightweight properties, longer lifespan, recyclability, reduced energy consumption, and corrosion resistance make them a valuable material for sustainable industry practices. As MMCs continue to be developed and new applications are identified, their potential to contribute to sustainable industry practices will continue to grow. By embracing these advanced materials, industries can take important steps towards reducing their environmental impact and improving their bottom line, while helping to create a more sustainable future for generations to come.

REFERENCES

[1]   Rajak, D.K.; Pagar, D.D.; Kumar, R.; Pruncu, C.I. Recent progress of reinforcement materials: A comprehensive overview of composite materials. *J. Mater. Res. Technol.,* **2019**, *8*(6), 6354-6374.
       [http://dx.doi.org/10.1016/j.jmrt.2019.09.068]

[2]   Åström, B.T. *Manufacturing of Polymer Composites*; Routledge, **2018**.
       [http://dx.doi.org/10.1201/9780203748169]

[3]   Nagarajan, T.; Sridewi, N.; Wong, W.P.; Walvekar, R.; Khanna, V.; Khalid, M. Synergistic performance evaluation of MoS_2–hBN hybrid nanoparticles as a tribological additive in diesel-based engine oil. *Sci. Rep.,* **2023**, *13*(1), 12559.
       [http://dx.doi.org/10.1038/s41598-023-39216-0] [PMID: 37532805]

[4]   Soutis, C.; Yi, X.; Bachmann, J. How green composite materials could benefit aircraft construction. *Sci. China Technol. Sci.,* **2019**, *62*(8), 1478-1480.
       [http://dx.doi.org/10.1007/s11431-018-9489-1]

[5]   Sreejith, M.; Rajeev, R.S. Fiber reinforced composites for aerospace and sports applications. In: *Fiber Reinforced Composites*; Elsevier, **2021**; pp. 821-859.
       [http://dx.doi.org/10.1016/B978-0-12-821090-1.00023-5]

[6]   Dahiya, M.; Khanna, V.; Bansal, S.A. Finite element analysis of the mechanical properties of graphene aluminium nanocomposite: Varying weight fractions, sizes and orientation. *Carbon Letters,* **2023**, *33*(6), 1601-1613.
       [http://dx.doi.org/10.1007/s42823-023-00543-x]

[7]   Balaji, V.; Sateesh, N.; Hussain, M.M. Manufacture of aluminium metal matrix composite (Al7075-SiC) by stir casting technique. *Mater. Today Proc.,* **2015**, *2*(4-5), 3403-3408.

[http://dx.doi.org/10.1016/j.matpr.2015.07.315]

[8]   Bansal, S.A.; Khanna, V.; Balakrishnan, N.; Gupta, P., Eds. *Diversity and Applications of New Age Nanoparticles*; IGI Global, **2023**.
[http://dx.doi.org/10.4018/978-1-6684-7358-0]

[9]   Yang, Y.; Boom, R.; Irion, B.; van Heerden, D.J.; Kuiper, P.; de Wit, H. Recycling of composite materials. *Chem. Eng. Process.*, **2012**, *51*, 53-68.
[http://dx.doi.org/10.1016/j.cep.2011.09.007]

[10]   Khanna, V.; Kumar, V.; Bansal, S.A. Mechanical properties of aluminium-graphene/carbon nanotubes (CNTs) metal matrix composites: Advancement, opportunities and perspective. *Mater. Res. Bull.*, **2021**, *54*(11), 111224.
[http://dx.doi.org/10.1016/j.materresbull.2021.111224]

[11]   Chawla, K.K. Metal matrix composites. In: *Composite Materials*; Springer New York: New York, NY, **2012**; pp. 197-248.
[http://dx.doi.org/10.1007/978-0-387-74365-3_6]

[12]   Khanna, V.; Kumar, V.; Bansal, S.A.; Prakash, C.; Ubaidullah, M.; Shaikh, S.F.; Pramanik, A.; Basak, A.; Shankar, S. Fabrication of efficient aluminium/graphene nanosheets (Al-GNP) composite by powder metallurgy for strength applications. *J. Mater. Res. Technol.*, **2023**, *22*, 3402-3412.
[http://dx.doi.org/10.1016/j.jmrt.2022.12.161]

[13]   Dahiya, M.; Khanna, V.; Anil Bansal, S. Aluminium-graphene metal matrix nanocomposites: Modelling, analysis, and simulation approach to estimate mechanical properties. *Mater. Today Proc.*, **2022**, (Nov)
[http://dx.doi.org/10.1016/j.matpr.2022.10.181]

[14]   Reddy Nagavally, R. Composite materials-history, types, fabrication techniques, advantages, and applications *Materials Science,* **2016**.

[15]   Kaczmar, J.W.; Pietrzak, K.; Włosiński, W. The production and application of metal matrix composite materials. *J. Mater. Process. Technol.*, **2000**, *106*(1-3), 58-67.
[http://dx.doi.org/10.1016/S0924-0136(00)00639-7]

[16]   Dahiya, M.; Khanna, V.; Anil Bansal, S. Effect of graphene size variation on mechanical properties of aluminium graphene nanocomposites: A modeling analysis. *Mater. Today Proc.*, **2022**, (Jul)
[http://dx.doi.org/10.1016/j.matpr.2022.07.259]

[17]   Gupta, P.; Ahamad, N.; Kumar, D.; Gupta, N.; Chaudhary, V.; Gupta, S.; Khanna, V.; Chaudhary, V. Synergetic effect of CeO_2 doping on structural and tribological behavior of $Fe-Al_2O_3$ metal matrix nanocomposites. *ECS J. Solid State Sci. Technol.*, **2022**, *11*(11), 117001.
[http://dx.doi.org/10.1149/2162-8777/ac9c92]

[18]   Singh, K.; Khanna, V. Current Developments in Machining of Titanium Based Alloys using Wire EDM. In: *Advances in Nonconventional Machining Processes*; Bentham Science Publishers, **2020**; pp. 103-119.
[http://dx.doi.org/10.2174/9789811483653120010009]

[19]   Metal matrix composites: Mechanisms and properties Available from: https://www.osti.gov/biblio/7282168 (accessed Aug, **2023**, 22).

[20]   Hutchings, I.M. *Tribological properties of metal matrix composites,* **2013**, *10*(6), 513-517.
[http://dx.doi.org/10.1179/mst.1994.10.6.513]

[21]   Alman, D.E. Properties of metal-matrix composites. *Composites,* **2001**, (Dec), 838-858.
[http://dx.doi.org/10.31399/asm.hb.v21.a0003448]

[22]   Kumar, A.; Vichare, O.; Debnath, K.; Paswan, M. Fabrication methods of metal matrix composites (MMCs). *Mater. Today Proc.*, **2021**, *46*, 6840-6846.
[http://dx.doi.org/10.1016/j.matpr.2021.04.432]

[23]   Sharma, D.K.; Mahant, D.; Upadhyay, G. Manufacturing of metal matrix composites: A state of review. *Mater. Today Proc.,* **2020**, *26*, 506-519.
[http://dx.doi.org/10.1016/j.matpr.2019.12.128]

[24]   Kainer, K.U. Metal matrix composites. *Metal Matrix composites: Custom-made materials for automotive and aerospace engineering,* **2006**, (Jun), 1-314.
[http://dx.doi.org/10.1002/3527608117]

[25]   Chawla, N.; Chawla, K.K. Metal-matrix composites in ground transportation. *J. Miner. Met. Mater. Soc.,* **2006**, *58*(11), 67-70.
[http://dx.doi.org/10.1007/s11837-006-0231-5]

[26]   Hunt, W.H., Jr; Miracle, D.B. Automotive applications of metal-matrix composites. *Composites,* **2001**, (Dec), 1029-1032.
[http://dx.doi.org/10.31399/asm.hb.v21.a0003484]

[27]   Rawal, S.P. Metal-matrix composites for space applications. *J. Miner. Met. Mater. Soc.,* **2001**, *53*(4), 14-17.
[http://dx.doi.org/10.1007/s11837-001-0139-z]

[28]   Jamwal, A.; Mittal, P.; Agrawal, R.; Gupta, S.; Kumar, D.; Sadasivuni, K. K.; Gupta, P. Towards sustainable copper matrix composites: Manufacturing routes with structural, mechanical, electrical and corrosion behaviour **2020**, *54*(19), 2635-2649.
[http://dx.doi.org/10.1177/0021998319900655]

[29]   Henry Ononiwu, N.; Akinlabi, E.T.; Ozoegwu, C.G. Sustainability in production and selection of reinforcement particles in aluminium alloy metal matrix composites: A review. *J. Phys. Conf. Ser.,* **2019**, *1378*(4), 042015.
[http://dx.doi.org/10.1088/1742-6596/1378/4/042015]

[30]   Bahrami, A.; Soltani, N.; Pech-Canul, M. I.; Gutiérrez, C. A. Development of metal-matrix composites from industrial/agricultural waste materials and their derivatives **2015**, *46*(2), 143-207.
[http://dx.doi.org/10.1080/10643389.2015.1077067]

[31]   Veeresh Kumar, G.B.; Pramod, R.; Rao, C.S.P.; Gouda, P.S.S. Artificial neural network prediction on wear of al6061 alloy metal matrix composites reinforced with -Al$_2$o$_3$. *Mater. Today Proc.,* **2018**, *5*(5), 11268-11276.
[http://dx.doi.org/10.1016/j.matpr.2018.02.093]

[32]   Banerjee, T.; Dey, S.; Sekhar, A.P.; Datta, S.; Das, D. Design of alumina reinforced aluminium alloy composites with improved tribo-mechanical properties: A machine learning approach. *Trans. Indian Inst. Met.,* **2020**, *73*(12), 3059-3069.
[http://dx.doi.org/10.1007/s12666-020-02108-2]

CHAPTER 2

Structure-Property Correlations in Metal Matrix Composites

Neeraj Kumar Sharma[1,*] and **Abhimanyu Singh Rana**[1]

[1] *Centre for Advanced Materials and Devices, School of Engineering & Technology, BML Munjal University, Sidhrawali, Gurgaon, Haryana, India*

Abstract: Metal matrix composites (MMCs) having particulate or laminate structure are extensively used in a wide range of applications including cutting tools, automotive vehicles, aircraft, and consumer electronics. In a composite material, two or more dissimilar materials are combined to form another material having superior properties. The matrix is a continuous phase in a composite material and is usually more ductile and less hard phase. In the matrix phase, aluminum, magnesium, titanium and copper are some of the metals widely used matrix materials. Compared with unreinforced metals, MMCs offer much better mechanical and thermal properties as well as the opportunity to tailor these properties for a particular application. In order to fabricate MMCs, various processing techniques have been evolved which can be categorized as liquid state method: Stir Casting, Infiltration, Gas Pressure Infiltration, Squeeze Casting Infiltration, Pressure Die Infiltration, solid state method: Diffusion bonding, Sintering and vapor state method: Electrolytic co-deposition, Spray co-deposition and Vapor co-deposition. The microstructure of MMCs such as orientation, distribution and aspect ratio of reinforced phase can effectively influence the properties of composite materials. The effective properties of MMCs can be predicted using the analytical or numerical methods. Analytical methods such as: Turner Model, Kerner Model, Schapery bonds, Hashin's bond and Rule-of-Mixtures are used widely for effective properties computation. However, analytical methods cannot take into account the material microstructure, and therefore, the finite element method has been used extensively to model the real microstructure of composites and to predict the deformation response and effective properties of composites.

Keywords: Hashin bound, Metal matrix composites, Shtrikman bound, Structural properties.

COMPOSITE MATERIALS

The ever-increasing demands from technology due to rapid advancements in the fields of aerospace, automotive, marine, sporting goods and aircraft have led to

* **Corresponding author Neeraj Kumar Sharma:** Centre for Advanced Materials and Devices, School of Engineering & Technology, BML Munjal University, Sidhrawali, Gurgaon, Haryana, India; E-mail: neeraj.sharma@bmu.edu.in

Virat Khanna, Prianka Sharma & Santosh Kumar (Eds.)

the development of high-performance composite materials [1]. Various processing techniques and advanced technologies have been developed to fabricate composite materials and this has made composites an attractive candidate and superior alternative to traditional materials. It is difficult for conventional metals and alloys to keep up with technological advancements.

Two or more materials, having considerably different properties, are combined to form the composite material. Therefore, a composite material can be defined as a material that consists of two or more materials having chemically distinct phases, microscopically heterogeneous but homogeneous macroscopically. Different constituents of a composite material do not dissolve or blend into each other and work together to yield much superior properties [1 - 5].

The matrix phase is the continuous phase of a composite material. In general, the matrix phase is a ductile material that helps to hold the dispersed phase. But this definition is not always applicable, for example, in the case of ceramic matrix composites, the matrix phase is harder and brittle. The phases embedded in the matrix in a discontinuous form are known as reinforcement. The reinforced phase should be uniformly dispersed in the matrix phase for better properties. The reinforced phase is usually stronger than the matrix. The properties of composites depend on the properties of their constituents, the bonding between constituents, and the size of reinforced particles. The distribution of particles: uniform or clustered also influences the effective material behavior. A strong bonding between the matrix and the reinforcement helps the matrix to transfer the load to the reinforcement phase [6 - 8].

CLASSIFICATION OF COMPOSITE MATERIALS

The classification of composites is based on: the type of matrix, type of reinforced, and fabrication techniques used [1]. The matrix phase could be metal, ceramic, and polymer (Fig. **1**).

METAL MATRIX COMPOSITES

Metal matrix composites are an important class of composites relevant to a wide variety of applications. Low-density metals, such as aluminum or magnesium are widely used as the matrix materials in these composites [6, 7]. The metal phase is usually reinforced with particulate or fibers of a ceramic material, such as silicon carbide or graphite. A much better combination of mechanical, thermal, and thermo-mechanical properties as well as the opportunity to tailor these properties for a particular application are offered by MMCs [8 - 15].

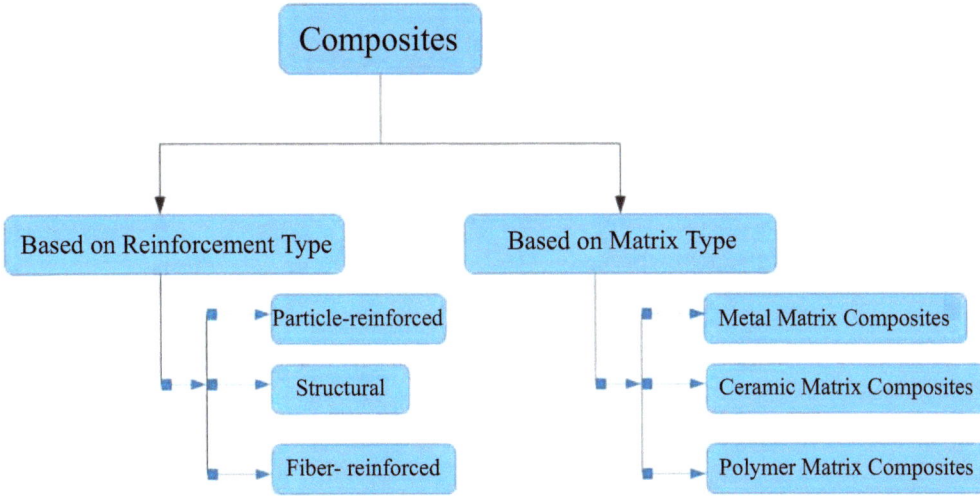

Fig. (1). Classification of composite materials.

Based on matrix material

a. Metal Matrix Composites (MMCs)

b. Ceramic Matrix Composites (CMCs)

c. Polymer Matrix Composites (PMCs)

Based on reinforcing material structure

a. Particle Reinforced Composites

b. Fibre Reinforced Composites

c. Laminated Composites

ANALYSIS OF COMPOSITE'S BEHAVIOR

The composite materials are fabricated by combining constituent materials having significantly different properties, and therefore computation of the effective properties of composites is a field of vital interest. Different types of analytical and numerical methods have been used by researchers to predict the effective properties of composites. Virtual simulation of deformation behavior using numerical methods such as finite element method (FEM), finite difference method (FDM) or atomistic simulations can be applied to understand the new materials behavior and their effective properties, and this reduces the expense on laboratory and experiments [16]. The deformation behaviour of composite materials subjected to mechanical, thermal, thermo-mechanical, or thermo-electrical loadings can be studied using the finite element method. The chief goal is to speed up the trial and error experimental testing and to be able to simulate the real phenomena that occur at the micro level of the composites. Three different approaches are utilized to know the material behavior and effective properties:

- Experimental Characterization
- Analytical Modeling
- Numerical Modeling

The increase in the computational capacity of computers has opened up the possibility of mathematical modeling and simulation. It raises the possibility that modern numerical methods can play a significant role in the analysis of heterogeneous microstructures [17 - 19].

These analytical or numerical modeling approaches for composite materials can also be categorized as either macroscopic or microscopic in nature, respectively. Macroscopic modeling of composites is often simple in application and can be used to predict the average or global response of a composite with minimal computational resources. Generally, in a macroscopic model, a uniform distribution of particles in a metal matrix is assumed and the volume fraction and properties of the individual phases are used to predict the effective properties of composites [20]. Macroscopic models ignore the reinforcement size, shape, arrangement, and orientation. The properties of the micro constituents and phenomena on the microscale significantly influence the macroscopic properties of all composite materials. A better understanding of the macroscopic behavior can be obtained by studying the description of the microstructural phenomena. However, predicting exact microstructure, especially in case of particle-reinforced composites, is quite complicated, so in general some statistical assumption has to be made.

The sample of composites obtained by applying these statistical assumptions is called the homogenized sample of material. These homogenized samples can be used effectively to predict the properties of effective composites and to study the stress-strain distribution. The homogenized sample of material of a volume element, is often called a representative volume element (RVE) [20]. The objective of homogenization is to extract sample data which can be used to find a material model for the computation of effective material properties. All macroscopic properties of the microheterogeneous material are supposed to be represented by the homogenized sample. In general, the homogenized material model is not assumed to be of the same type as the model used for the micro constituents, since it significantly complicates the search for an effective material model. The random distribution of particles in RVEs is important to simulate.

Until the development of computers, the determination of effective material parameters for homogenization was only possible by either performing experiments or tests with the existing material sample or by making use of semi-analytical methods. In semi-analytical methods, rather strong assumptions on the mechanical field variables or on the microstructure of the material are used, and therefore obtaining accurate results from semi-analytical methods is difficult. Especially in case of metal matrix composites, the metal behavior is influenced by elastoplastic deformation, and the determination of effective material parameters with the commonly used semi-analytical methods leads to considerable deviation in results from reality.

Fabrication and Experimental Characterization of MMCs

Different processing routes such as liquid state, solid state, and vapor state processing have been used for the fabrication of MMCs. In liquid state methods, the metal matrix is heated to the molten state and the preheated reinforced phase is added to the molten metal. Liquid-state processing of MMCs is inexpensive and more efficient compared to the solid or vapor state processing. Some of the commonly used liquid state methods are: stir casting, gas pressure infiltration, squeeze casting, and pressure die infiltration. In solid-state methods, both the metal and the reinforcement phase are processed in the powder form. The powder-form constituents are compressed at an elevated temperature and a bonding between the matrix and the reinforced phase is obtained either by diffusion or sintering. A comparatively low temperature in solid-state processing than the liquid-state processing eliminates the undesirable chemical reactions at the interface of matric and reinforcement phases. Vapor state methods such as electrolytic co-deposition, spray co-deposition and vapor co-deposition are used for the fabrication of MMCs. In spray deposition, the atomized molten metal is sprayed on a substrate and the dispersed reinforced particles are supplied from a separate container for deposition.

Fibers, continuous or discontinuous, whiskers, or particulates are some of the common reinforcement materials for MMCs to tailor their stiffness, strength, thermal conductivity, thermal expansion coefficient (CTE), and wear resistance. Various experimental techniques are used to measure the effective properties of these composites.

Tensile tests are conducted to measure the effective elastic modulus, yield strength and the ultimate tensile or compressive strength of MMCs using the universal testing machines (UTM). The composite specimens are held between the two cross heads subjected to the uniaxial load. The resistance of the sample is tested against the compressive or the tensile load. The cross-head moves with a certain velocity which is called 'strain rate'. It is important that the material behaves differently at different strain rates. The load-displacement data recorded during the experiment can be used to calculate the engineering stress and strain. The stress-strain plots provide important information about the strain-hardening behavior of composite materials.

The thermal expansion behavior of MMCs can be studied using the differential thermal analyzers (DTA). As the sample is heated in a furnace, the change in length is measured using a linear variable displacement transducer. The thermocouples located next to the heating element measure the applied temperature differences. The measured thermal strain against the temperature changes provides the CTE of composite samples. For thermal conductivity measurement, the guarded heat flow meter method is used. In this method, the composite samples are subjected to a heat source at one end and the other end is subjected to a coolant. In order to ensure the heat flow is one-directional, a ring guard heater is used at its lateral surface to eliminate any temperature difference and consequent heat loss from the lateral surface. The thermocouples at both ends measure the temperature and effective thermal conductivity can be measured by applying Fourier's law of heat transfer.

Analytical Models for Composite Analysis

Apart from finite element methods, analytical models are also used extensively to predict the effective material's properties. Various analytical models are reported by many researchers for the analysis of the mechanical and thermal behavior of particle-reinforced metal matrix composites [21 - 27]. Analytical models are simplistic in nature and are derived using the variational principle of linear elasticity. An explicit assumption that the material is homogeneous within itself is a common feature in mathematical formulations of the mechanical behavior of solid materials.

However, in real materials, the presence of heterogeneities and discontinuities is evident at the level of grain clusters and/or microcracks due to their discrete structure. In this macroscopic approach to the modeling of composites, the complete microstructural details such as the shape, size, orientation, and concentration of reinforced particles, and interface bonding among constituents cannot be taken into account for analysis. Analytical models predict the average or global response of a composite material by considering only individual properties and volume fraction of constituents.

Hashin and Shtrikman bound

Hashin presented a model in 1962 considering the spherical inclusions embedded in a continuous matrix phase. The volume fractions can be used to obtain the constant ratio of radii. A uniform distribution and filling of all gaps between composite spheres require sizes reduced to infinitesimal. A single representative composite sphere has to be analyzed by applying the homogeneous volumetric stress on the surface of the sphere [21 - 24].

Hashin's bounds were obtained by applying the variational principles in the linear elasticity. The elastic polarization tensor, has been obtained to derive the upper and lower bounds of elastic modulus for quasi-isotropic multiphase materials of arbitrary phase geometry. The bounds derived are close enough to provide a good estimate for the effective moduli in case the ratios between the different phase moduli are not too large. The bulk modulus and shear modulus are obtained as:

$$K_{upper} = K_p + \left(1 - V_p\right) \times \left[\frac{1}{\left(K_m - K_p\right)} + \frac{3V_p}{\left(3K_p + 4G_p\right)} \right]^{-1} \tag{1}$$

$$K_{lower} = K_m + V_p \times \left[\frac{1}{\left(K_p - K_m\right)} + \frac{3\left(1 - V_p\right)}{\left(3.K_m + 4.G_m\right)} \right]^{-1} \tag{2}$$

$$G_{upper} = G_p + \left(1 - V_p\right) \times \left[\frac{1}{\left(G_m - G_p\right)} + \frac{6V_p\left(K_p + 2G_p\right)}{5G_p\left(3K_p + 4G_p\right)} \right]^{-1} \tag{3}$$

$$G_{lower} = G_m + V_p \left[\frac{1}{\left(G_p - G_m\right)} + \frac{6\left(1 - V_p\right)\left(K_m + 2G_m\right)}{5G_m\left(3K_m + 4G_m\right)} \right]^{-1} \tag{4}$$

where G and K are the shear and bulk moduli, Vp denotes the volume fraction of the reinforcement phase, and 'm' and 'p' denote the matrix and the particles, respectively. The overall bounds of Young's modulus are then obtained by substituting the modulus values in the following equation:

$$E = \frac{9KG}{\left(G + 3K\right)} \tag{5}$$

Upper Bound:

$$E_{upper} = \frac{9 K_{upper} G_{upper}}{\left(G_{upper} + 3 K_{upper} \right)} \tag{6}$$

Lower Bound:

$$E_{lower} = \frac{9 K_{lower} G_{lower}}{\left(G_{lower} + 3 K_{lower} \right)} \tag{7}$$

Analytical Thermo-Elastic Model

Analytical thermo-elastic models are used to study the temperature-dependent material properties. Ceramic-reinforced metal matrix composites are used extensively in high-temperature applications because of the sustainability of mechanical properties by composites at elevated temperatures. Particle-reinforced metal matrix composites having a high reinforcement volume fraction (>50 vol.%) are used for thermal management applications such as electronic packaging. The application of high temperature causes the expansion of composite materials quantified by the coefficient of linear thermal expansion (CTE). Thermal expansion results in thermal stresses and significantly affects the material's deformation behavior. It is difficult to predict the CTE of the MMCs precisely because it is influenced by several factors including plasticity and the internal structure of the composite. In subsequent sections, the thermo-elastic models used in the present work, for the prediction of effective CTE of composite materials, are discussed.

Kerner Model

The Kerner model [26] predicts the effective CTE considering the reinforcement is spherical and discontinuous and is surrounded by a uniform matrix layer. The model considers the composite, a volume element, in which the spherical-shaped reinforced particle is surrounded by a shell of a matrix with their respective volume fractions in the composite. Assuming a macroscopically isotropic and homogeneous distribution, the CTE of composites using this model can be given as:

$$\alpha_c = \alpha_p V_p + \alpha_m V_m + (\alpha_p - \alpha_m) V_p \frac{V_m (K_p - K_m)}{V_p K_p + V_m K_m + 0.75 \frac{K_p K_m}{G_m}}, \tag{8}$$

where 'V' represents the volume fraction, and α the CTE of the constituents. The subscripts 'm', 'p', and 'c' refer to the matrix phase, particle phase (reinforcement) and composite, respectively. 'K' is the bulk modulus of the components of the composite, which is related to Young's modulus 'E' and Poisson's ratio μ of isotropic materials by:

$$E = 3K(1-2\mu) \tag{9}$$

'G' is the shear modulus, which is related to the Young's modulus 'E' of isotropic materials by:

$$E = 2G(1+\mu) \tag{10}$$

Schapery Bounds

Schapery obtained the bounds on effective CTEs of an isotropic and anisotropic composite materials composites by applying the extremum principles of thermo-elasticity coupled with Hashin's bounds [25]. Hashin's bounds are used to obtain the upper and lower values of Bulk modulus. The upper bound and lower bound of CTE for a two-phase isotropic, particulate reinforced composites are obtained as:

$$\alpha_c = \alpha_p V_p + \alpha_m V_m + \left[\frac{4G_m}{K_c}\right]\left[\frac{(K_c - K_p)(\alpha_m - \alpha_p)V_p}{4G_m + 3K_p}\right] \tag{11}$$

$$\alpha_c = \alpha_p V_p + \alpha_m V_m + \left[\frac{4G_p}{K_c}\right]\left[\frac{(K_c - K_m)(\alpha_p - \alpha_m)V_m}{4G_p + 3K_m}\right] \tag{12}$$

The effective CTE (α_c) depends on the volume fraction and geometry of constituents only through their effect on the bulk modulus. In these expressions, K_c denotes the average of upper and lower value of Bulk modulus that can be obtained by Hashin's bound and is given by equations 4.3 and 4.4 as:

$$K_{upper} = K_p + (1 - V_p) \times \left[\frac{1}{(K_m - K_p)} + \frac{3 V_p}{(3 K_p + 4 G_p)} \right]^{-1}$$

$$K_{lower} = K_m + V_p \times \left[\frac{1}{(K_p - K_m)} + \frac{3(1 - V_p)}{(3. K_m + 4. G_m)} \right]^{-1}$$

It is important to see that the bounds are derived considering only the elastic deformation, therefore if the plastic deformation of metal matrix is present, the effective CTE of composite derived through these bounds will deviate with experimental results.

Turner Model

Turner model [27] is a relatively simple model used for the estimation of CTEs of composites. Turner model is based on the assumption that only homogeneous strains exist throughout the composite. The stresses present in the microstructure is assumed as uniform hydrostatic stresses and also the stresses are insufficient to disrupt the microstructure of the composite. The presence of isostrain conditions among constituents causes to change the dimensions with a temperature change for each constituent at the same rate. The shear deformation is assumed to be negligible. Turner derived the coefficient of thermal expansion for a two-phase composite by applying the balance of internal average stresses as [88]:

$$\alpha_c = \frac{\alpha_p V_p K_p + \alpha_m V_m K_m}{V_p K_p + V_m K_m} \tag{13}$$

Where subscripts 'p' and 'm' refer to the particle phase and matrix phase, respectively and α is the linear CTE of constituents, 'V' is the volume fractions and 'K' is the shear modulus. Interpenetrating composites, having one phase continuous in the matrix material should be well described by the Turner model because it is expected that these types of composites approach the assumption of the same dimension change in the average (expansion and shrinkage, respectively) of each component of the composite. It is important to note that the stronger the interphase bonding among constituents, the greater would be the possibility of isostrain conditions among constituents.

Finite Element Methods (FEM)

Finite element methods are extensively used to simulate the composite's behavior. In FEM, a complex geometry is discretized into smaller elements. Complex structured composites are difficult to analyze satisfactorily using analytical methods. A composite microstructure requires the following parameters to be considered for analysis [28 - 30]:

- Shape and size of reinforced particles
- Concentration of particles
- Orientation of particles
- Presence of voids
- Interface bonding among constituents

The microstructure modeling of composites considering all above parameters is possible using finite element methods. Once, a satisfactory representation of microstructure is obtained, the discretization and meshing is the next step. The discretization of a composite microstructure in finite elements has several advantages:

- It leads to accurate representation of complex geometry
- Easy representation of the total solution
- Distribution of various fields can be captured at the microscopic level

The effects of various microstructural parameters are required to be properly understood on the composite's behavior. The properties of composite materials are influenced by:

- Natural parameters
- Geometrical parameters

Natural parameters include the bulk properties of the constituent phases and the interface behavior. In geometric parameters, the volume fraction, size, shape of the reinforced particles, and the placement of the anisometric phases with respect to the loading direction are considered. Finite element method is one of the powerful tools for such an analysis. FEM methods are computationally intensive, but have the advantage of being able to readily incorporate the microstructural details of composites.

A discrete model composed of a set of piecewise continuous functions (polynomials) for any continuous quantity, such as temperature, pressure, or displacement defined over a finite number of subdomains or elements is used in FEM. Different elements are connected through sharing nodes. Energy functional is developed using piecewise continuous functions with unknown coefficients. The energy functional will be minimized in relation to the unknown leads to the system of equations and the unknown coefficients are determined by functional minimization. A continuous physical problem is transformed into a discretized finite element problem with unknown nodal values. Two features of the FEM are worth to be mentioned:

1) Good precision can be obtained using piecewise approximation of physical fields on finite elements. In order to obtain better precision, either the number of elements should be increased or the degree of approximating polynomials should be high.

2) Different types of elements, linear for 1D, triangular, quadrangular for 2D problems, and tetrahedron, hexahedron, wedge, *etc.* can be used for 3D problems.

Modeling the composite's structure is an important step in FEM and on the basis of this, different types of FEM approaches are used. Broadly, all the finite element methods can be categorized into 2-dimensional and 3-dimensional FEM. Unit cell modeling approach is used extensively by researchers in both 2D and 3D finite element analysis [18 - 21]. Two dimensional finite element methods utilize plane stress or plane strain hypothesis to make the analysis of composites behavior. Many problems in elasticity may be treated satisfactorily by this two-dimensional or plane theory of elasticity [30]. These two types can be defined by applying certain assumptions and restrictions on geometry. In case of geometries, having one dimension much smaller than the others, as that of a plate, and also subjected to loading uniformly over the thickness of the plate, plane stress theory is applicable. In plane strain, one deals with a situation in which the dimension of the structure in one direction is much larger in comparison with the dimensions of the structure in the other two directions, and the structure is subjected to transverse loading.

The advantages of two-dimensional finite element analysis include a much shorter computation time, lesser computing resources, and computation cost. Object Oriented Finite Element Method (OOFEM) [28, 30] is one such method that performs the two-dimensional analysis using the scanning electron microscopic

(SEM) images of composites. SEM images of composites are used to generate the microstructure of composites as shown in Fig. (**2**).

(a)   (b)

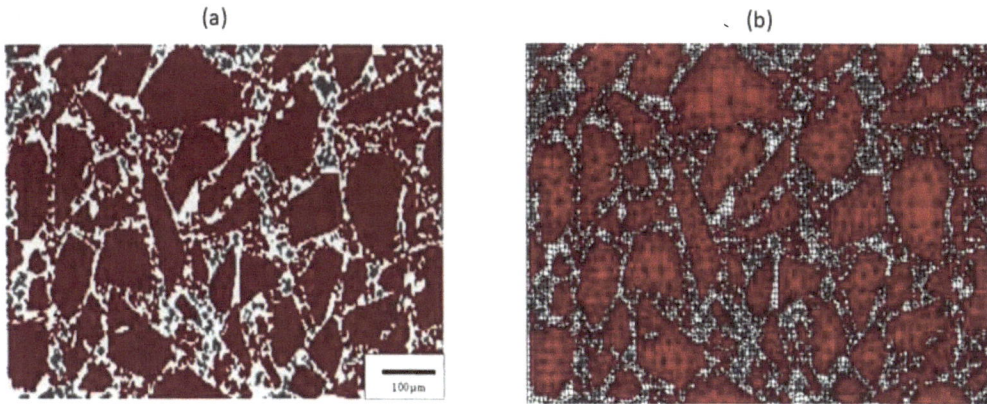

Fig. (2). Two-dimensional finite element analysis.

2D-FEA USING OOFEM

The composites having a complex spatial geometry can be analyzed conveniently for their mechanical and thermal behavior using finite element analysis. The most finite element packages require a numerical representation of the model geometry in terms of simple building blocks as input. The boundaries of domains are described by straight lines, planes or simple curves. However, in the analysis of composites, the domains have complex spatial geometry and it is difficult to model the microstructures. This difficulty can be avoided if the SEM images are used for mesh generation, simultaneously identifying boundaries in the image and bringing the finite element mesh into correspondence with these boundaries. The National Institute of Standards and Technology (NIST) has developed an object-oriented finite-element software (OOFEM) [32] that can be used for two-dimensional finite-element analysis.

OOFEM is the image-based analysis software that uses SEM images of composites for mesh generation and this way, the explicit mathematical description of the image boundaries is not required, and is not explicitly constructed. OOFEM is used to conduct the two dimensional analysis of composites considering the linear relationship between stress and strain in this chapter. The steps involved in finite element mesh generation using OOFEM is described in forthcoming paragraphs.

Microstructure Modeling Using OOFEM

The first task while working with OOFEM is to define the microstructure of composite using the SEM image of the composite material. The pixel group has to be created for the matrix and reinforcement phases. The OOFEM program includes a number of tools for selecting pixels belonging to the matrix and reinforced materials of composites. Equally important is that all the pixels have to be selected, otherwise the composite will behave as a porous material that can significantly affect the properties. Properties like colors that are purely decorative were assigned to differentiate the two types of pixels conveniently. Once the Microstructure is created, the materials are assigned to each pixel in the microstructure to make the microstructure meaningful. Materials are defined as sets of properties.

Finite Element Mesh Generation

Mesh generation in OOFEM is carried out with the skeleton of the mesh. The node positions and element edges of a finite element mesh are specified in the skeleton. The skeleton doesn't contain any information about the interpolation function to be used in finite element analysis. A microstructure can have many skeletons [32].

The adaptive skeleton option facilitates discretization of the image such that it conforms to the pixel boundary. Two properties control the conformation to the microstructure boundary: (i) homogeneity and (ii) the shape of the element (see Fig. **3**). The homogeneity energy represents the degree of pixel boundary conformation by mesh elements and the total energy of the mesh can be defined as:

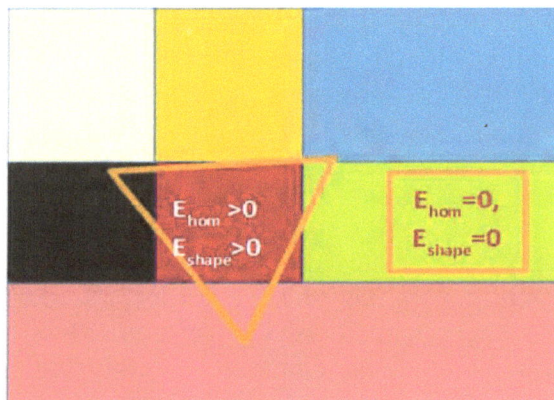

Fig. (3). E_{hom} and E_{shape} for square and triangular elements during mesh generation [33].

$$E = \alpha\, E_{hom} + (1-\alpha)\, E_{shape} \qquad\qquad (14)$$

where the parameter α has a value between 0 and 1, and E_{hom} and E_{shape} are the functions that depend on the element's homogeneity and shape, respectively [32].

The finite element mesh introduces the interpolation functions to the skeleton. After creating a Mesh, the fields were defined and activated. OOFEM currently analyzed problems related to displacement, temperature and voltage field [31, 32].

Once the boundary condition is assigned, the conventional force balance equation can be solved by the conjugate gradient method using a linear driver.

Nonlinear Analysis Using OOFEM

When a material is subjected to external loading, the initial deformation follows a linear relationship between stress and strain and is called elastic deformation. After elastic deformation, the material starts yielding and the plastic deformation starts to take place. Plastic deformation is influenced by the hardening of materials. It is observed that in case of ductile materials like metal, the plastic deformation significantly influences the material behavior. OOFEM can be used only to simulate the elastic deformation of composites. In order to simulate the elasto-plastic deformation of metal matrix composites, the commercial software ABAQUS is used. Fig. (4) explains the steps for nonlinear analysis using OOFEM [31].

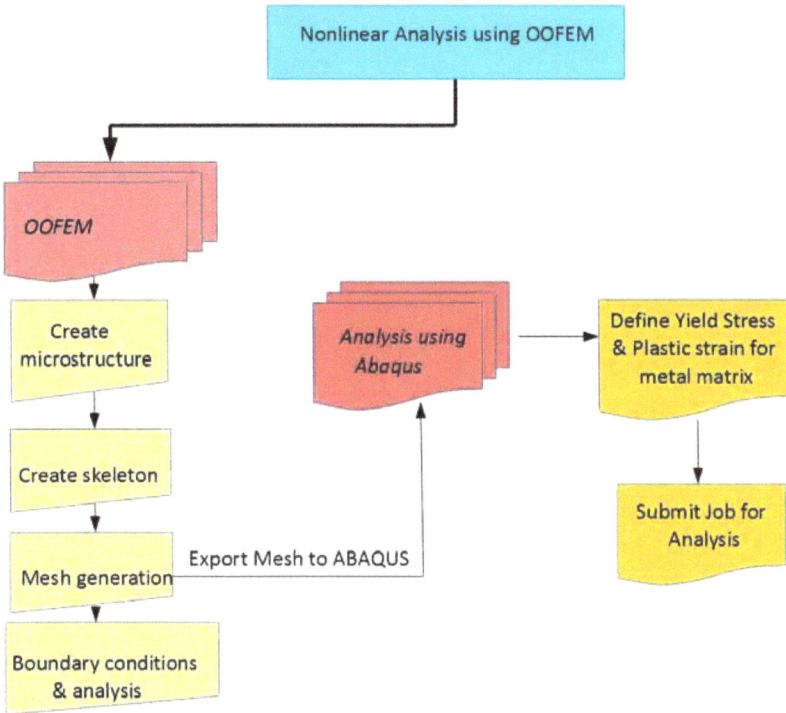

Fig. (4). Nonlinear analysis using OOFEM.

OOFEM Analysis for Ni-Alumina Composite

The two-dimensional finite element analysis using OOFEM has been applied to analyze Ni-Alumina composites having particulate and interpenetrating phase structures [30]. The effective CTE of composites has been predicted using FEA and the predicted results are compared with the reported experimental values [6]. Three composite samples, containing 40, 60, and 80 vol% alumina, each from the 'particle reinforced' and 'interpenetrating phase' are analyzed. In order to model the microstructure of the composites, two pixel groups 'Ni' and 'Alumina' were created. The pixels belonging to these two phases were added from the SEM images reported by the author [5]. The properties obtained from the literature [5, 6] were added to the microstructure.

Once the material microstructure was modeled successfully, the skeleton for the finite element mesh was generated. A homogeneity index above 98% was obtained for all skeletons using various skeleton modification operations. The skeletons were meshed using isoparametric 2-noded edge elements, isoparametric 3-noded triangular elements, and isoparametric 4-noded quadrilateral elements. After completion of the mesh generation process, the boundary conditions are assigned to solve for the thermal strain and the effective CTEs of the composites. In order to conduct the thermal expansion analysis, the Dirichlet boundary conditions were applied considering the temperature and displacement fields. The conjugate gradient method will be used to solve the force balance equation with a linear driver.

Thermal-expansion of Interpenetrating Phase Composites

The effective CTEs of 40 vol%, 60 vol% and 80 vol% interpenetrating phase composites computed using OOF have been compared with the experimental values in Fig. (5). For the OOF, predictions are in good agreement with experimental values for 40 vol% and 80 vol% composites and for 60 vol%, the OOF predicted values are slightly lower (Fig. 5). A steeper rise in the measured CTE compared to that of OOF, from 200°C to 500°C could be attributed to the relaxation in elastic stresses due to a sharp mismatch in CTE of constituents. The elastic modulus decreases with temperature and hence yields strength. This causes higher plastic deformation in a high-temperature range [30].

THREE-DIMENSIONAL FINITE ELEMENT ANALYSIS

In order to conduct the 3D finite element analysis of composites, the representative volume elements are generated using the homogenization approach. The macroscopic properties of all composite materials are strongly influenced and determined by the properties of the micro constituents and phenomena on the microscale. However, predicting the microstructure is very difficult. The exact microstructure of composites is not known and therefore, some statistical assumption has to be made [29].

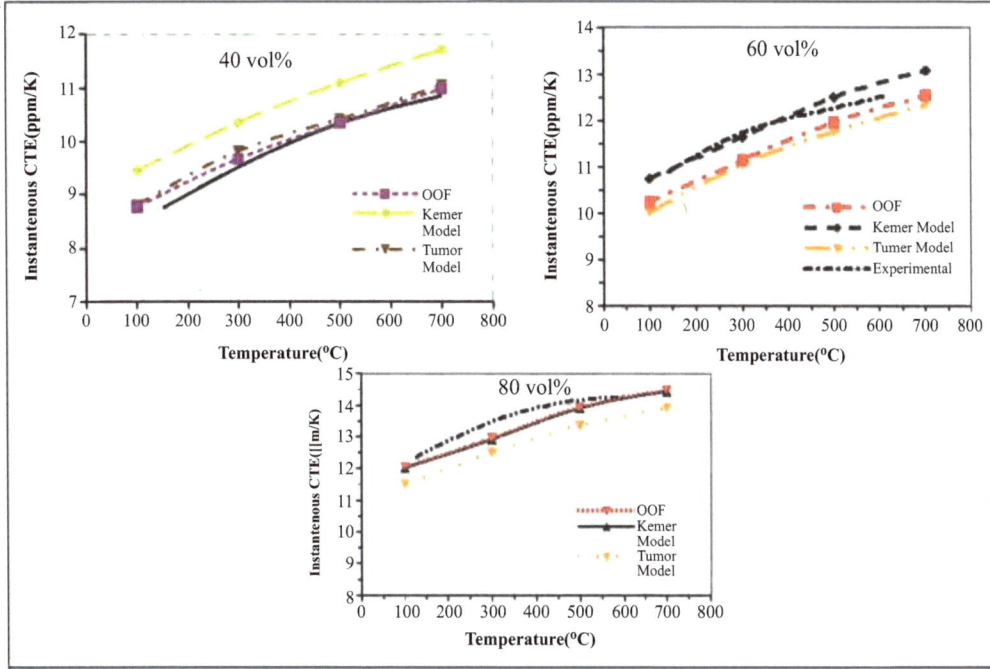

Fig. (5). Comparison of effective CTE of IPCs using OOFEM, Kerner Model, Turner Model with experimental values **(a)** 40 vol% IPC **(b)** 60 vol% IPC **(c)** 80 vol% IPC [30].

In order to find out the macroscopic properties, a homogenization process approach is used. The effective stresses and strains acting on the effective, homogenized sample of material can be obtained from this. The sample of material used for simulation is called a representative volume element (RVE) [30]. The goal of the homogenization approach is to generate data for effective material modeling and to identify the parameters introduced in this material model. The effective material properties represent all macroscopic properties of the micro heterogeneous material.

In order to generate the RVEs of composites, the random distribution of inclusion particles in a cube representing the matrix phase can be made using random sequential adsorption algorithm. For the inclusion phase, the choice of various shaped particles like ellipsoid, icosahedron, sphero-cylinder, prism and sphere have been made by the researcher (Fig. **6**). The inclusion phase is defined using the volume fraction or the mass fraction in the composite. The next task is to apply the boundary conditions. Two types of boundary conditions, either the Periodic or the Dirichlet can be used. Periodic boundary conditions can be implemented only on the periodic microstructure.

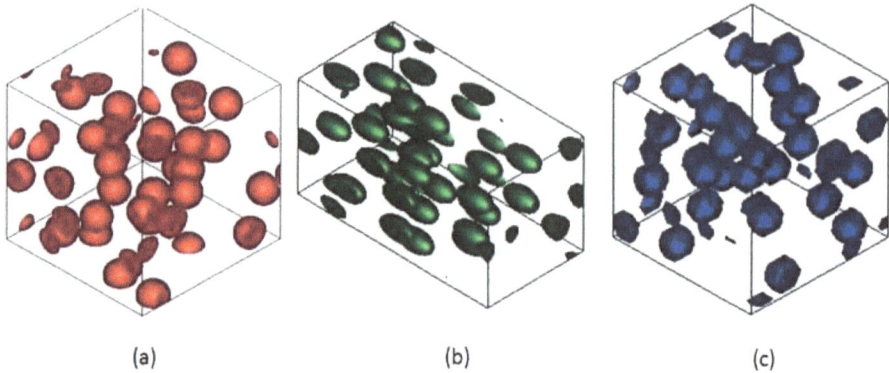

(a)　　　　　　　　　　　　(b)　　　　　　　　　　　　(c)

Fig. (6). Microstructure modeling in 3D-FEM **(a)** Spherical inclusions **(b)** Ellipsoid inclusion **(c)** Icosahedron [33].

Apart from predicting the effective material properties, the local stress, strain, plastic deformation and other fields can be studied by plotting filled contours [31]. These contours provide vital information about the effect of microstructural parameters, such as the shape, numbers, orientation, aspect ratio and the distribution of particles, on the effective material's properties and failure criteria.

3D Finite Element Modeling of Al–B$_4$C Composites

In this section, three-dimensional finite element modeling has been applied to model the deformation behavior of boron carbide reinforced aluminum matrix composite [20]. Since the reinforced particles have a complex shape, therefore, icosahedron particles have been used for microstructure generation. The authors have studied the 4, 8 and 12 vol% particle reinforced composites considering the random orientation and periodic distribution of icosahedron particles with size varying from 0.01 to 0.15 μm, The random sequential adsorption algorithm was used for the particle distribution (Fig. **7**, pictorial representation). A perfect bonding at the interface of matrix and particles was assumed.

Meshing and boundary conditions

The modeled microstructure was meshed using the commercial software Abaqus using the 10-node tetrahedral elements (C3D10 M) with hourglass control. Tetrahedral elements are preferred for automatic mesh generation. The 2nd order elements can correctly capture the stress distribution especially in case of elastoplastic deformation of the material. The optimum RVE size was determined by performing a mesh and RVE sensitivity analysis to attain objective mesh-independent results. For this, a convergence study was performed considering the RVE side length of 0.6, 0.8, 1, 1.2 and 1.4 mm. Similarly mesh seed size of 0.03, 0.04, 0.05 and 0.06 was used for the mesh sensitivity analysis. The aluminum metal matrix was modeled as the linear elastic material as well as the elasto-plastic material with temperature-dependent properties. From the mesh and RVE convergence study, the RVE of 1 mm side length and mesh of seed size of 0.05 was obtained.

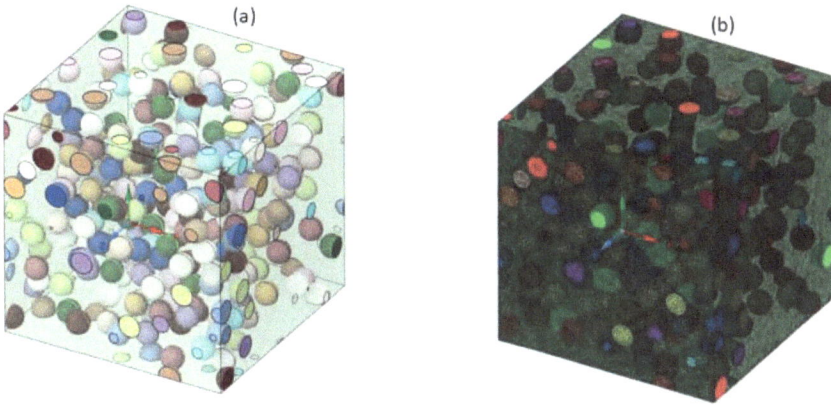

Fig. (7). Three-dimensional RVE generation and meshing of 12 vol% composite microstructures.

Effective Elastic Modulus of Composites

A mismatch between the coefficients of thermal expansion gives rise to thermal residual stresses. The influence of residual stresses was studied on the effective elastic modulus [20]. First, the thermal residual stresses were generated by cooling the composites from processing temperature (600°C) to room temperature, and then the RVEs with existing residual stresses were subjected to periodic boundary conditions, and the effective elastic moduli were predicted. (Fig. **8**) shows the effective elastic modulus of 4, 8 and 12 vol% composites considering linear elastic as well as elastoplastic deformation behavior of aluminum. The predicted results show that the effective elastic modulus is high for linear elastic deformation. This is due to the fact that the matrix phase experiences a significant plastic deformation. The predicted elastic moduli are not much influenced by residual stresses for linear elastic deformation behavior, though, it decreases by 0.35%, 0.78%, and 3.5% in case of elastoplastic deformation behavior. The predicted elastic modulus considering elastoplastic deformation matches well with the experimentally measured values. The presence of voids can also influence the effective modulus.

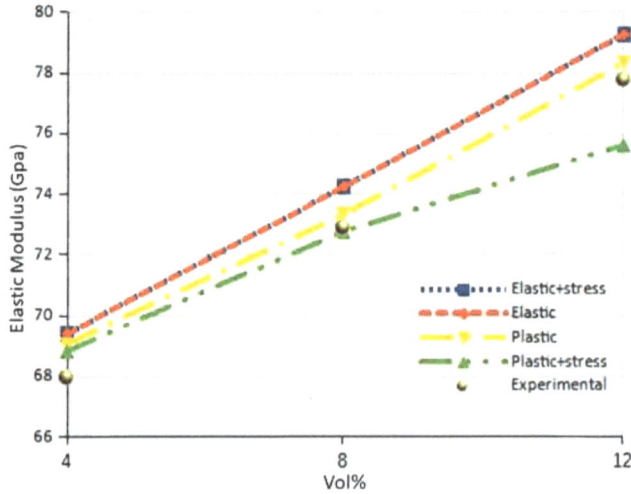

Fig. (8). Effective elastic modulus predicted using 3D FEA.

Coefficient of Thermal Expansion of Composites

Thermal expansion behavior of MMCs is another important design parameter for various application. Densely packed ceramic-reinforced MMCs are used in various electronic applications due to their stability at high temperatures. In this section, the linear coefficient of thermal expansion is measured using 3D FEA [20]. Al matrix was modeled as elastic as well as elastoplastic deformation behavior with temperature-dependent properties. The RVEs with existing residual stresses were heated from room temperature to 500°C and the volumetric average of thermal strain was computed. Thermal strain per unit temperature rise yielded the linear CTE.

CTE values for 4 vol% composite show that composites with thermal residual stresses exhibit lower CTE than that of the composites with no residual stresses. The maximum difference between predicted CTEs with and without thermal residual stresses for all three RVEs is observed under 100°C temperature only. Therefore, the effect of residual stress subsided as the composites were heated. Fig. (9) shows that the FE-FE-predicted results are in good agreement with the measured experimental values.

The Effect of Voids

During the processing of MMCs, the voids get generated due to the large difference in CTS of constituents and the presence of voids can influence the effective properties. The authors measured the void volume fraction of composites using the SEM images of composites. The image processing in OOF2 software was used to capture void volume fraction. A void volume fraction of 0.65%, 0.56%, and 1.63% was obtained for 4 vol%, 8 vol% and 12 vol% composite, respectively. These voids can grow further and nucleate as the composite is subjected to tensile stresses during the uniaxial tensile test. The void growth and nucleation were simulated using the Gurson-Tveerguard constitutive model.

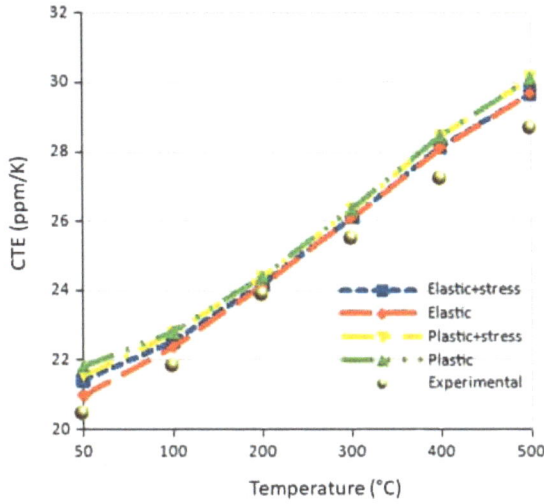

Fig. (9). Effective CTEs of composites with and without thermal residual stresses, for elastic and elastoplastic deformation behavior for 4 vol% composite.

Next, the effective composite properties were measured by modeling the RVEs with voids incorporating the void volume fraction obtained from the GTN constitutive model. A very low elastic modulus (0.00001 GPa) and high a Poisson's ratio (0.48) were assigned to the void phase in RVEs to simulate it as a very soft material. The aluminum matrix was modeled as elastoplastic material and the RVEs were assumed stress-free at room temperature. The predicted effective modulus and the effective CTE both decrease. The predicted effective CTE decreased at room temperature. At high temperatures due to the increased thermal expansion of the metal phase, the voids get filled up and the effective thermal strain reduces. In case of 4 and 8 vol% composites, the predicted elastic modulus is in good agreement with the experimental values, however, in case of 12 vol% composite, it under predicts.

CONCLUSION

• The thermal and thermo-mechanical behavior of particle-reinforced metal matrix composites is influenced by microstructural morphologies, *i.e.*, shape, size, orientation, distribution, and concentration of reinforced particles, presence of voids, presence of thermal residual stresses, temperature and elasto-plastic deformation of the metal matrix. A realistic distribution of local stress and strain can be obtained from 3D FEM which provides vital information about the deformation behavior of composites. However, the computation cost and time increase manifold as compared to that of 2D FEM. FE analysis revealed that the local stress distribution, the effective elastic moduli and the effective CTE are significantly influenced by the metal matrix behavior: linear elastic or elasto-plastic deformation.

- The analytical models cannot take into account the effects induced due to microstructural morphologies, thermal residual stresses, elasto-plastic deformation and the presence of voids. Therefore the precise prediction of effective material properties using analytical models is difficult. However, in the absence of computational resources, the effective CTE and the effective elastic moduli can be predicted using Schapery and Hashin's bounds, respectively.

- We conclude that finite element modeling can be used as an effective tool for modeling the microstructure of composites. It is observed that the thermo-mechanical behavior of composites is significantly influenced by processing conditions, microstructure, and thermal residual stresses.

REFERENCES

[1]   Trumper, R.L. Metal matrix composites: Application and prospects. *Met. Mater.*, **1987**, *3*, 662-667.

[2]   Wildan, M.; Edrees, H.J.; Hendry, A. Ceramic matrix composites of zirconia reinforced with metal particles. *Mater. Chem. Phys.*, **2002**, *75*(1-3), 276-283.
[http://dx.doi.org/10.1016/S0254-0584(02)00076-7]

[3]   Aldrich, D.E.; Fan, Z. Microstructural characterisation of interpenetrating nickel/alumina composites. *Mater. Charact.*, **2001**, *47*(3-4), 167-173.
[http://dx.doi.org/10.1016/S1044-5803(01)00183-8]

[4]   Kawai, C.; Park, J-J. Jin-Joo Park, Mechanical and thermal properties of Al-Si$_3$N$_4$ composites fabricated by the infiltration of molten Al into a porous Si$_3$N$_4$ ceramic with network. *J. Mater. Sci. Lett.*, **2001**, *20*(4), 385-388.
[http://dx.doi.org/10.1023/A:1006735130117]

[5]   Bruck, H.A.; Rabin, B.H. Evaluating microstructural and damage effects in rule-of-mixtures predictions of the mechanical properties of Ni-Al$_2$O$_3$ composites. *J. Mater. Sci.*, **1999**, *34*(9), 2241-2251.
[http://dx.doi.org/10.1023/A:1004509220648]

[6]   Bruck, H.A.; Rabin, B.H. Evaluation of rule-of-mixtures predictions of thermal expansion in powder-processed Ni-Al$_2$O$_3$ composites. *J. Am. Ceram. Soc.*, **1999**, *82*(10), 2927-2930.
[http://dx.doi.org/10.1111/j.1151-2916.1999.tb02182.x]

[7]   Huda, M.D.; Hashmi, M.S.J.; El-Baradie, M.A. MMCs: Materials, manufacturing and mechanical properties. *Key Eng. Mater.*, **1995**, *104-107*, 37-64.
[http://dx.doi.org/10.4028/www.scientific.net/KEM.104-107.37]

[8]   Wang, X.J.; Wang, N.Z.; Wang, L.Y.; Hu, X.S.; Wu, K.; Wang, Y.Q.; Huang, Y.D. Processing, microstructure and mechanical properties of micro-SiC particles reinforced magnesium matrix composites fabricated by stir casting assisted by ultrasonic treatment processing. *Mater. Des.*, **2014**, *57*, 638-645.

[http://dx.doi.org/10.1016/j.matdes.2014.01.022]

[9]   Kevorkijan, V. Mg AZ80/SiC composite bars fabricated by infiltration of porous ceramic preforms. *Metall. Mater. Trans., A Phys. Metall. Mater. Sci.,* **2004**, *35*(2), 707-715.
        [http://dx.doi.org/10.1007/s11661-004-0381-4]

[10]   Bahraini, M.; Schlenther, E.; Kriegesmann, J.; Graule, T.; Kuebler, J. Influence of atmosphere and carbon contamination on activated pressureless infiltration of alumina–steel composites. *Compos., Part A Appl. Sci. Manuf.,* **2010**, *41*(10), 1511-1515.
        [http://dx.doi.org/10.1016/j.compositesa.2010.06.013]

[11]   Sutherland, T.J.; Gibeling, J.C. Fatigue crack propagation in dispersion strengthened aluminium alloy metal matrix composite (MMC). *Met. Powder Rep.,* **1992**, *47*, 57-58.
        [http://dx.doi.org/10.1016/0026-0657(92)91512-I]

[12]   Gopalakrishnan, S.; Murugan, N. Production and wear characterisation of AA 6061 matrix titanium carbide particulate reinforced composite by enhanced stir casting method. *Compos., Part B Eng.,* **2012**, *43*(2), 302-308.
        [http://dx.doi.org/10.1016/j.compositesb.2011.08.049]

[13]   Sajjadi, S.A.; Torabi Parizi, M.; Ezatpour, H.R.; Sedghi, A. Fabrication of A356 composite reinforced with micro and nano Al2O3 particles by a developed compocasting method and study of its properties. *J. Alloys Compd.,* **2012**, *511*(1), 226-231.
        [http://dx.doi.org/10.1016/j.jallcom.2011.08.105]

[14]   Dobrzański, L.A.; Włodarczyk, A.; Adamiak, M. The structure and properties of PM composite materials based on EN AW-2124 aluminum alloy reinforced with the BN or Al_2O_3 ceramic particles. *J. Mater. Process. Technol.,* **2006**, *175*(1-3), 186-191.
        [http://dx.doi.org/10.1016/j.jmatprotec.2005.04.031]

[15]   Torralba, J.M.; da Costa, C.E.; Velasco, F. P/M aluminum matrix composites: An overview. *J. Mater. Process. Technol.,* **2003**, *133*(1-2), 203-206.
        [http://dx.doi.org/10.1016/S0924-0136(02)00234-0]

[16]   Mazen, A.A.; Ahmed, A.Y. Mechanical behavior of Al-Al_2O_3 MMC manufactured by PM techniques part I—scheme I processing parameters. *J. Mater. Eng. Perform.,* **1998**, *7*(3), 393-401.
        [http://dx.doi.org/10.1361/105994998770347846]

[17]   Rahimian, M.; Ehsani, N.; Parvin, N.; Baharvandi, H. The effect of particle size, sintering temperature and sintering time on the properties of Al–Al_2O_3 composites, made by powder metallurgy. *J. Mater. Process. Technol.,* **2009**, *209*(14), 5387-5393.
        [http://dx.doi.org/10.1016/j.jmatprotec.2009.04.007]

[18]   Rahimian, M.; Parvin, N.; Ehsani, N. Investigation of particle size and amount of alumina on microstructure and mechanical properties of Al matrix composite made by powder metallurgy, *Materials Science and Engineering A. Structures,* **2010**, *527*, 1031-1038.

[19]   Lloyd, D.J. Particle reinforced aluminium and magnesium matrix composites. *Int. Mater. Rev.,* **1994**, *39*(1), 1-23.
        [http://dx.doi.org/10.1179/imr.1994.39.1.1]

[20]   Sharma, N.K.; Misra, R.K.; Sharma, S. Finite element modeling of effective thermomechanical properties of Al–B4C metal matrix composites. *J. Mater. Sci.,* **2017**, *52*(3), 1416-1431.
        [http://dx.doi.org/10.1007/s10853-016-0435-1]

[21]   Hashin, Z. The moduli of an elastic solid, containing spherical particles of another elastic material *Proceedings IUTAM Symposium on Non-homogeneity in Elasticity and Plasticity,* Warsaw, Poland, **1959**, pp. 463-478.

[22]   Hashin, Z. The elastic moduli of heterogeneous materials. *J. Appl. Mech.,* **1962**, *29*(1), 143-150.
        [http://dx.doi.org/10.1115/1.3636446]

[23]   Hashin, Z. Theory of mechanical behavior of heterogeneous media. *Appl. Mech. Rev.,* **1964**, *17*, 1-9.

[24]   Hashin, Z.; Shtrikman, S. A variational approach to the theory of the elastic behaviour of multiphase materials. *J. Mech. Phys. Solids,* **1963**, *11*(2), 127-140.
[http://dx.doi.org/10.1016/0022-5096(63)90060-7]

[25]   Schapery, R.A. Thermal expansion coefficients of composite materials based on energy principles. *J. Compos. Mater.,* **1968**, *2*(3), 380-404.
[http://dx.doi.org/10.1177/002199836800200308]

[26]   Turner, P.S. Thermal-expansion stresses in reinforced plastics. *J. Res. Natl. Bur. Stand.,* **1946**, *37*(4), 239-250.
[http://dx.doi.org/10.6028/jres.037.015]

[27]   Kerner, E.H. The elastic and thermo-elastic properties of composite media. *Proc. Phys. Soc. B,* **1956**, *69*(8), 808-813.
[http://dx.doi.org/10.1088/0370-1301/69/8/305]

[28]   Neeraj, Kr. Sharma, RK Misra & Satpal Sharma, Experimental characterization and numerical modeling of thermomechanical. *Ceram. Int.,* **2017**, *43*, 513-522.

[29]   Neeraj, Kr. Modeling of thermal expansion behavior of densely packed Al/SiC composites. *International Journal of Solids and Structures,* **2016**, *77-88*, 102-103.

[30]   Neeraj Kumar Sharma, R.K. Misra & Satpal sharma, Thermal expansion behavior of Ni-Al$_2$O$_3$ composites with particulate and interpenetrating phase structures: An analysis using object oriented finite element method. *Comput. Mater. Sci.,* **2014**, *90C*, 130-136.
[http://dx.doi.org/10.1016/j.commatsci.2014.04.008]

[31]   Neeraj Kumar Sharma, R.K. Misra & Satpal sharma, 3D micromechanical analysis of thermal expansion behavior of Al/Al$_2$O$_3$ composites. *Comput. Mater. Sci.,* **2016**, *115*, 192-201.
[http://dx.doi.org/10.1016/j.commatsci.2015.12.051]

[32]   Reid, A.C.E.; Langer, S.A.; Lua, R.C.; Coffman, V.R.; Haan, S.I.; García, R.E. Image-based finite element mesh construction for material microstructures. *Comput. Mater. Sci.,* **2008**, *43*(4), 989-999.
[http://dx.doi.org/10.1016/j.commatsci.2008.02.016]

[33]   Sharma, ; Kumar, N. Synthesis and Characterization of Boron Carbide Reinforced MMCs By FEA. *Gautam Buddha University.,* **2017**.
[http://hdl.handle.net/10603/370374]

CHAPTER 3

A Critical Review of Fabrication Techniques and Possible Interfacial Reactions of Silicon Carbide Reinforced Aluminium Metal Matrix Composites

Jatinder Kumar[1,*], **Dilbag Singh**[2] and **Nirmal S. Kalsi**[3]

[1] *Department of Mechanical Engineering, Modern Group of Colleges, Mukerian, Punjab, India*

[2] *Department of Mechanical Engineering, Sardar Beant Singh State University, Gurdaspur, India*

[3] *Department of Mechanical Engineering, Shri Vishwakarma Skill University, Palwal,Haryana, India*

Abstract: In this review article, the current status of and recent developments in fabrication techniques for all types of Silicon Carbide reinforced Aluminium Metal Matrix Composites (SiC-AMMCs) have been elaborately discussed. The comparative studies on fabrication methods have also been reported in this article. Furthermore, the possible interfacial reactions between aluminium and silicon carbide that have been presented by researchers were also explored and their causes and remedies have been discussed. The entire discussion in this review article reveals that liquid fabrication processes (especially stir casting) are used effectively for mass production, intricate shapes, a variety of products, nano-composites, *etc*. The solid-state processes are performed below the melting temperature of matrices, resulting in the least possible interfacial reactions leading to unwanted compounds' formation. The literature on interfacial reactions reveals that the Al_4C_3 compound is mostly formed as a result of the reactions between aluminium and silicon carbide and exhibits a deleterious effect on the composite properties.

Keywords: Aluminium, Fabrication, Interfacial reactions, Metal matrix composite, Review, Silicon carbide.

INTRODUCTION

A few decades ago, conventional materials were used to produce aerospace, automotive, marines, defence, sports equipment, *etc*. but these materials have high density, low corrosion and wear resistance, low strength to weight ratio and high

---
* **Corresponding author Jatinder Kumar:** Department of Mechanical Engineering, Modern Group of Colleges, Mukerian, Punjab, India; E-mail: jatinderbahal@gmail.com

Virat Khanna, Prianka Sharma & Santosh Kumar (Eds.)

cost. At present, researchers are focusing their attention on overcoming these shortcomings by replacing conventional materials with composite materials [1]. The superior qualities of the composite, which replace conventional monolithic materials and their alloys, extend their applications in automobile, defense, marine, sports and recreation industries [2, 3].

Composite is a combination of matrix and reinforcements. The matrix is a base or mother metal/alloy, in which reinforcement particles are embedded. The matrix is always a light metal, which supports the reinforced particles within the grain structure and the newly developed strong grain structure can hold more external loads. So, it is desirable to have strong interfacial bonding and wettability between the matrix and reinforced particles to get the material with improved properties [4].

Matrix Materials

Light metals, such as aluminium, titanium, magnesium, and zinc and their alloys are used as a matrix in MMCs. Other materials like copper, nickel, lead, iron, and tungsten are also used in some particular applications. Moreover, cobalt and Co-Ni alloys are used as a base matrix in particular applications, where the materials are subjected to high temperature [5 - 9]. In the last one-two decades, the use of Aluminium Matrix Composites (AMCs) has increased rapidly, particularly in automotive, recreation and aerospace applications due to their lower density, lower coefficient of thermal expansion, higher wear resistance, lower costs, availability, higher strength/weight ratio, and higher resistance to corrosion and lower processing temperature requirement than competitive materials [10].

Reinforcement Materials

Reinforcements in composite materials are second-phase materials, which are added in the matrix. A reinforcement prevents deformation and improves desirable properties such as strength, hardness, stiffness, wear resistance, *etc.* of the base metal/alloy.

Before 1970, mono-filaments of tungsten (W), Be, Al_2O_3, *etc.* were used as reinforcements. At the beginning of 1970, the interest was directed towards the use of SiC, Al_2O_3 and C multi-filaments and whiskers [11]. In the last decade, the use of particulate reinforcements has attracted the interest of researchers due to their availability at low cost, well-established fabrication processes, *etc.* Also, the discontinuous particulate reinforcements-based composite exhibits the least interfacial reactions, superior properties, high strength-to-weight ratio, relative ductility, *etc* [12, 13].

Particulate-reinforced materials can generally be classified as, (1) ceramics/metallic materials, (2) agro waste materials, and (3) industrial waste materials. The ceramics mostly used as reinforcement are borides, carbides, nitrides, and oxides of the metallic materials *e.g.* SiC, B_4C, Al_2O_3, Ni_3Al, Al_4N_3, TiB_2, and $ZrSiO_4$ [14, 15]. Apart from these, other reinforcements are MWCNT, WC, diamond, AlO, Si_3N_4, *etc.* Among these materials, SiC particulate ceramic is widely used in different composites. The addition of SiC in the aluminium matrix enhances wear resistance, hardness value and tensile strength, but with the loss of machinability and toughness resistance [16]. Industrial waste materials are the second category of reinforcements, which are used in AMCs. These materials include fly ash (FA) and red mud. Several authors have reported their extensive work regarding the utilization of fly ash as reinforcement in mono as well as hybrid MMCs [17 - 19]. However, limited research work is reported on the use of red mud as reinforcement in the AMCs. The third category of reinforcement is related to agro waste materials. These materials include BLA (bamboos leaf-ash), eggshell waste, RHA (rice husk-ash), sugarcane bagasse-ash, maize stalks waste-ash, CCA (corn cob-ash), *etc.* No doubt, the single agro waste reinforced AMCs have improved properties than the unreinforced ones, but exhibit inferior properties than synthetic reinforced composites. The main motive of using such materials is to reduce materials cost by maintaining the desired level of essential properties [20 - 24].

Of these, the Silicon Carbide reinforced Aluminium Metal Matrix Composites (SiC-AMMCs) are ones being the most popular composite materials used these days. Such lightweight composites are widely used in automotive and aircraft where the weight is the main concern. With the addition of SiC (in different ratios) in the aluminium matrix, the properties like hardness, tensile strength, wear resistance, density, *etc.* are increased, but with the loss of ductility and toughness [16]. Moreover, these composites exhibit heterogeneous and anisotropic properties [25, 26], which are not desirable because of internal voids, irregular grain structure, variable density, and porosity that occur in such composites. However, the machinability of SiC-AMCs can be improved by using secondary reinforcements (like Graphite) with SiC in aluminium matrices [27]. Also, the secondary processing of novel composites (like extrusion, rolling, forging, and heat treatment processes) can remove these defects. A small amount of Mg may be added to improve the interfacial bonding between the constituents [28]. Also, the coating of copper and CNT on SiC improves the wettability of particles and strengthens metallurgical bonding [29]. So the basic awareness of Al-SiC composite materials and their properties can provide a solid foundation for widespread industrial applications.

To summarize, the applications of SiC-MMCs are increasing day by day. These materials possess superior properties than monolithic materials. The interfacial reaction between aluminium and SiC during fabrication results in the formation of uneven compounds. These compounds produce a deleterious effect on several properties of the novel composites. So from the application point of view, the knowledge of appropriate fabrication techniques should be required to produce composite materials with superior quality at minimum cost. Generally, the mechanical properties *viz* the strength, and hardness of MMCs vary proportionally w.r.t. the contents of SiC [30, 31]. However, with the addition of secondary reinforcements, the mechanical properties are influenced significantly. In this review article, various fabrication techniques, used for the processing of SiC-AMCs are elaborately discussed. Also, comparative studies on various processing routes are presented. The articles also explored the possible interfacial reactions between constituents during processing and discussed possible remedies to prevent the formation of unwanted compounds.

SCOPE OF THE REVIEW ARTICLE

Abundant research on aluminium matrix composite has already been reported and many review articles on fabrication, properties, characterization and machining of AMCs have been presented by the various researchers [2, 11, 12, 27]. This review article focused on the specific area in relation to fabrication methods that have been used by researchers and possible interfacial reactions of SiC-AMMCs.

In the first section of this article, different fabrication techniques used by numerous researchers in the processing of SiC-AMMCs have been elaborately reviewed and discussed. The comparative study of these techniques has also been explored. The second section describes possible interfacial reactions between silicon carbide and aluminium matrices during fabrication and secondary phase formations resulting from these reactions. The outcomes of these reactions and possible preventive measures used to overcome the possibility of unwanted compound formations are also presented and elaborately discussed. The concluding section explained the comparative study on fabrication techniques used for SiC-AMMCs, the formation of secondary phases during the fabrication of these composites, and their preventive measures.

FABRICATION OF SIC-AMMC

The fabrication technique is the main concern while producing composite materials because each fabrication technique alters the composite properties in its own ways [32]. Generally, the MMCs processing techniques are classified into two categories; *Ex-situ* and *In-situ*. In case of the *In-situ* technique, the oxides and carbides of constituents are formed within the molten matrix due to chemical reactions, which take place among the ingredients. For example, Si and C react with each other within the aluminium matrix melt to form SiC [33]. On the other hand, in *Ex-situ*, the already formed reinforcements are introduced into the molten matrix. The *Ex-situ* processes are further classified according to the state of constituent particles, such as (1) Liquid states processing, (2) Solid states processing, (3) Semi-solid states processing, and (4) Gaseous states processing [7, 19, 34 - 37]. Fig. (**1**) indicates the block diagram of techniques used by researchers for the production of silicon carbide reinforced mono and hybrid aluminium matrix composites.

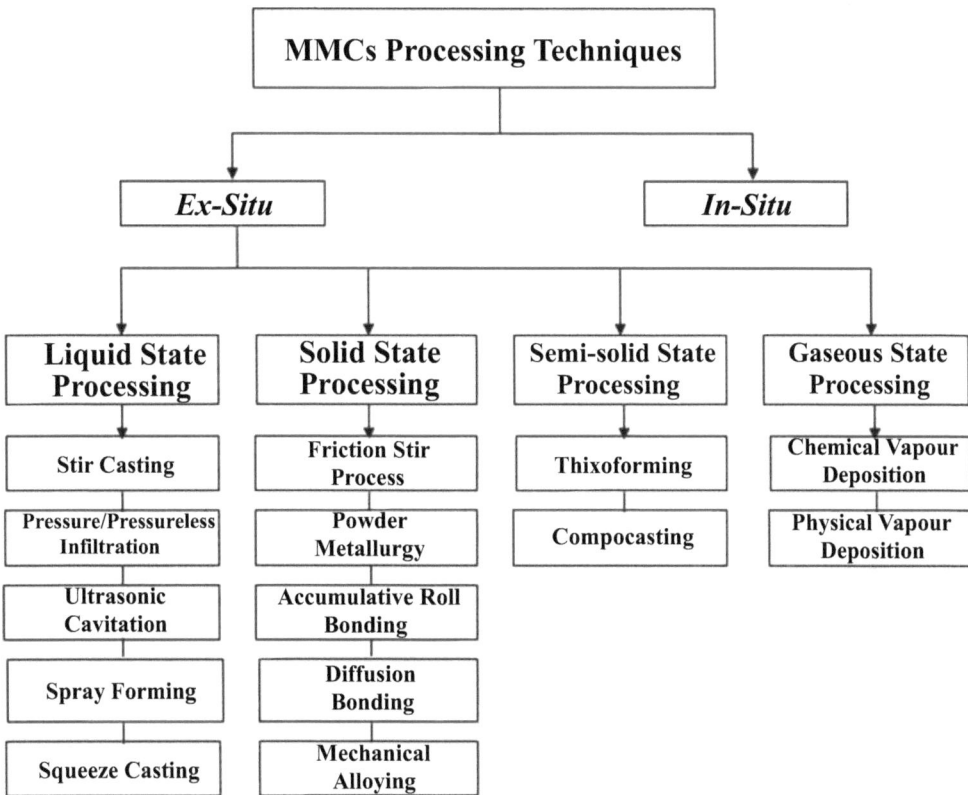

Fig. (1). Block Diagram of SiC-MMC Fabrication Methods.

Liquid States Processing

Liquid state processing is widely used for the production of SiC-AMCs. In this technique, initially, the liquid phase of the matrix is produced then solid reinforcement(s) are introduced in it. These techniques possess specific features like a simple setup, less time-consuming, cheaper and easily understandable [38].

But the poor wettability of the SiC with the molten aluminium matrix is the drawback of this process [35]. Also, the requirement of using high temperatures for composites triggers interface reactions between constituents. However, these drawbacks can be eliminated with the addition of Mg in the melt or by using CNT/Cu/Si coatings on the SiC particles that enhance the wettability [39, 67]. Numerous researchers have reported their valuable work in the production of Al-SiC composites using liquid fabrication techniques as discussed as follows.

Stir Casting Process

The production of SiC-AMCs through the stir casting process attracts great interest in industries as it is relatively simple, inexpensive and most effectively used to produce such composites. This technique is best suitable for mass production and large-size castings for a variety of matrix materials. The proper selection of process parameters ensures regular distribution of undamaged reinforcement constituents within the matrix grain structure [27, 40 - 44]. Generally, in this technique, the reinforcement(s) (ceramics, metallic, agro wastes or industrial wastes) is/are mixed with liquid matrix metal and stirred mechanically under controlled condition [45]. The schematic setup of the stir-casting process is shown in Fig. (2).

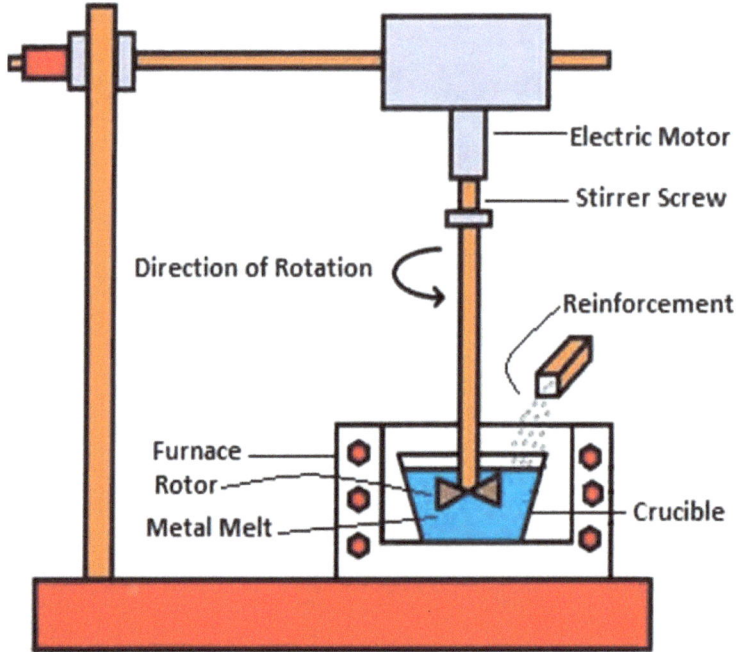

Fig. (2). Illustration of stir casting process.

To achieve composites with optimum properties, the reinforcements should be dispersed uniformly within the matrix, and should possess high wettability and interface bonding along with minimum porosity level and reactions between constituent particles. The factors affecting the performance of this technique include the stirring time and speed, the shape, size and material of the stirrer and its position in molten metal, preheating temperature and the time of reinforcements, stirrer, and mould [46 - 48]. Many researchers have followed stir casting technique to develop SiC-AMCs and a few of them are discussed as follows:

Rahman and Rashed [31] successfully stir-cast Al-SiC composite materials with varying weight fractions of SiC (*viz.* 0, 5, 10 and 20%). They melted the aluminium alloy at 750°C using an electrical resistance furnace, followed by the incorporation of SiC (preheated at 800°C) and stirred at 500 rpm for 10 minutes using a graphite stirrer. In the end, the semi-solid slurry was introduced into the mould at 670°C and allowed to solidify at room temperature. Dwivedi *et al.* [36] have cast Al-(5-15%) SiC composites with the help of an electromagnetic stir casting process. The steps involved the melting of the A356 alloy at 650°C in a graphite crucible using a muffle furnace, followed by the addition of SiC (25µm size) particles in the molten melt. Then they stirred the mixture using a three-phase induction motor at 210 rpm for 7 minutes under the envelope of the electromagnetic field. The mixture was then discharged in a preheated mould. Vanarotti *et al.* [39] prepared similar composites using the stir-casting route. The authors used copper-coated (5, 10, and 15 wt.%) SiC particles to enhance metallurgical bonding between A356 and SiC. They melted the base matrix at 450°C temperature for one hour, added SiC particles in the melt and stirred for about 7 minutes at an average speed of 350 rpm and then poured the slurry into the preheated mould. Tonythomas *et al.* [45] developed new feeding and stirring mechanisms to develop LM6-SiC_p AMCs. The study focused on the design and fabrication of different stirrers and feeders to ensure uniform and controlled spray of SiC particulates in the melts. The results of the newly designed feeder and stirrer ensured the uniform dispersion of SiC particles in the matrix grain structure. In another research, Palanikumar *et al.* [49] cast A356-20SiC_p composite and optimized machining parameters in terms of minimum surface roughness. They cleaned the A356 ingot with acetone and melted at 800°C. The flux and hexachloroethane were added to remove impurities and degassing. A small amount of Mg was also introduced to enhance the wetting property and interface bonding of constituents. SiC particles (56-185µm) were heated at 600°C and poured into the melt. The melt was stirred with a mechanical stirrer, poured into the preheated mould and allowed to cool to get the desired shape. Lokesh and Mallikarjun [50] have developed SiC and Gr-reinforced mono/hybrid Al6061 alloy matrix composites. During fabrication, SiC varied from 2 to 10 wt.%. The

Gr particles varied from 2 to 8 wt.% for mono composites and 3 wt.% (constant) for hybrid composites. Kumar *et al.* [51] produced Al (LM25) alloys matrix based composites reinforced with silicon carbide (SiC) and molybdenum (Mo) using the stir casting route. SEM results revealed that SiC and Mo dispersed uniformly within the matrix grain structure. Lakshmipathy and Kulendran [52] prepared SiC-reinforced Al7075-T6 alloy matrix composites. Fabrication steps involved melting of 2.7 kg matrix in a graphite crucible at 820°C and stirring with a graphite stirrer at 800 rpm. Then, 300 g SiC particles (heated at 800°C) were introduced into the melt and stirred continuously for 8-12 minutes. In last, the mixed slurry was discharged in the heated green sand mould. A similar fabrication route was followed by Kumar *et al.* [53] for the preparation of aluminium-based LM25/10SiC$_p$/xCr$_p$ hybrid MMC. In another research, Patel *et al.* [54] fabricated SiC$_p$ (5 and 10 wt.%) reinforced aluminium composites by using different particle sizes (50 and 150μm) of SiC. The objective of the research was to analyze the effect of weight fraction and grain size of SiC on the machinability of the base matrix. A similar casting route was adopted by Pawar and Utpat [55] to develop Aluminium-based composites using SiC (2.5-10% with steps of 2.5.) as the reinforced material.

The stir casting is also used economically in the processing of hybrid AMCs, because it provides good particles bonding between aluminium metals/alloys and reinforcements. Padmavathi *et al.* [1] prepared AA6061 alloy-based composites, which were reinforced with SiC (15% constant) and MWCNT (0.5 and 1.0vol.%). Authors melted AA6061 alloy at 750°C and SiC (preheated at 620°C) was introduced in the melt and stirred at an average speed of 450 rpm for 5 minutes using alumina coated mild steel impeller. During stirring, impeller blades were placed at 20mm above the crucible base. Johny-James *et al.* [16] developed SiC and TiB$_2$ reinforced hybrid Al6061-T6 matrix composites. The authors kept a contribution of the reinforcements as 10% (SiC+TiB$_2$) by weight and TiB$_2$ varied from 0 to 5%. The processing steps involved, melting of Al alloy at 750°C temperature using electric furnace. SiC and TiB$_2$ particles were heated at 1000°C and 200°C, respectively and included in the molten matrix. 2g Mg was also added to enhance the wetting property of the constituents and the mixture was stirred at 350rpm for about15 minutes. Then, a semisolid mixture was poured in heated moulds to give the desired shapes. Prasad *et al.* [40] also used a similar route to manufacture Al-SiC-RHA hybrid composite materials with reinforcement (SiC+RHA) composition varying from 0-8%. During their fabrication, both reinforcements were used in an equal mass ratio (1:1). Kumar and Chauhan [42] produced Al7075-10SiC$_p$ composite and Al7075-7SiC$_p$-3Gr$_p$ hybrid composites bottom with a stir casting setup. In another research, double stir casting route was followed by Alaneme *et al.* [43] to develop Al6063 alloy-SiC-RHA hybrid composites. The authors used 5 to 10wt.% reinforcements (SiC+RHA) contents with mixing ratios of SiC with RHA varied as 1:0, 1:1, 1:3, 3:1, 0:1. Rajmohan and Palanikumar [44] prepared aluminium alloy (Al356) based composites using SiC (25µm) and mica (45µm) particles as reinforcing phases, where the composition of mica as 3wt.% (constant) and SiC from 2.5 to 5wt.%. The production steps involved melting of the Al356 alloy in a silica crucible at 750°C, followed by the inclusion of preheated SiC and mica particles in the melt. Then, the mixture was stirred at 500rpm for 5-7 minutes and discharged in a metallic mould. In another research, Muthukrishnan *et al.* [47] produced hybrid Al356 alloy-10%SiC-5%B_4C composite using a conventional stir-casting method. Umanath *et al.* [48] developed stir-cast Aluminium (AA6061-T6) Reinforced SiC and Al_2O_3 particulate hybrid composites. During fabrication, an equal volume fraction of SiC and Al_2O_3 was used with a grain size of 25 and 45µm, respectively. Venkatesan *et al.* [56] followed a similar route during the fabrication of Al (356 alloy)-SiC- B_4C hybrid MMCs. During fabrication, the composition of SiC varied (*viz.* 5, 10 and 15wt.%), while B_4C kept constant(5wt.%). Initially, Al alloy Al356 was cleaned with acetone and melted at 700°C. The impurities, from the base metal alloy, were removed by using coverall powder in the melt. The SiC (10-20µm) and B_4C (30-70µm) were heated at 650°C and added continuously in the melt. Mg was added to improve wetting characteristics. The mixture was stirred at 350rpm for 15 minutes and an inert gas atmosphere was created in the vicinity of the melt in order to prevent the formation of oxides by reacting with atmospheric air. After stirring, the semi-solid mixture was poured into the mould. Shoba *et al.* [57] stir cast Aluminium matrix-based SiC and RHA-reinforced hybrid MMCs. During the experiments, the percent composition of SiC and RHA varied as 0, 2, 4, 6 and 8% by weight with a ratio of 1:1 of both the reinforced materials. Narasimha *et al.* [58] fabricated hybrid MMCs using AlMg1SiCu alloy (Al6061) as a matrix material and SiC (9%) and graphite (3 to 9%) particulates as reinforcement phase with a vortex (stir casting) method. The steps involved melting of Al 6061 alloy to maintain its liquidus temperature. After degassing with hexachloroethane, SiC and Gr particles were heated at 600°C and 1100°C, respectively. Mg was added in a small fraction to increase the wettability of Al 6061 alloy and reinforcements. The total weight of the reinforced particles was added in three steps and stirred before and after the addition of a zirconia-coated steel impeller at 300rpm. The semisolid mixture was again heated to the liquid temperature for 10 minutes and stirred at 500rpm followed by pouring in CI mould at 730°C. Boopathi *et al.* [59] fabricated the Al (Al2024) alloy based AMCs using SiC, fly ash (FA) and SiC+FA with different concentrations of the reinforcements. The MMCs were developed with a two-step stir casting route. Mg was introduced to improve the wettability of reinforcement particles. Alaneme *et al.* [60] used bamboo leaf ash (BLA) as a reinforced material along with SiC in AlMgSi alloy-based HMMCs. During fabrication, aluminium alloy was melted at

750°C, followed by the addition of heated SiC and BLA particulates along with 0.1wt% of Mg. The mixture was stirred continuously at 400rpm for 10 minutes. The mixed slurry was then introduced in a sand mould to give the desired shape. Tiruvenkadam *et al.* [61] used a bottom pouring stir casting process to produce Al(6061)-SiC-ZrO₂-Gr hybrid MMCs. The authors have selected nanoparticles of ZrO₂ and graphite and micro-particles of SiC. During the fabrication of HMMCs, the reinforcement phase varied from 0 to 3% with steps of 0.75%. Radha and Vijayakumar [62] produced Al (AA6061) alloy matrix-based hybrid composites, which were reinforced with nanoparticles of SiC and graphene. The authors kept the constant composition of SiC (10%), while the graphene varied as 0, 0.3, 0.5 and 0.7%. Similar MMCs were stir cast by several authors to evaluate various machining aspects (*viz.* machinability, forces, surface roughness *etc*)., with different Aluminium alloys and percent contributions of SiC and Gr reinforcements [63, 64]. The same route was followed by Rajmohan *et al.* [65] during the processing of Al356 alloy - (5-15%) SiC - (3%) mica hybrid composites. In another research, Fatile *et al.* [66] developed Al alloy (AlMgSi) based hybrid composites reinforced with SiC and corn cob ash (CCA). During fabrication, the authors kept the constant composition of the reinforcements, while the ratio of SiC and CCA varied as 10:0, 9:1, 8:2, 7:3, and 6:4 by weight. Sangeetha *et al.* [67] produced aluminium matrix composites. 10% SiC particles coated with MWCNT using sonication method were used as reinforcement. The results revealed that the composite reinforced with coated SiC exhibited better interfacial bonding and smaller porosity than the uncoated one.

Pressureless/pressure Infiltration Process

The pressureless/pressure infiltration (PI) process is one type of liquid fabrication technique, which involves the infiltration of liquid metal/ alloys in the preformed porous bed of reinforcement particles with or without any external pressure [68]. This is a cost-effective route to fabricate composite materials with a large volume percentage of reinforced particles among others. This process was generally performed under an inert gas atmosphere and the capillarity action between preformed reinforcements and matrix causes infiltration [69, 70]. The effectiveness of this process depends upon capillarity action, a flow of molten matrix over the preformed bed, internal reaction between constituents and solidification of composites [71, 72]. Fig. (**3**) presents the illustration of pressure less/pressure infiltration process.

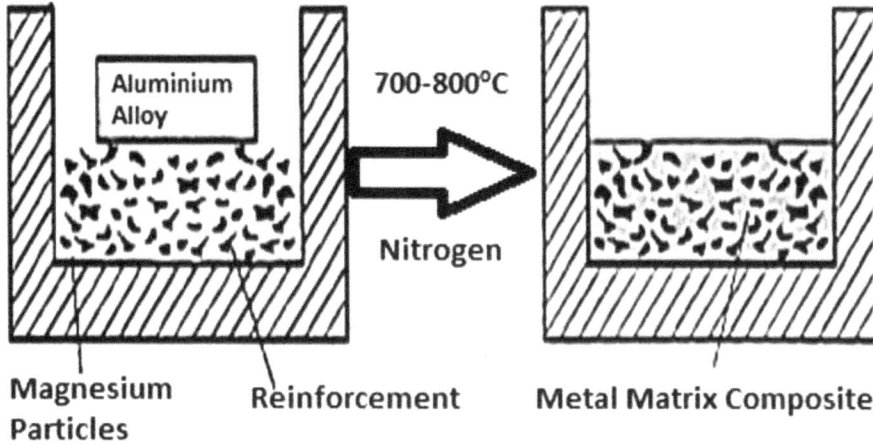

Fig. (3). Illustration of pressure less/pressure infiltration process.

The composite materials containing a high volume percent of reinforcements exhibit significantly improved properties like excellent wear resistance higher modulus value, low coefficient of thermal expansion (CTE), superior strength and hardness, *etc*. The presence of superior properties increases their applications in various sectors such as aerospace, automobile industries electronic industries, *etc* [73]. Yan *et al.* [74] prepared SiC (55-57%) reinforced AMCs. The fabrication steps involved the preparation of a powder bed of densely packed abrasive graded SiC particles and placing loose particles of SiC over the preformed bed, followed by packing with vibrations for 10 minutes to achieve reinforcement particle density equal to 55 to 57%. Then, the matrix was located over the densely packed reinforcement bed and the whole setup inside the refractory container. The refractory container was then heated at 790-810°C temperature for 2-12 hours as per bed thickness of SiC powder. A continuous N_2 atmosphere was maintained and the composite material was cooled down up to 500°C in the same environment. Xu *et al.* [75] developed Al3Si2Mg alloy-55wt.%SiC$_p$ composites. The authors used 15 and 30μm sized SiC particles and fabrication was performed at a temperature of 850°C. Lee *et al.* [76] developed AA6061/SiC$_p$ MMC using this route. During fabrication, the infiltration of molten aluminium alloy over the SiC powder bed was carried out at 800°C temperature for 60 minutes under the N_2 gas atmosphere. Yang and Zhang [77] successfully fabricated the SiC$_p$/Cu–Al alloy composite using a high percent contribution of SiC (72.2 vol.%). Authors have studied the effects of infiltration temperature, moulding pressure, and infiltration time on the depth of infiltration. The results revealed that the composite structure was more dense and homogeneous with the least porosity when infiltration was performed at 850°C temperature and 10 MPa moulding pressure applied for 120 minutes. Also, Hemambar *et al.* [78] have followed the same route to develop new Al-SiC MMCs. The novel composite was the replacement of Kovar material, which is used in microwave-integrated circuits. This new material with controlled CTE exhibited additional advantages like higher thermal conductivity and stiffness, lower density and desirable electrical conductivity to maintain dimensional tolerances. Qiang *et al.* [79] have developed an AlMgSi alloy-SiC composite. Before use, the composite was pre-oxidized at 1000°C temperature for 60 minutes. Results revealed that SiC was dispersed uniformly inside the matrix grain structure. Chen and Chung [80] fabricated Al/SiC$_w$ composites using the same route. During fabrication, Al-5Mg, Al-10Mg and Al-5Si-5Mg alloys were used as matrices in order to make pressure infiltration possible, SiC$_w$ was coated with nickel metal by using electroless plating process. The steps involved placing a matrix alloy on a preformed SiC bed in the graphite crucible mould, followed by heating at various infiltration temperatures ranging from 800 to 950°C for 6 hours under the N$_2$ atmosphere. Cui *et al.* [81] produced Al-55%SiC MMC using the pressureless infiltration method. It was observed that hardness increased up to 242% than base Al-Mg-Si alloy, while as-cast composite exhibited only 22% improvement. Salim *et al* [82] cast Al-Si alloy-based composites reinforced with SiC$_p$. The authors used polystyrene as an external binder that distributed the SiC uniformly within the base alloy.

Ultrasonic Cavitation Process

Ultrasonic Cavitation (UC) process is a modified liquid stir casting process. In this technique, an ultrasonic wave passes through the melt for the continuous dispersion of reinforced particles in the matrix phase. In this process, an ultrasonic cavitation produces transient micro "hot spots" for a period of nanoseconds and a temperature of around 5000°C and pressures about 1000 atmospheric with high heating and cooling rates. Thus, the ultrasonic wave of frequency 17-20 kHz is generated with alternate compression and expansion cycle. The strong ultrasonic wave with extremely high local temperature potentially cleans the particles' surfaces, makes small bubble growth and breaks the nanoparticle clusters, thus preventing cluster formations. This wave produces no linear effect (like transient cavitation and acoustic streaming) in the melts and is responsible for refining microstructure, degassing of melts, reduction in porosity level, and homogeneity in reinforcement distribution in matrix metals/alloys. Fig. (**4**) represents a schematic set-up of the ultrasonic cavitation process for AMCs.

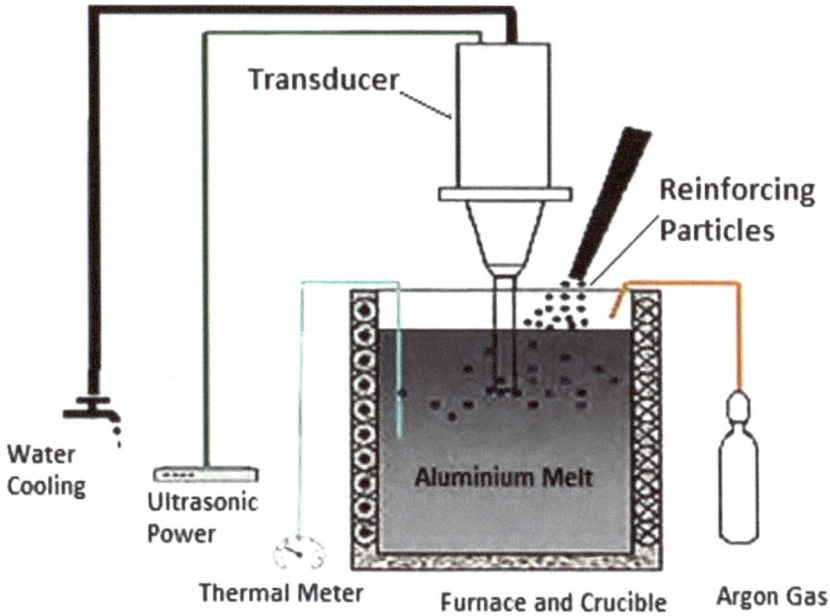

Fig. (4). Schematic set-up of ultrasonic cavitation process [83].

As the nanoparticle inside the cluster is loosely packed with each other, the air traps in the empty space of the cluster and forms a nucleus for the cavitation process. Moreover, the cluster size varies from nano to micrometer due to poor interfacial bonds and the existence of attraction force in matrix metal/alloy and reinforced nanoparticles [73 - 77]. Yang and Li [83] reported that ultrasonic cavitation was an effective route to develop bulk Aluminium (A356) based nanocomposites and provided a homogeneous dispersion of reinforcement particles with negligible cluster formation. In another research, Poovazhagan *et al.* [84] fabricated SiC and B_4C ceramics reinforced Al6061 alloy composites. During fabrication, nano-SiC (0.5-1.5% with steps of 0.5) and nano B_4C (0.5%) particulates were used. The fabrication steps involved melting of Al alloy in a stainless steel crucible at 680°C temperature for 10 minutes under the inert gas atmosphere. The titanium alloy-based ultrasonic probe connected with the horn at one end was dipped inside the melt at about 30mm height and another end connected to the transducer. In the first stage, the reinforcement particles were introduced in the molten melt from the top side and stirred mechanically to produce the uniform dispersion of reinforced particles in the matrix melt. In the second stage, the stirrer was removed and the melt was processed ultrasonically for one hour. Then, the slurry was reheated at 850°C temperature to enhance the flowability. In the end, melt slurry was poured into the heated steel mould. In another research, Rana *et al.* [85] used an ultrasonically assisted stir-casting process to fabricate SiC-reinforced Al (AA5083) alloy composites. During fabrication, the alloy was heated above its melting temperature and stirred continuously at 450rpm stainless steel stirrer to form a vortex of the melt. SiC was heated at 400°C prior to introducing into the melt. The mixture was then stirred for five minutes, to uniformly disperse the reinforcement particles into the matrix. Then, the ultrasonic probe of 21.21 kHz with a vibration amplitude of 40μm and 1.2kW power was introduced in molten alloy for around five minutes. N_2 gas was used for degassing of the composite melt. In the end, the melted slurry was heated again up to liquidus temperature and then discharged in the mild steel mould. The same fabrication route was used by Gopalakannan and Senthilvelan [86] to develop Al alloy (7075) matrix-based composite material reinforced with nano-particulates of SiC and B_4C of 50nm average particle sizes. During fabrication, the niobium alloy ultrasonic probe was used to generate ultrasonic waves, as this metal does not react with Al melt. The results of the high-resolution field emission scanning electron micrograph (FESEM) and scanning electron micrograph (SEM) revealed the uniform distribution of the reinforcement's nanoparticles in the base metal. Murthy *et al.* [87] successfully developed Al 2219 alloy-based composites reinforced with SiC using a similar fabrication route. The authors varied the contents of nano-SiC from 0.5 to 2 wt.% with steps of 0.5. The particle size of SiC used during fabrication was 50 and 150nm. The results of SEM reported the homogeneous distribution of the reinforced particle in the base aluminium alloy.

Squeeze Casting Process

Squeeze (SQC) casting process combines the large pressure die-casting and forging processes. This process can be carried out under high pressure, within the dies, during the time of melt solidification. This process includes melting the metal alloy, mixing reinforcements, pouring in preheated dies (mould), and squeezing with external pressure [88] The applied pressure may alter the melting temperature of liquid melt during solidification time, and thus improves the solidification rate [89, 90]. Fig. (**5**) shows the illustration of the squeeze casting process.

This process refines the macro and microstructures of the composites with negligible shrinkage and porosity. It also improves the flowability and castability of alloy, eliminates the segregation of reinforced particles and the requirement of gates and risers, and reduces materials' consumption. Moreover, common wrought alloys and cast alloys can be fabricated with this process [91, 92]. The common factors affecting this process are matrix melting temperature and their properties, squeeze die pressure, solidification rate, die temperature, reinforcements particle size, porosity, *etc.* Squeeze casting can effectively be used for developing AMCs. In this process, the application of applied pressure improves interface bonding between aluminium and SiC particles, enhances wettability, reduces porosity, and prevents the nucleation of gas bubbles [93]. According to Singh *et al.* [94], ultrasonic-assisted squeeze casting can produce composite materials with superior properties along with refining grain size than ultrasonic-assisted stir casting, under the same conditions. Taufik and Sulaiman [95] have developed a thermal expansion model for Al-based composites reinforced with SiC using squeeze casting. The authors used Al6061 and LM25 alloys during modeling. The results of the model revealed that the coefficient of thermal expansion was dependent on thermal stresses and interfacial interactions between the constituents. Kalkanli and Yilmaz [96] have developed Al (Al7075)-SiC MMC. The fabrication steps involved, the melting of Al alloy at a temperature of about 750°C to 780°C into the induction furnace. After the formation of homogeneous mixtures of the elementary particles, magnesium particles were added followed by SiC and stirred continuously until the highly viscous slurry was obtained. The semi-solid mixture was then moulded into the desired shape by pouring the slurry into a die, followed by squeeze casting with the application of 80 MPa pressure on the movable die. Karnezis *et al.* [97] cast Al (A356)-SiC particulate composite materials. The composition of SiC varied from 0 to 20wt.% with steps of 10. The fabricated composites were further subjected to solution treatment at 540°C temperature for 12 hours and then quenched with water. Experimental results revealed that the composites exhibited refined microstructures as compared to the base matrix and the result of the treatment revealed higher ductility than untreated ones.

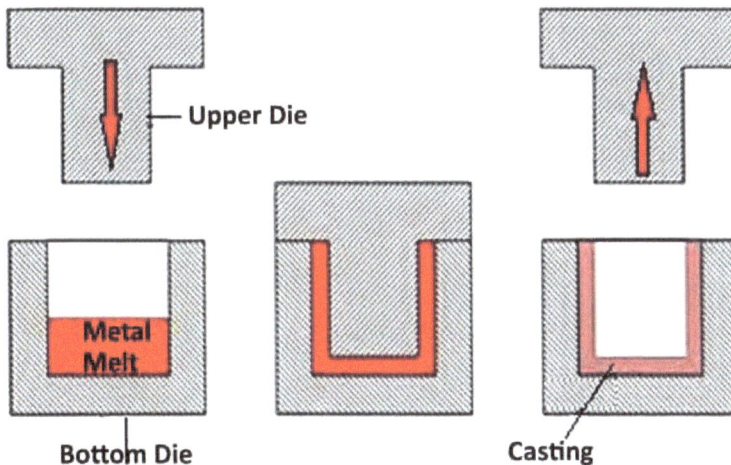

Fig. (5). Illustration of the squeeze-casting process.

Spray Forming Process

Spray Forming Process (SFP) is a liquid-state innovative technology used for producing aluminium alloys and their composites. This is one of the fast solidification methods, in which the molten metal melt atomizes in a pre-existed inert gas environment. The atomized liquid metal in the form of droplets, sprayed onto the substrate of predefined geometry, solidified rapidly and gave a near net shape. The schematic set-up of the spray-forming process is shown in Fig. (**6**).

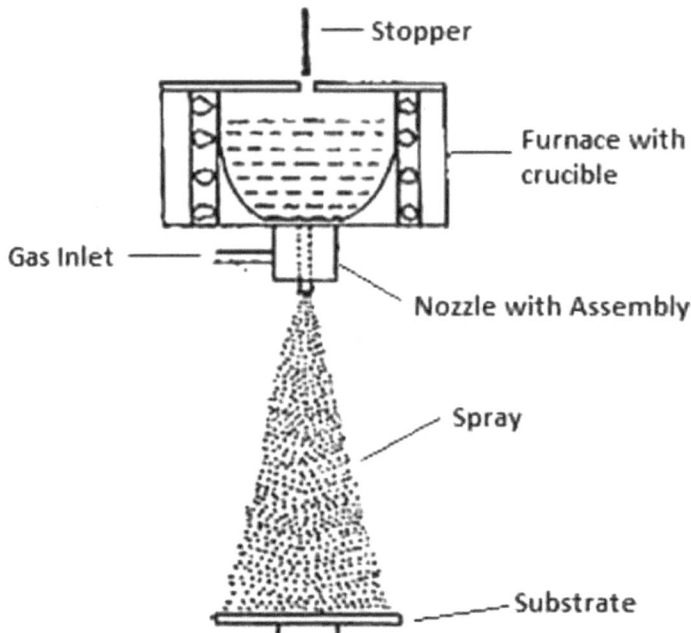

Fig. (6). Schematic set-up of the spray forming process..

The reinforced particles are either mixed in the liquid matrix melt by stirring and spraying combined with matrix melt or co-sprayed instantaneously along with the matrix melt spray. The components thus produced had fine grain size and mechanical properties. The microstructure refining of the composites depends upon the various process parameters like the flow rate of liquid metal, the distance between the spray nozzle and substrate, gas to melt ratio, the pressure of atomizing gas, and substrate orientation. These parameters can be independently controlled to fabricate materials with improved properties. This technique can be used for the fabrication of ferrous and nonferrous materials with near net shape components like discs, strips, tubes, billets, *etc.* This process possesses some drawbacks like loss of materials due to overspray of droplets, bounce off the impacting droplets, machining losses due to being out of specifications and low processing efficiency [98]. Gui *et al.* [99] fabricated SiC_p-based AMCs using the plasma spraying route. The authors used 55 and 75% volume fractions of SiC powder during fabrication. The sprayed feedstock was deposited on the graphite substrate. The authors observed negligible interface reactions between the constituents. The impurities from ball mixing media, zirconium oxide and steel were reacted with Al and Si and produced different compounds. In another research conducted by Srivastava and Ojha [100], SiC_p reinforced Al-4.5Cu-1-i-0.5Mg AMCs were fabricated by using this route. N_2 gas was used for atomization of the molten melt at 1MPa pressure. SiC particulates of different sizes (ranging from 6 to 45µm) were used during fabrication. Authors observed uniform particle distribution of large-size reinforced particles, whereas, in small sizes, clusters are formed near grain junctions and boundaries. The authors also reported that the electrical conductivity of the composites decreased with increasing vol.% as well as the grain size of SiC.

Solid States Processing

In solid fabrication techniques, the matrix is used in powder or foil form whereas reinforcements are in powder form only. During fabrication, the constituents are mixed with each other, followed by sintering under pressure. In solid fabrication techniques, extra energy is required to develop powders/foils and also to sinter the blends.

Friction Stir Process

The Friction Stir Process (FSP) works on the principle of friction stir welding. This method uses a revolving tool having a pin and shoulder, moves down to the face of the matrix plate, and with stirring action, produces a highly plastic sheared zone [101]. The illustration of the friction stir process is shown in Fig. (7).

The problem that arises in liquid state fabrication techniques is the critical control over reinforcement distribution, unwanted particle reactions in reinforced materials and matrix metals/alloys, detrimental phase generation, critical control of the process parameters, *etc.* [102]. The above limitations can be eliminated in FSP because in this process, the temperatures are below the melting temperatures of matrix metals/alloys, eco-friendly, energy-proficient, and flexible [103]. Refined microstructure is achieved due to the extensive plastic deformation. Because of better homogeneity and densification of the composite, the FSP process is more favourable than others [104].

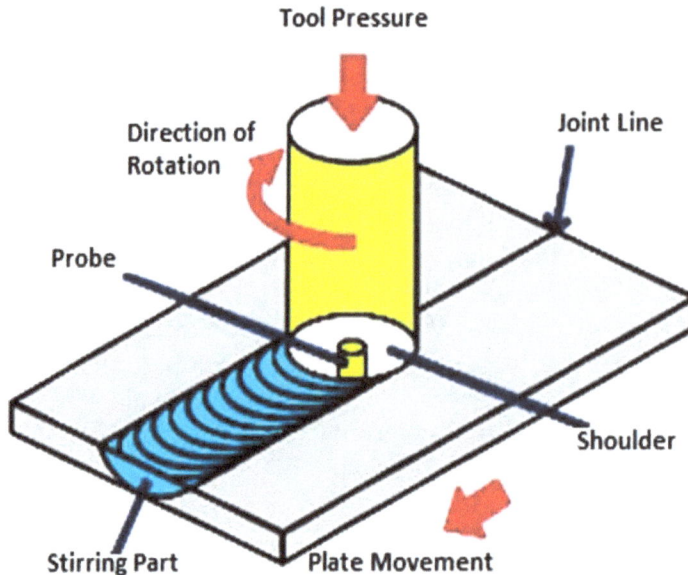

Fig. (7). Illustration of the friction stir process.

Researchers have effectively used the friction stir process for the development of SiC-AMCs. Dhayalan *et al.* [103] have produced SiC and Gr-Gr-reinforced Al (Al6063) alloy composites using this route. Authors developed three types of composites *viz.* Al/0.80SiC, Al/0.80Gr, and Al/0.40SiC/0.40Gr by altering the volume fraction of reinforcements. The rotational speed of 1000rpm, transverse speed of 30mm/min and vertical load as 10 kN were selected as process parameters during fabrication. A similar fabrication route was used by Mahmoud *et al.* [105] in the development of aluminium 1050/H24 alloy-(Al_2O_3-SiC) HMMCs with 20% Al_2O_3 and 80% SiC. The authors observed superior hardness and wear resistance than the base metal/alloy. Devaraju *et al.* [106] followed the FSP route to fabricate SiC and Al_2O_3/Gr reinforced Al (6061-T6) alloy composites. An aluminium plate of 4mm thickness and a total of 12 vol.% of SiC and Al_2O_3/Gr reinforcements were selected in a 2:1 ratio for all composites. The groove size of 3x3 mm was contrived in the center of the matrix specimen. H13 grade steel was used as a tool material threaded with a tapered probe having a shoulder size of 24mm, pin of 8mm, and pin length of about 3.5mm. A reinforcement mixture with specific ratios was packed inside the groove. The vertical milling machine with 5kN axial load, 900rpm, 40mm/minute transverse speed and 2.5° tilt angle was used during fabrication. Puviyarasan and Senthilkumar [107] optimized FSP process parameters for the development of Al(AA6061)/SiC$_p$ composites. The parameters selected for the experimentation were tool rotational speed (1200-1800rpm), transverse tool speed (36-72mm/min) and tool tilt angle (ranging from 1 to 3°). The high-speed steel tool with 18mm diameter flat shoulder, screwed cylindrical pin with size 6mm and height 5.9mm was chosen during processing. SiC (99.9% pure) was selected as a reinforced material with 3μm size. The authors reported that tool rotational speed and transverse speed were the most significant parameters, while the tool tilt angle is insignificant to maximize the tensile strength of the composites. Also, the maximum strength was achieved when the tool rotated at 1800rpm, the transverse tool moved up to 58mm/min when the tool was tilted at 2°. Similar research has been conducted by Sahraeinejad *et al.* [108] during the optimization of FSP parameters (*viz.* a number of passes, groove depth, shoulder diameter and pin length). The authors evaluated the effect of grain size of SiC, Al_2O_3, and B_4C on the properties of Al 5059 alloy MMCs. Experimental results revealed that superior properties were achieved at 2mm groove depth with 3 passes of FSP. Also, the homogeneity in the distribution of reinforcing particles varied proportionally w.r.t. friction stir process that passes in upper as well as in lower portions of stir zone of the material.

Powder Metallurgy Process

Due to the low processing temperature, the powder metallurgy (PM) process is effectively used in the production of Al-SiC composite materials. This method generally involves three steps *viz.* blending/mixing, compacting and sintering. Fig. (**8**) represents the block diagram of the powder metallurgy process for SiC-MMCs.

The Al-SiC composites with improved properties (*viz.* hardness, homogeneous particle distributed) are prepared with this process. In the PM process, the formation of detrimental intermetallic phases and undesirable interfacial reactions are lesser than liquid cast composites [109 - 111]. Mosisa *et al.* [91] fabricated Al-SiC-Mg hybrid composites using the PM process. The results illustrated that the optimum value for SiC and Mg varied from 12 to 16% and 1 to 1.5% by weight respectively for better mechanical properties. Puhan *et al.* [109] prepared Al-SiC$_p$ MMC using the same process by altering two material parameters *viz.* wt.% and mesh size of SiC in the composites. The fabrication steps involved heating of SiC particulates at 700°C temperature for one hour to avoid the formation of unwanted reactions between constituents as a very small thickness layer of SiO_2 was formed on the SiC particles. The mixing of the matrix and reinforcement powder was carried out in a ball planetary machine at 600rpm for a time period of 30 minutes. The mixture was then placed in the metallic die and compacted in a cold uniaxial pressing machine. Then, the samples were baked inside the horizontal tubular furnace below the melting temperature of the major constituent under a controlled atmosphere of argon gas at 1 bar pressure. The compact structure was then heated at 580°C for one hour followed by cooling to room temperature. Hao *et al.* [110] optimized the powder metallurgy route for Al2024-SiC$_p$ composites to obtain better mechanical properties and microstructure. The composites were developed by choosing different temperatures (520° to 600°), and sintered for 2 to 4 hours under sintering pressures ranging from 45 to 90 MPa. Authors reported that the mechanical properties and density of the composites increased with increasing hot pressing temperature. Jaykumar *et al.* [112] successfully developed Al (356 alloys)-SiC AMCs using the PM route. The authors varied SiC (1μm size) from 0 to 20 vol.% in matrix powder with steps of 5. Alves *et al.* [113] successfully developed Al$_2$O$_3$ and SiC-reinforced Al (AA2124) alloy matrix composites using the PM route. During experimentation, both Al$_2$O$_3$ and SiC varied from 5 to 15% by mass with the steps of 5. Abhik *et al.* [114] used the PM route to develop Al alloy-based composite materials, for brake pads, which are reinforced with SiC. Two different vol.% of SiC (10% & 20%) were selected during preparation. Experimental steps were involved in the mixing of Al-2014 alloy powder with SiC powder by using a ball mill mixer to ensure homogeneous powder mixing. The mixture of particles was then compacted within dies under the load of 150kN applied for 20 minutes with the help of a universal testing machine. After that, sintering was performed with a microwave furnace at 550°C temperature for 30 minutes. The results indicated that the sample with 20% reinforced material exhibited high wear rate as well as hardness due to its lower density. Lee *et al.* [115] have manufactured Al-Si/SiC$_p$ MMC as piston material used in automotive industries. The authors fabricated composites by using the hot vacuum pressing PM route. After fabrication, the composites were subjected to extrusion with an 8:1 ratio at 500°C temperature. Results indicated that the mechanical properties of base metal/alloy, as well as composite materials, increased due reduction in grain size during hot extrusion and forging processes. Kwok and Lim [116] used the PM route to develop Al/4.5Cu-13SiC$_p$ MMCs. During processing, Al (150μm), Cu (150μm), and SiC (50μm) were selected as constituent particles. The steps involved mixing of specified particle ratios in a ball milling machine at 86 rpm and 163rpm for 10 hours under steady-state conditions. The powder mixture was then passed through the powder metallurgical route with a compact pressure of 420MPa followed by sintering at 590°C temperature up to 5 hours. Bodukuri *et al.* [117] have prepared Al-SiC-B$_4$C hybrid MMCs using the PM route. The authors selected three compositions *viz.* 90Al-8SiC-2B$_4$C, 90Al-5SiC-5B$_4$C, and 90Al-3SiC-7B$_4$C and evaluated their density and hardness values. Karabulut *et al.* [118] have performed a comparative study of Al (AA7039)-10 wt.% SiC/Al$_2$O$_3$/B$_4$C composites fabricated by the PM method. The fabrication of composites was followed by hot extrusion. The composites were compared on the basis of mechanical properties, drilling properties, and microstructure. Authors observed that Al-Al$_2$O$_3$ MMC has better hardness, rupture strength, elongation and drilling performance as compared to the base matrix alloy as well as other AMCs. Kumar and Rajadurai [119] have fabricated Al-SiC-TiO$_2$ hybrid MMCs using the PM route. The authors kept a constant composition of SiC (15%), while TiO$_2$ (rutile) varied from 0 to 12% mass fraction with steps of 4. With the addition of TiO$_2$ contents, superior tribological and mechanical properties were obtained. In another research, Li *et al.* [120] prepared CNT-coated SiC reinforced MMCs with aluminium as a base matrix. After fabrication, the composites were subjected to extrusion. The authors have observed the uniform distribution of reinforcement particles along with significant improvements in mechanical properties. Ghasali *et al.* [121] performed a comparative study during sintering (conventional and microwave methods) of Al-15wt%SiC-7wt%TiC hybrid composite. The authors used pure Al and Al-1056 alloy as matrix powders during fabrication. They observed that microwave sintering produced better results as compared to conventional ones. In another research, Ghasali *et al.* [122] have fabricated similar composites using plasma and conventional sintering methods with 5% SiC and 5% TiC weight fractions as reinforcement in pure Al matrix powder. The authors observed superior mechanical properties with plasma sintering than the conventional one. In another research, Hu *et al.* [123] used vacuum hot pressing during the fabrication of Al (2024)-SiC-Gr hybrid MMCs. SiC varied as 0, 5, 10 vol.% while graphite as 0, 3, 6 vol.% during fabrication. After fabrication, the composites were followed by hot extrusion. The authors observed a negative effect of reinforcement particles on the strength and elongation. Moreover, graphite indicated a higher negative effect than SiC. Madeira *et al.* [124] produced Al-SiC$_p$ MMCs using hot pressing PMP route. During fabrication, authors have used three different particle sizes of SiC *viz.* 13, 38.8 and 118μm. They concluded that the damping capacity and Young's modulus can significantly be improved with suitable selection of reinforcement particles. Montalba *et al.* [125] produced AlMg5-SiC-PLZT hybrid MMCs using the PM route followed by hot extrusion. The authors kept a constant percentage of SiC (1wt.%) and PLZT (piezoelectric lead lanthanum zirconate titanate) varied from 0 to 15wt% with the steps of 5. The authors evaluated the microstructure and damping behaviour of the newly developed hybrid MMCs. Carvalho *et al.* [126] have performed a comparative study on the mechanical behaviour of Al-5%Si--2%CNT hybrid composite, Al-5%SiC composite, Al-2%CNT and base matrix. The composites were developed by PMP. The authors observed that hybrid MMC indicated better overall results than other ones.

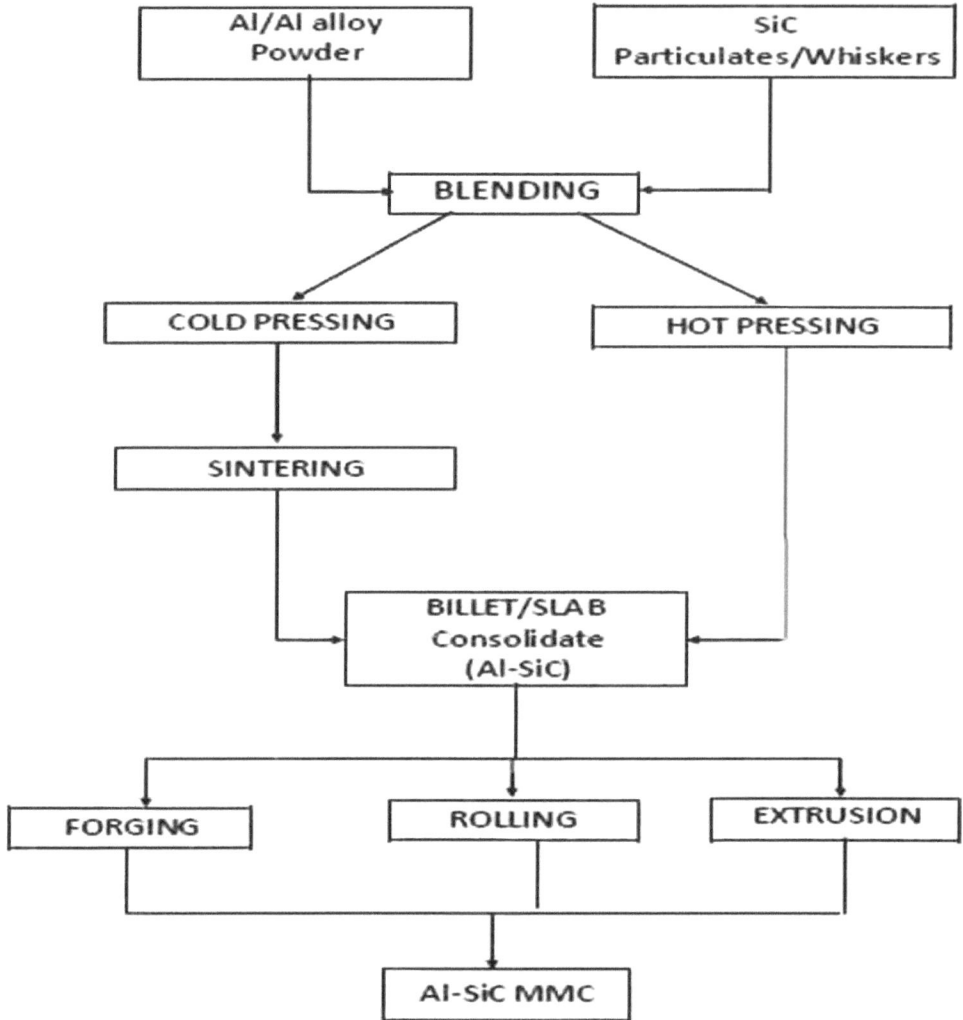

Fig. (8). Block Diagram of the powder metallurgy process for SiC-MMCs.

Accumulative Roll Bonding Process

The accumulative roll bonding (ARB) process is a newly developed process that has attracted the interest of many industrial applications. This process was invented in 1998 by Saito *et al*. [127] and was followed by the principle of plastic deformation of the matrix materials. The ARB process involves the cyclic phenomena of cutting, surface preparation, adding particles, stacking, and roll bonding with a 50% reduction in thickness and again repeating the same [128]. Fig. (**9**) presents the illustration of the accumulative roll bonding process.

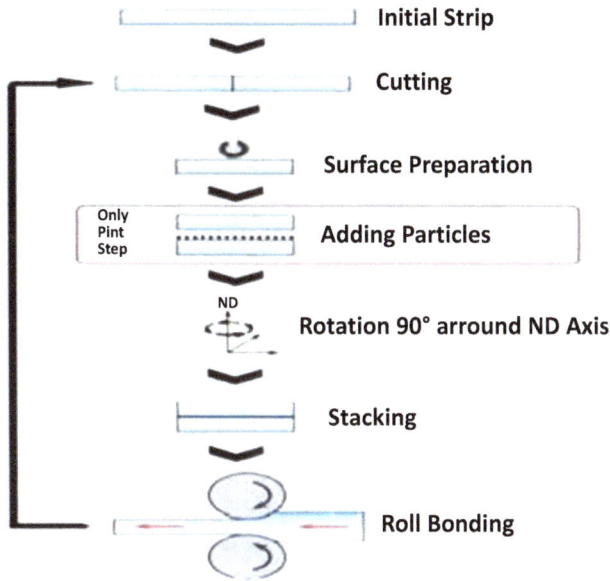

Fig. (9). Illustration of the accumulative roll bonding process [128].

Many researchers have used this route for the fabrication of Al-SiC MMCs. Naseri *et al.* [128] have performed a comparative study of cross ARB and linear ARB processes during the fabrication of Al-2wt.% SiC-2wt.% B_4C hybrid MMCs. The authors observed that the cross ARB process was performed better in particle distribution and exhibited superior mechanical properties without forming additional phases. These improvements were attributed to variations in strain path during rolling. Fattah-alhosseini *et al.* [129] used nine repeated cycles of the ARB process to produce Al(1050)-1%B_4C-2.5%SiC hybrid composites. The results indicated uniform dispersion of reinforced particles with negligible porosity. In another research, Baazamat *et al.* [130] produced Al(1050)-SiC-WO_3 hybrid nanocomposites using the same route. The authors reported significant improvements in specific properties of nanocomposites with the increasing number of passes of the ARB route. The same fabrication route was followed by Ardakani *et al.* [131] to prepare Al-2vol.% SiC_p nano-composite and after eight cycles of CARB process, the uniform distribution of reinforcements with the least porosity was observed. In another work, Hosseini *et al.* [132] prepared Al (AA1050)-SiC_p nanocomposites using a cross ARB route. The authors observed better properties of the newly developed nanocomposites than the base matrix. Darmiani *et al.* [133] fabricated Al(1050)-SiC nanocomposites using the same route in order to evaluate corrosion resistance of new AMC. Prior to use, authors have annealed the matrix material at 380°C temperature for 2hours. Alizadeh *et al.* [134] have produced Al-7%SiC nanocomposite using the ARB process. A sheet of the matrix was also subjected to the same ARB route in order to compare it with the newly developed composite. The results indicated better properties of the composite as compared to the monolithic base matrix. Ahmadi *et al.* [135] have followed this route in the production of Al-(0.5-2vol.%) SiC-1.6vol.% Al$_2$O$_3$ hybrid composites. The authors observed that reinforced particles dispersed uniformly within the matrix without porosity. This was due to the fracture phenomena of Al$_2$O$_3$ with the increasing number of cycles. In another research, a similar hybrid composite of Al(AA1050)-Al$_2$O$_3$-SiC with 2 vol.% of reinforced particles (SiC+Al$_2$O$_3$) was prepared by Reihanian *et al.* [136] using the same route. Amirkhanlou *et al.* [137] have used continual annealing and roll bonding processes during the preparation of the Al-8vol.%SiC composite. Pure Al strips and SiCp powder of 40μm average particle size were selected during processing. A similar route has been followed by Alizadeh *et al.* [138] and Hosseini *et al.* [139] during the development of Al-SiC$_p$ composites using repeated roll bonding processes. The authors observed the continuous dispersion of reinforced particles with reduced porosity and significant variations in properties of the composite material with the increasing number of cycles.

Semi-Solid States Processes

The semi-solid process is the modified die-casting method that exhibits the advantages of casting as well as forging processes. With this method, the reinforced particles are entrapped by semi-solid matrix alloy and thus prevent settling down agglomeration and flotation of these particles. In this method, extensive shearing is applied to the semi-solid alloys within the solidification range.

Thixomoulding Process

Thixomoulding (also known as thixoforming) is a newly developed amongst the semi-solid state fabrication process, which is suitable for non-ferrous matrices like Al, Mg, Cu *etc*. In this process, the benefits of the liquid state casting process and solid-state powder metallurgy process are combined. This process is based on thixotropy property of the fluid which allows the fluid material to shear off when flows and thickens when standing [140]. This process is simple and produces composites with negligible porosity, intricate parts with net shape, accurate tolerance, excellent mechanical properties, least shrinkage, uniformly microstructure and suitable for thin wall materials [141, 142]. A few kinds of literature, related to use of this process for the production of SiC-AMCs, is available. Zhang *et al.* [143] followed the thixoforming process to fabricate Al (6061)-based MMCs reinforced with 10 vol.% SiC. The authors have evaluated the consequences of solution treatment on strengthening mechanism and tensile strength. The fabrication steps involved mixing of Al and SiC with ball milling at 100rpm for 40minutes, followed by cold compaction at 150MPa. The specimen first heated at 660°C and then forged in a thixoforming mould at 160MPa pressure. Fig. (**10**) indicates the schematic illustration of the thixoforming process.

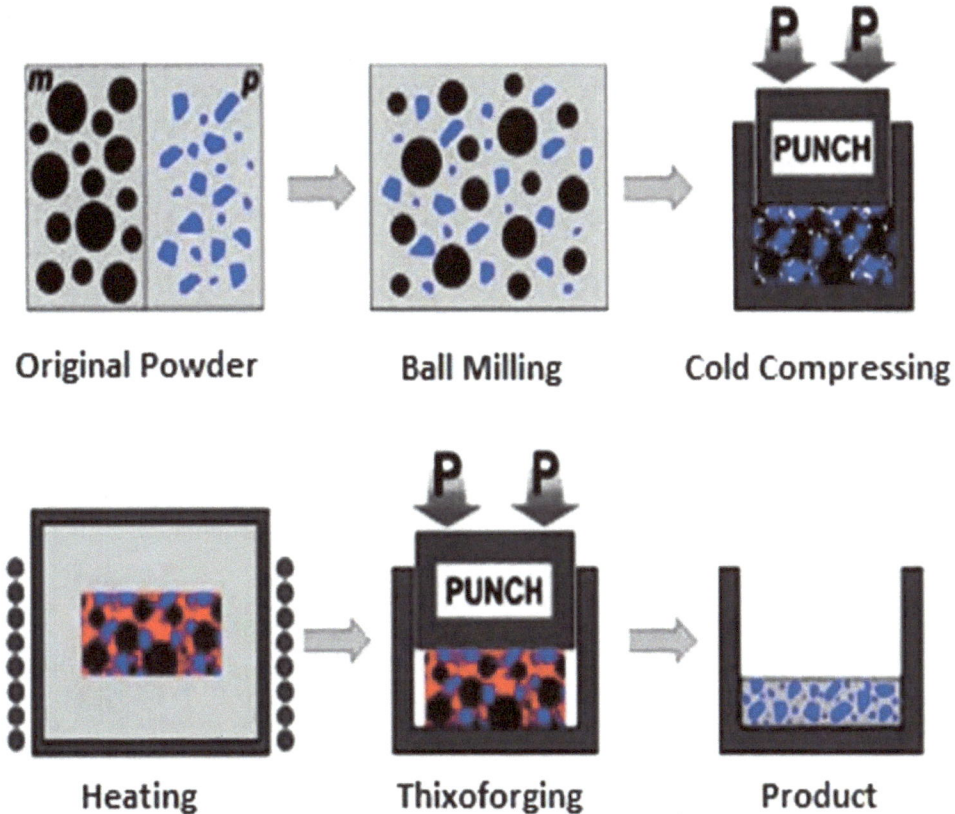

Fig. (10). Schematic illustration of the thixoforming process.

Ozyurek *et al*. [144] fabricated Al(A356)/ (5-20vol.%) SiC composites with the thixomoulding method. The fabrication process was carried out at two different temperatures (590°C and 600°C). The fabrication steps involved the mixing of Aluminium A356 matrix particles with reinforcement particles for 30minutes, followed by pre-shaping of the powder under cold press with pressure around 800 MPa to ensured continuous dispersion of reinforcing particle in the aluminium metals/alloys. Then, the pre-shaped composites were thixomoulded at 590°C and 600°C temperature (below the melting point of the main constituent). Authors observed that with process temperature, the grain size of the composites was improved, but with the loss of hardness value. Kandemir *et al*. [145] have thixoformed Al(A356)-SiC and Al(A356)-TiB$_2$ nanocomposites by combining ultrasonic cavitation of molten alloy and green compaction of nanoparticles. The authors used SiC and TiB$_2$ nanoparticles with an average diameter ranging from 20 to 30nm during the experiments. The steps involved the addition of reinforcement particles into green compacts using powder powder-forming process followed by the melting of alloy and then treated ultrasonically. Guo *et al.* [146] have developed high SiC-based AMCs used in electronics assemblies by using the thixomoulding route. The results of microstructure and properties indicated gradient distribution formation and phenomena of particles segregation exhibited between SiC and the liquid metal alloy.

Compocasting Process

This process is basically performed at lower freezing (in between liquidus and solidus) temperature of matrices. The compo-casting (Rheocasting), casting steps involve melting the matrix above liquidus temperature, allowing it to cool down slowly to the semi-solid state, stirring continuously to form the vortex, and the introduction of reinforcements and discharge to preheated mould. The advantages of this process are that it reduces the possibility of interfacial reactions and interface destruction at low process temperatures during mixing [147]. The high viscosity of melt slurry prevents the segregation of constituents as in the case of liquid casting processes [38].

Little literature is available on the utilization of this process in the production of SiC-AMCs. Selvam *et al.* [30] have developed Al-SiC-FA hybrid composite materials using the compo-casting route. Authors used AA6061-T6 Al alloy with different wt.% of SiC (7.5 & 10%) and a constant wt.% Flyash (7.5%). The fabrication steps involved melting 1500g Al alloy at 920°C temperature, followed by stirring to form a vortex of the melt. It was then cooled down to liquidus temperature followed by the introduction of heated SiC and Flyash (preheated at 900°C for 90 minutes) slowly and continuously in the melt. 1% of Mg by weight was introduced to enhance the wetting characteristics of reinforcements with Al alloy. An inert atmosphere (argon gas) was maintained over the surface of the molten metal to prevent atmospheric contamination. Two stages of stirring were performed during casting to ensure uniform dispersion of reinforcements in matrix melt. Rahmani Fard *et al.* [198] have fabricated Al (A356) matrix-based AMCs, in which different particle sizes of SiC (*viz.* 38, 62 and 82μm) were embedded. The age-cast composites were passed through extrusion (18:1 ratio) at 450, 500 and 550°C extrusion temperatures. Authors observed that the extruded composites possessed more refined microstructure along with lesser porosity than the base matrix and as-cast composites.

In-Situ Processing Technique

The *ex-situ* process exhibits several limitations like porosity, unequal dispersion of reinforcements, improper wettability, unwanted interfacial reaction, segregation, costly equipment and reinforcement, gas entrapment, *etc.* To overcome these limitations, the researchers developed an alternate method for reinforcement additions In the *in-situ* process, unlike *ex-situ* processes, the reinforcement is formed within the molten matrix in which, constituent particles chemically react with Al melt, nucleate and grow up during the processing [148]. This is a recently developed fabrication method that is relatively economical and cost-effective in the development of MMCs. The major benefits of this method include a homogeneous dispersion of the reinforcements, high compatibility in the matrix and reinforcing particles, low cost, *etc.* However, this process is not equally acceptable in all systems. These processes are still in the developing stage [149 - 151].

During reinforcement formation reactions, the first reacting element is generally a particle of matrix metal/alloy, whereas the second element is either externally added solid powder or any other gaseous phase. The reinforced particles are thus formed uniformly within the matrix with excellent bonding between matrix and reinforcement particles. However, this process requires great screening attention during the reaction system while forming the reinforcements. The researchers must have knowledge about favorable thermodynamics related to the anticipated reaction [152, 153]. Solid-state *in-situ* synthesis of reinforcement phases is a technique for fabricating composite materials that involves synthesizing the reinforcing phase inside the matrix. As opposed to *ex-situ* composites, where the reinforcing phase is first synthesized differently and then incorporated into the matrix during a subsequent process like infiltration or powder processing, the former uses a single manufacturing step. Various solid-state *in-situ* synthesis processes used to develop composites are as under:

Displacement reactions

This is a reaction between two or more phases that results in the formation of a completely novel phase. To create TiC-reinforced $MoSi_2$ composites, for instance, Si and TiC can react to yield $MoSi_2$.

Internal oxidation

This route involves the oxidation of metal to create a ceramic phase. For example, the oxidation of aluminum to form aluminum oxide (Al_2O_3) can be used to synthesize Al_2O_3-reinforced aluminum composites.

Reactive milling

This route uses powders of the matrix and reinforcements that are milled together under a liquid or gas atmosphere resulting in a reaction among the powders and forming the reinforcing phase.

Cryomilling

In this process, the powders are milled at cryogenic temperatures which prevent the development of unwanted phases.

Goudarzi and Akhlaghi [154] have used *in-situ* powder metallurgy route to develop Al-SiC nanocomposites. The fabrication steps involved the mixing of nanoparticles of SiC of different sizes (average size of 60 nm) with micron size SiC particles (250-600μm) followed by preheating and incorporating in Al molten melt. The mixture was then stirred with an impeller and the kinetic energy of the impeller was transferred to the melt. This transformation resulted in the disintegration of non-wetting SiC particles. The liquid droplets were created and solidified to produce Al powders and SiC nanoparticles mixture. This mixture was then passed through the powder metallurgy process route to develop composites.

Solid-state *in-situ* synthesis, however, is not without its difficulties. The possibility of delayed reaction kinetics, which can make it challenging to obtain a complete reaction, is one problem. Another difficulty is that unfavorable phase development may occur alongside the process. Solid-state *in-situ* synthesis is a potential method for creating high-performance composites despite these difficulties. This procedure is adaptable and can be used to make composites with a variety of characteristics.

Spark Plasma Sintering

In the Spark Plasma Sintering (SPS) process, uniaxial pressure and pulsed electric current are combined to consolidate powdered materials. The powder particles and the die are heated by the pulsed current, which causes the particles to sinter together. High pressure helps in preventing cracking or deformation of the particles during sintering. The SPS is based on the principle of electrical spark discharge phenomena. The powder produces momentary spark plasma at high localized temperatures, from several to ten thousand degrees Celsius, when a high energy, low voltage spark pulse current is given to it. A dense and robust sintered material is produced as a result of the powder particles melting and fusing together in this plasma. When compared to traditional sintering methods, the heating rate in SPS is often substantially faster. This is because, as opposed to depending on conduction through the die, the pulsed current warms the powder particles directly. SPS can be used to sinter materials like ceramics and composites that are challenging to sinter using traditional techniques due to the high heating rate. SPS is superior to alternative sintering methods in a variety of ways, including:

- No need for a vacuum or inert environment
- Shorter sintering time
- Higher sintered density
- Better mechanical characteristics
- Reduced grain formation
- Reduced residual stresses

A wide range of materials, including ceramics, metals, composites, and electronic materials, are fabricated using SPS. When producing materials with intricate shapes or demanding high-performance standards, it is especially helpful.

INTERFACIAL REACTIONS IN SIC-AMMC

Numerous kinds of literature are available on the utilization of SiC (particulates, whiskers, fibers) as reinforcement in aluminium metal/alloys. Aluminium has a high affinity toward oxygen and forms an oxide layer on its surface when reacting with it under specific conditions. For example, aluminium, when heated at a 450°C temperature for 4 hours, forms a 50nm oxide layer on its surface [155]. This layer impedes the penetration of reinforcements in the liquid metal and leads to a reduction in the wettability of constituents [156]. Also, the ceramics are surrounded by a gas film and when discharged in molten metal, especially from the top side, this film leads to the prevention of reinforcement's contact with molten metal [157 - 159]. In liquid casting techniques, when molten matrix melt comes in contact with SiC particles at a high temperature, direct unwanted reactions occur, which cause weakening of the composite properties. For example, in SiC-AMCs, when Al reacts with SiC, a brittle intermetallic aluminium carbide (Al_4C_3) compound is formed [160] as per the heterogeneous reaction shown in equation 1.

$$4Al + 3SiC = Al_4C_3 + 3Si \qquad (1)$$

This Al_4C_3 compound reacts with atmospheric moisture and produces a deleterious effect on the properties of composites [161, 162] like a reduction in the overall thermal conductivity of the material [163] and elastic modulus [164]. Furthermore, the formation of Si after the reaction changes the base matrix composition [165]. The reactions between Al and SiC include chemical reactions between constituents, diffusion of Si and C in the Al matrix, compound formation until the concentrations of Al and C reach the equilibrium concentration of Al_4C_3. The level of these reactions can be estimated by measuring the intensity levels of Al, SiC, and Al_4C_3 peaks [166]

Many researchers have reported the different techniques to prevent the Al_4C_3 formation. The control of processing temperature and holding time can prevent unwanted compounds' formation up to some extent. The enhancement of Si contents in base metal prevents the dissolution of SiC by retarding the chemical kinetics and thus the formation of Al_4C_3 [167, 168]. Further, the coating of Cu, CNT, Si, *etc.* on SiC particles, pre-oxidizing of SiC, using an Al alloy containing high Si contents can eliminate unwanted compound formations [40, 120, 169]. As per quantum mechanics calculations by Li *et al.* [168], the interfacial bonding strength of Al-SiC for a particular plane crystalline structure is 50% higher than that of the Al-Al structure. Kimoto *et al.* [170] observed the negligible formation of compounds (Al_4C_3 and Si) when a novel Al-33%SiC composite was prepared with a friction stir powder process at 1500rpm rotational speed. The absence of these compounds was attributed to the transverse of the rotational tool, which prevented the local rise of temperature.

Unnecessary interfacial reactions and compound formation can be eliminated by modifying matrix composition, controlling process parameters of production method, surface modification of reinforcements with coatings, and oxidizing reinforcements [171]. Also, preheating of SiC particles at 700-1200°C temperature for a few hours forms an amorphous silica (SiO_2) layer on the particle surface. This formed SiO_2 layer reacts with aluminium and counteracts Al_4C_3 compound formation along with the liberation of excess Si contents as shown in equation 2 [171 - 175]. This Improvement in Si contents improves wettability among the constituents [156].

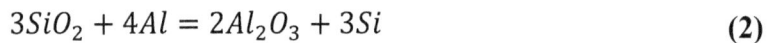

$$3SiO_2 + 4Al = 2Al_2O_3 + 3Si \qquad (2)$$

The formation of amorphous SiO_2 shows parabolic tendency and can be controlled by controlling process parameters like processing temperature, holding time, *etc* [173, 174].

KEY OBSERVATIONS

The comparative study of various processing methods of SiC-AMCs is presented in Table **1**. From the literature, it is observed that liquid state, solid state, semi-solid state, and *in-situ* techniques are successfully used by the researchers to develop SiC-AMCs. Fig. (**11**) illustrates the percentage contribution of different fabrication methods, which are used by researchers in the production of SiC-AMCs in the last two decades.

Table 1. Comparative studies on fabrication techniques for SiC-AMCs

1	Main Features	Cost Aspects	Applications	Ref(s).
Stir Casting	Best suited for particulate reinforced AMC, requires less sophisticated equipment, suitable for bulk size (up to 500 kg) and mass production, restricts to reinforcements ratio up to 30% due to cluster formation, and the possibility of defects like voids and shrinkage.	Least expensive	A commercial technique for developing aluminium-based alloys and composites, used in automotive and aerospace industries.	[38 - 67]
Pressureless Infiltration	Best suitable for filament type reinforcements, best for high wt./vol.% of reinforcements, and provides good strength and stiffness with lower density.	Moderate/ Expensive	Production of structural beams and shapes, rods, tubes, *etc.*	[68 - 82]
Ultrasonic Cavitation	Best suitable for nano size reinforcement particles and complex structure components, with better reinforcement distribution.	Expensive	Net shape and intricate parts fabrication of complex structural components, and Mass production.	[83 - 87]
Squeeze casting	Suitable for all categories of reinforcement, for mass production, low shrinkage and porosity, no blow holes, produces near net shape Intricate parts, rapid solidification, and restricted to reinforcement's ratio up to 45%.	Moderate	Used for components like connecting rods, engine blocks, brake discs, rack housing, fuel pipes, pistons, cylinder heads, rocker arms, *etc.* in automobile and aeronautics industries.	[88 - 97]
Spray Forming (casting)	Best for particulate reinforcement, suitable for full density materials, near net shape components, minimum processing steps and time, intricate shapes can be produced, suitable for ferrous and nonferrous materials, and low processing efficiency.	Moderate	Automotive, aerospace industries, electronic packaging, cutting and grinding tools, electrical brushes and contacts, tubes, billets, discs, strips, *etc.*	[98 - 100]

1	Main Features	Cost Aspects	Applications	Ref(s).
Friction Stir Process	Used as surface and microstructure modification route, operated below the melting temperature of particles, so least possibility of chemical reactions and formation of unfavorable phases. An increase in micro hardness of the surface, and significantly improves wear resistance, with less distortion and porosity.	Moderate/ Expensive	In Automotive and Aerospace applications.	[101 - 108]
Powder Metallurgy	Both phases are used in powder form. Reinforcements may be long/short fibers or particulates, best control on particle distribution, low temperature during fabrication, relatively less possibility of chemical reaction among constituents and different reinforcements can be used in one composite, suitable for high volume production, high density, reduced cost and tedious second processing due to net shape production, and suitable for a high fraction of reinforcements.	Moderate	Best suitable for small circular components like valves, pistons, bolts, high heat resistant and high strength components. Major applications in sports automobile, defense aerospace and appliance industries.	[109 - 126]
Accumulative Roll Bonding	Best suitable for nano particulate composites, and uniform dispersion of reinforced particles,	Moderate/ Expensive	Automotive parts, aerospace and transportation industries.	[127 - 139]
Thixomoulding	Combines the advantages of liquid processing and solid (powder metallurgy) processing methods, simplicity and less time consuming process, good wear characteristics, modest composite strength, lack of control on the microstructure, and different ceramics can be dispersed into the melt.	Moderate	Defence, aerospace and automobile industries.	[140 - 146]
Compo casting	Best suitable for discontinuous fibers, particularly for particulate reinforcement, and low porosity.	Least expensive	Used in manufacturing, aerospace and automobile industries.	[30, 147]
In-situ process (reactive method)	Homogeneous dispersion of the reinforcements, unoxidized interfacial bonding in matrix material, and reinforcements with high strength and wettability.	Expensive	Automotive and aerospace industries.	[148 - 153]

(Table 1) cont.....

1	Main Features	Cost Aspects	Applications	Ref(s).
Diffusion bonding	Handles foils or sheets of matrix and filaments of the reinforcing element.	Expensive	Used for structural components, sheets, vane, shafts, blades, *etc*.	[176 - 190]
Vapour deposition	Physical Vapour Deposition (PVD) coatings are more corrosion-resistant and harder than the coatings applied by the electroplating process.	Moderate	Surgical/Medical Dies, aircraft industries, thin films for food packaging, Cutting tools, optics watches, firearms, and automotive.	

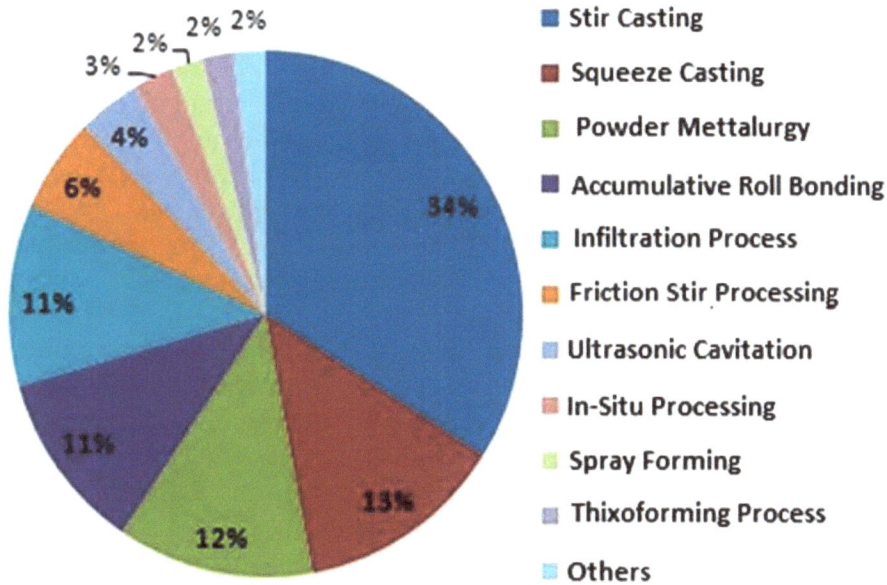

Fig. (11). Fabrication techniques of SiC-AMCs followed by researchers in last one decade as per search engine of Scopus.

The literature revealed that numerous researchers followed lthe iquid stir casting route during the processing of SiC-AMCs, especially in hybrid composites. This is due to its simplicity, flexibility, low processing cost, the requirement of less sophisticated equipment, suitability for bulk size and mass production, and casting of intricate shapes. Furthermore, hybrid AMCs can easily be processed without any modification. The casting cost of SiC-AMCs with this process is about one-third to half that of other competitive processes and for bulk production, it may rise up to one-tenth [178 - 183].

However, uneven reinforcement distributions, unwanted interfacial reactions, and lack of wettability are the big challenges of this route. Homogeneous distribution of reinforcement depends upon the state of the molten matrix and SiC slurry, stirring time and speed, the preheating temperature of reinforcements, stirrer, mould materials and shapes of equipment used, reinforcement wettability, interfacial reactions, reinforcement settlement due to higher density, the viscosity of molten matrix, cooling rate, *etc*. Moreover, double stirring, electromagnetic stirring, ultrasonic cavitation, and squeeze casting are the modified stir casting processes, which ensure uniform particle distribution along with superior properties. For example, the use of external pressure in squeeze casting ensures an appreciable reduction in porosity level and internal defects. However, these modifications in conventional stir casting increase processing costs, due to additional equipment requirements and more time-consumption. The infiltration method is best for nano-sized reinforcement particles, and high-volume fraction AMCs that provide good strength and stiffness with the least porosity but is more expensive than other ones. Powder metallurgy is another important route for the fabrication of SiC-AMCs. This route ensures homogeneous reinforcement dispersion due to best control during powders mixing, the least possibility of chemical reactions due to lower processing temperature. However, this method is more expensive and requires more sophisticated equipment. The friction stir process is best for surface modification, but it requires more costly equipment. Thixomoulding is a newly developed semi-solid method that combines the advantages of liquid processing technique and solid processing technique (powder metallurgy). This process is simple and less time-consuming, but subjected to a lack of control over the microstructure and is most expensive. The *In-situ* route ensures a homogeneous dispersion of the reinforcements, unoxidized interfacial bonding in the matrix, and reinforcements with higher strength and wettability. However, this method requires high concentration during fabrication and is an expensive process [177 - 185].

The affinity of Aluminium toward oxygen formed an oxide layer on its surface, which restricted the penetration of SiC in the liquid metal and reduced wettability. The gas film around SiC prevented its contact with molten metal. Also, the liquid aluminium melted when came in contact with SiC particles at high temperatures, and the Al_4C_3 compound formed, which reacted with atmospheric moisture resulting in deleterious effects on composite properties and alteration of the base matrix composition. This uneven reaction can be eliminated by increasing Si contents in base metal pre-oxidizing of SiC, using Al alloy containing high Si contents, parameters of the processing method, preheating of SiC particles, surface modification of SiC particles with different coatings of Cu, CNT, Si, *etc* [186 - 190].

CONCLUDING REMARKS

In this review article, the recent scenario of SiC-AMCs processing techniques has been discussed. The interfacial reactions between liquid Al melt and SiC have also been explored along with their possible remedies. In general, the above literature discussion can be summarized as follows:

- Liquid fabrication processes (especially stir casting) for SiC-AMCs are used effectively for mass production, intricate shapes, a variety of secondary reinforcements (metallic/ceramics, agro waste, industrial waste), nano-composites, *etc.* but up to 30% of reinforcement contents, because the difference in densities of constituents caused agglomeration and clusters formation. This drawback can be overcome using solid-state processing techniques (especially powder metallurgy) which provide a more uniform dispersion of SiC contents than liquid processing techniques. Also, these processes are performed below melting temperature of matrices, resulting in least possible interfacial reactions which means unwanted compounds formation. The semi-solid techniques are the combination of the liquid state and solid state processing and carried out below liquidus temperature of matrix, which result in least interfacial reactions. The *in-situ* fabrication process is used to overcome several drawbacks of *ex-situ* processes such as porosity and unequal dispersion of reinforcements, improper wettability, unwanted interfacial reaction, segregation, costly equipment and reinforcement, gas entrapment, *etc.*
- The formation of an oxide layer on the aluminium surface, gas film around SiC particles prevented the contact of SiC with molten aluminium matrix, which causes an exothermic reaction between constituents resulting in reduced wettability. The outcome of this reaction formed an Al_4C_3 compound, which produced a deleterious effect on composite properties. The formation of this compound can be prevented by increasing Si contents in base metal, pre-oxidizing SiC, using Al alloy containing high Si contents, optimizing parameters of the processing method, preheating of SiC particles, modifying SiC particles with different coatings like Cu, CNT, Si, *etc.*

LIST OF ABBREVIATIONS

Al   Aluminum

MMCs   Metal Matrix Composites

AMCs   Aluminum Matrix Composites

SiC   Silicon Carbide

TiB₂   Titanium Diboride

MWCNT Multi Wall Carbon Nano Tube

RHA   Rice Husk Ash

FA	Flyash
CCA	Corn Cob Ash
FSP	Friction Stir Process
SC	Stir Casting
PM	Powder Metallurgy
SQC	Squeeze Casting
B$_4$C	Boron Carbide
Al$_2$O$_3$	Aluminum Oxide
ARB	Accumulative Roll Bonding
Al$_4$C$_3$	Aluminium Carbide
Gr	Graphite
BLA	Bamboo Leaf Ash

ACKNOWLEDGEMENTS

The authors will like to acknowledge IKG Punjab Technical University, Kapurthala, Punjab, India for providing an opportunity to write this review article.

REFERENCES

[1]   Kumar, J.; Singh, D.; Kalsi, N.S.; Sharma, S. Influence of reinforcement contents and turning parameters on the machining behaviour of Al/SiC/Cr hybrid aluminium matrix composites. In: *Additive and Subtractive Manufacturing of Composites*; Springer: Singapore, **2021**; pp. 33-51. [http://dx.doi.org/10.1007/978-981-16-3184-9_2]

[2]   Kala, H.; Mer, K.K.S.; Kumar, S. A review on mechanical and tribological behaviors of stir cast aluminum matrix composites. *Procedia Mater Sci.,* **2014**, *6*, 1951-1960.

[3]   Dhavamani, C.; Alwarsamy, T. Optimization of machining parameters for aluminum and silicon carbide composite using genetic algorithm. *Procedia Eng.,* **2012**, *38*, 1994-2004. [http://dx.doi.org/10.1016/j.proeng.2012.06.241]

[4]   Kumar, J.; Kumar, V.; Sharma, S.; Chohan, J.; Kumar, R.; Singh, S.; Obaid, A.J.; Akram, S.V. Optimizations of reinforcing particulates and processing parameters for stir casting of aluminium metal matrix composites for sustainable properties. *Mater. Today Proc.,* **2022**, *68*, 1172-1179. [http://dx.doi.org/10.1016/j.matpr.2022.10.109]

[5]   Miklaszewski, A. Effect of starting material character and its sintering temperature on microstructure and mechanical properties of super hard Ti/TiB metal matrix composites. *Int. J. Refract. Hard Met.,* **2015**, *53*, 56-60. [http://dx.doi.org/10.1016/j.ijrmhm.2015.04.011]

[6]   Uhlmann, E.; Bergmann, A.; Gridin, W. Investigation on additive manufacturing of tungsten carbide-cobalt by selective laser melting. *Procedia CIRP,* **2015**, *35*, 8-15. [http://dx.doi.org/10.1016/j.procir.2015.08.060]

[7]   Kumar, J.; Sharma, S.; Singh, J.; Singh, S.; Singh, G. Optimization of Wire-EDM process parameters for Al-Mg-0.6Si-0.35Fe/15%RHA/5%Cu hybrid metal matrix composite using TOPSIS: Processing and characterizations. *J. Manuf. Mater. Process.,* **2022**, *6*(6), 150. [http://dx.doi.org/10.3390/jmmp6060150]

[8]   Cao, H.; Qian, Z.; Zhang, L.; Xiao, J.; Zhou, K. Tribological behavior of Cu matrix composites containing graphite and tungsten disulfide. *Tribol. Trans.,* **2014**, *57*(6), 1037-1043.
[http://dx.doi.org/10.1080/10402004.2014.931499]

[9]   Tiwary, A.K.; Singh, S.; Kumar, R.; Sharma, K.; Chohan, J.S.; Sharma, S.; Singh, J.; Kumar, J.; Deifalla, A.F. Comparative study on the behavior of reinforced concrete beam retrofitted with CFRP strengthening techniques. *Polymers,* **2022**, *14*(19), 4024.
[http://dx.doi.org/10.3390/polym14194024] [PMID: 36235971]

[10]   Alaneme, K.K.; Aluko, A.O. Fracture toughness () and tensile properties of as-cast and age-hardened aluminium (6063)–silicon carbide particulate composites. *Sci. Iran.,* **2012**, *19*(4), 992-996.
[http://dx.doi.org/10.1016/j.scient.2012.06.001]

[11]   Delannay, F.; Froyen, L.; Deruyttere, A. The wetting of solids by molten metals and its relation to the preparation of metal-matrix composites composites. *J. Mater. Sci.,* **1987**, *22*(1), 1-16.
[http://dx.doi.org/10.1007/BF01160545]

[12]   Kumar, J.; Singh, D.; Kalsi, N.S. Tribological, physical and microstructural characterization of silicon carbide reinforced aluminium matrix composites: A review. *Mater. Today Proc.,* **2019**, *18*, 3218-3232.
[http://dx.doi.org/10.1016/j.matpr.2019.07.198]

[13]   Singh, K.; Khanna, V.; Rosenkranz, A. Panorama of physico-mechanical engineering of graphene-reinforced copper composites for sustainable applications. *Materials Today Sustainability.,* **2023**, *100560*, 100560.
[http://dx.doi.org/10.1007/978-981-16-3184-9_2]

[14]   Shirvanimoghaddam, K.; Khayyam, H.; Abdizadeh, H.; Karbalaei Akbari, M.; Pakseresht, A.H.; Abdi, F.; Abbasi, A.; Naebe, M. Effect of B4C, TiB2 and ZrSiO4 ceramic particles on mechanical properties of aluminium matrix composites: Experimental investigation and predictive modelling. *Ceram. Int.,* **2016**, *42*(5), 6206-6220.
[http://dx.doi.org/10.1016/j.ceramint.2015.12.181]

[15]   Karabulut, Ş. Optimization of surface roughness and cutting force during AA7039/Al₂O₃ metal matrix composites milling using neural networks and Taguchi method. *Measurement,* **2015**, *66*, 139-149.
[http://dx.doi.org/10.1016/j.measurement.2015.01.027]

[16]   James, S.J.; Venkatesan, K.; Kuppan, P.; Ramanujam, R. Hybrid aluminium metal matrix composite reinforced with SiC and TiB2. *Procedia Eng.,* **2014**, *97*, 1018-1026.
[http://dx.doi.org/10.1016/j.proeng.2014.12.379]

[17]   Chandrashekara, K.N.; Murthy, B.N.; Sreenivasa, K.; Krupakara, P.V. Corrosion properties of advanced materials like aluminium 6013 metal matrix composites reinforced with red mud particulates in acid chloride medium. *Asian J. Chem.,* **2016**, *28*(8), 1770-1772.
[http://dx.doi.org/10.14233/ajchem.2016.19817]

[18]   Mahendra, K.V.; Radhakrishna, K. Fabrication of Al-4.5% Cu alloy with fly ash metal matrix composites and its characterization. *materials science-Poland,* **2007**, *25*(1), 57-68.

[19]   Geetha, B.; Ganesan, K. Optimization of tensile characteristics of Al 356 alloy reinforced with volume fraction of red mud metal matrix composite. *Procedia Eng.,* **2014**, *97*, 614-624.
[http://dx.doi.org/10.1016/j.proeng.2014.12.290]

[20]   Hassan, S.B.; Aigbodion, V.S. Effects of eggshell on the microstructures and properties of Al–Cu–Mg/eggshell particulate composites. *J. King Saud Univ. Eng. Sci. ,* **2015**, *27*(1), 49-56.
[http://dx.doi.org/10.1016/j.jksues.2013.03.001]

[21]   Oghenevweta, J.E.; Aigbodion, V.S.; Nyior, G.B.; Asuke, F. Mechanical properties and microstructural analysis of Al–Si–Mg/carbonized maize stalk waste particulate composites. *J. King Saud Univ. Eng. Sci. ,* **2016**, *28*(2), 222-229.
[http://dx.doi.org/10.1016/j.jksues.2014.03.009]

[22]   Atuanya, C.U.; Ibhadode, A.O.A.; Dagwa, I.M. Effects of breadfruit seed hull ash on the

microstructures and properties of Al–Si–Fe alloy/breadfruit seed hull ash particulate composites. *Results Phys.,* **2012**, *2*, 142-149.
[http://dx.doi.org/10.1016/j.rinp.2012.09.003]

[23]   Bello, S.A.; Raheem, I.A.; Raji, N.K. Study of tensile properties, fractography and morphology of aluminium (1xxx)/coconut shell micro particle composites. *J. King Saud Univ. Eng. Sci. ,* **2017**, *29*(3), 269-277.
[http://dx.doi.org/10.1016/j.jksues.2015.10.001]

[24]   Kumar, J.; Singh, R. Investigating the effect of vulcanization in tread rubber applications. *Mater. Sci. Forum,* **2013**, *751*, 1-7. Trans Tech Publications Ltd.
[http://dx.doi.org/10.4028/www.scientific.net/MSF.751.1]

[25]   Rajmohan, T.; Palanikumar, K. Application of the central composite design in optimization of machining parameters in drilling hybrid metal matrix composites. *Measurement,* **2013**, *46*(4), 1470-1481.
[http://dx.doi.org/10.1016/j.measurement.2012.11.034]

[26]   Ozben, T.; Kilickap, E.; Çakır, O. Investigation of mechanical and machinability properties of SiC particle reinforced Al-MMC. *J. Mater. Process. Technol.,* **2008**, *198*(1-3), 220-225.
[http://dx.doi.org/10.1016/j.jmatprotec.2007.06.082]

[27]   Sultan, U.; Kumar, J.; Dadra, S.; Kumar, S. Experimental investigations on the tribological behaviour of advanced aluminium metal matrix composites using grey relational analysis. *Mater. Today Proc.,* **2022**.
[http://dx.doi.org/10.1016/j.matpr.2022.12.171]

[28]   Saha, P.; Tarafdar, D.; Pal, S.K.; Saha, P.; Srivastava, A.K.; Das, K. Multi-objective optimization in wire-electro-discharge machining of TiC reinforced composite through Neuro-Genetic technique. *Appl. Soft Comput.,* **2013**, *13*(4), 2065-2074.
[http://dx.doi.org/10.1016/j.asoc.2012.11.008]

[29]   Gowda, B.U.; Ravindra, H.V.; Ullas, M.; Prakash, G.N.; Ugrasen, G. Estimation of circularity, cylindricity and surface roughness in drilling Al-Si3N4 metal matrix composites using artificial neural network. *Procedia. Mat. Sci.,* **2014**, *6*, 1780-1787.

[30]   David Raja Selvam, J.; Robinson Smart, D.S.; Dinaharan, I. Synthesis and characterization of Al6061-Fly Ashp-SiCp composites by stir casting and compocasting methods. *Energy Procedia,* **2013**, *34*, 637-646.
[http://dx.doi.org/10.1016/j.egypro.2013.06.795]

[31]   Rahman, M.H.; Rashed, H.M.M.A. Characterization of silicon carbide reinforced aluminum matrix composites. *Procedia Eng.,* **2014**, *90*, 103-109.
[http://dx.doi.org/10.1016/j.proeng.2014.11.821]

[32]   Mishra, A.K.; Sheokand, R.; Srivastava, R.K. Tribological behaviour of Al-6061/SiC metal matrix composite by Taguchi's techniques. *International Journal of Scientific and Research Publications,* **2012**, *2*(10), 1-8.
[http://dx.doi.org/10.15373/22778179/OCT2013/37]

[33]   Shivalingappa, D. *In-Situ* Magnesium Based Composites-Development and Tribological Behavior **2007**,

[34]   Singh, G.; Singh, H.; Kumar, J. Effect of coating thickness with carbide tool in hard turning of AISID3 cold work steel. *Int. Res. J. Eng. Technol.,* **2017**, *4*(04), 436-440. [IRJET].

[35]   Nishida, Y. *Introduction to metal matrix composites: Fabrication and recycling*; Springer Science & Business Media, **2013**.
[http://dx.doi.org/10.1007/978-4-431-54237-7]

[36]   Dwivedi, S.P.; Sharma, S.; Mishra, R.K. RETRACTED: Microstructure and mechanical properties of A356/SiC composites fabricated by electromagnetic stir casting. *Procedia. Mater. Sci.,* **2014**, *6*, 1524-

1532.
[http://dx.doi.org/10.1016/j.mspro.2014.07.133]

[37]   Naebe, M.; Shirvanimoghaddam, K. Functionally graded materials: A review of fabrication and properties. *Appl. Mater. Today,* **2016**, *5*, 223-245.
[http://dx.doi.org/10.1016/j.apmt.2016.10.001]

[38]   Shirvanimoghaddam, K.; Hamim, S.U.; Karbalaei Akbari, M.; Fakhrhoseini, S.M.; Khayyam, H.; Pakseresht, A.H.; Ghasali, E.; Zabet, M.; Munir, K.S.; Jia, S.; Davim, J.P.; Naebe, M. Carbon fiber reinforced metal matrix composites: Fabrication processes and properties. *Compos., Part A Appl. Sci. Manuf.,* **2017**, *92*, 70-96.
[http://dx.doi.org/10.1016/j.compositesa.2016.10.032]

[39]   Vanarotti, M.; Shrishail, P.; Sridhar, B.R.; Venkateswarlu, K.; Kori, S.A. Study of mechanical properties & residual stresses on post wear samples of A356-SiC metal matrix composites. *Procedia. Mater. Sci.,* **2014**, *5*, 873-882.
[http://dx.doi.org/10.1016/j.mspro.2014.07.374]

[40]   Shorowordi, K.M.; Laoui, T.; Haseeb, A.S.M.A.; Celis, J.P.; Froyen, L. Microstructure and interface characteristics of B4C, SiC and Al_2O_3 reinforced Al matrix composites: A comparative study. *J. Mater. Process. Technol.,* **2003**, *142*(3), 738-743.
[http://dx.doi.org/10.1016/S0924-0136(03)00815-X]

[41]   Prasad, D.S.; Shoba, C.; Ramanaiah, N. Investigations on mechanical properties of aluminum hybrid composites. *J. Mater. Res. Technol.,* **2014**, *3*(1), 79-85.
[http://dx.doi.org/10.1016/j.jmrt.2013.11.002]

[42]   Kumar, R.; Chauhan, S. Study on surface roughness measurement for turning of Al 7075/10/SiCp and Al 7075 hybrid composites by using response surface methodology (RSM) and artificial neural networking (ANN). *Measurement,* **2015**, *65*, 166-180.
[http://dx.doi.org/10.1016/j.measurement.2015.01.003]

[43]   Alaneme, K.K.; Adewale, T.M.; Olubambi, P.A. Corrosion and wear behaviour of Al–Mg–Si alloy matrix hybrid composites reinforced with rice husk ash and silicon carbide. *J. Mater. Res. Technol.,* **2014**, *3*(1), 9-16.
[http://dx.doi.org/10.1016/j.jmrt.2013.10.008]

[44]   Rajmohan, T.; Palanikumar, K. Experimental investigation and analysis of thrust force in drilling hybrid metal matrix composites by coated carbide drills. *Mater. Manuf. Process.,* **2011**, *26*(8), 961-968.
[http://dx.doi.org/10.1080/10426914.2010.523915]

[45]   Thomas, A.T.; Parameshwaran, R.; Muthukrishnan, A.; Kumaran, M.A. Development of feeding & stirring mechanisms for stir casting of aluminium matrix composites. *Procedia. Mater. Sci.,* **2014**, *5*, 1182-1191.
[http://dx.doi.org/10.1016/j.mspro.2014.07.415]

[46]   Ambhai, K.G. *Study on machinability of Al-SiC particulate metal matrix composite ,* **2007**,

[47]   Muthukrishnan, N.; Babu, T.S.M.; Ramanujam, R. Fabrication and turning of Al/SiC/B $_4$ C hybrid metal matrix composites optimization using desirability analysis. *J Chin Inst Ind Eng,* **2012**, *29*(8), 515-525.
[http://dx.doi.org/10.1080/10170669.2012.728540]

[48]   Umanath, K.; Selvamani, S.T.; Palanikumar, K.; Sabarikreeshwaran, R. Dry sliding wear behaviour of AA6061-T6 reinforced SiC and Al_2O_3 particulate hybrid composites. *Procedia Eng.,* **2014**, *97*, 694-702.
[http://dx.doi.org/10.1016/j.proeng.2014.12.299]

[49]   Palanikumar, K.; Muthukrishnan, N.; Hariprasad, K.S. Surface roughness parameters optimization in machining a356/sic/20 P metal matrix composites by pcd tool using response surface methodology and desirability function. *Mach. Sci. Technol.,* **2008**, *12*(4), 529-545.

[http://dx.doi.org/10.1080/10910340802518850]

[50] Lokesh, T.; Mallikarjun, U.S. Mechanical and morphological studies of Al6061-Gr-SiC hybrid metal matrix composites. *Appl. Mech. Mater.,* **2015**, *813-814*, 195-202. Trans Tech Publications Ltd. [http://dx.doi.org/10.4028/www.scientific.net/AMM.813-814.195]

[51] Kumar, J.; Singh, D.; Kalsi, N.S.; Sharma, S.; Mia, M.; Singh, J.; Rahman, M.A.; Khan, A.M.; Rao, K.V. Investigation on the mechanical, tribological, morphological and machinability behavior of stir-casted Al/SiC/Mo reinforced MMCs. *J. Mater. Res. Technol.,* **2021**, *12*, 930-946. [http://dx.doi.org/10.1016/j.jmrt.2021.03.034]

[52] Singh, B.; Kumar, J.; Kumar, S. Investigation of the tool wear rate in tungsten powder-mixed electric discharge machining of AA6061/10% SiCp composite. *Mater. Manuf. Process.,* **2016**, *31*(4), 456-466. [http://dx.doi.org/10.1080/10426914.2015.1025965]

[53] Kumar, J.; Singh, D.; Kalsi, N.S.; Sharma, S.; Pruncu, C.I.; Pimenov, D.Y.; Rao, K.V.; Kapłonek, W. Comparative study on the mechanical, tribological, morphological and structural properties of vortex casting processed, Al–SiC–Cr hybrid metal matrix composites for high strength wear-resistant applications: Fabrication and characterizations. *J. Mater. Res. Technol.,* **2020**, *9*(6), 13607-13615. [http://dx.doi.org/10.1016/j.jmrt.2020.10.001]

[54] Patel, K.M.; Pandey, P.M.; Rao, P.V. Understanding the role of weight percentage and size of silicon carbide particulate reinforcement on electro-discharge machining of aluminium-based composites. *Mater. Manuf. Process.,* **2008**, *23*(7), 665-673. [http://dx.doi.org/10.1080/15560350802316702]

[55] Pawar, P.B.; Utpat, A.A. Development of aluminium based silicon carbide particulate metal matrix composite for spur gear. *Procedia Mater Sci,* **2014**, *6*, 1150-1156.

[56] Venkatesan, K.; Ramanujam, R.; Joel, J.; Jeyapandiarajan, P.; Vignesh, M.; Tolia, D.J.; Krishna, R.V. Study of cutting force and surface roughness in machining of Al alloy hybrid composite and optimized using response surface methodology. *Procedia Eng.,* **2014**, *97*, 677-686. [http://dx.doi.org/10.1016/j.proeng.2014.12.297]

[57] Shoba, C.; Ramanaiah, N.; Rao, D.N. Effect of reinforcement on the cutting forces while machining metal matrix composites–an experimental approach *Eng. Sci. Technol. Int. J.,* **2015**, *18*(4), 658-663.

[58] Narasimha, G.B.; Krishna, M.V.; Sindhu, R. Prediction of wear behaviour of almg1sicu hybrid MMC using taguchi with grey rational analysis. *Procedia Eng.,* **2014**, *97*, 555-562. [http://dx.doi.org/10.1016/j.proeng.2014.12.283]

[59] Boopathi, M.M.; Arulshri, K.P.; Iyandurai, N. Evaluation of mechanical properties of aluminium alloy 2024 reinforced with silicon carbide and fly ash hybrid metal matrix composites. *Am. J. Appl. Sci.,* **2013**, *10*(3), 219-229. [http://dx.doi.org/10.3844/ajassp.2013.219.229]

[60] Alaneme, K.K.; Ademilua, B.O. Bodunrin. MO,(2013)," Mechanical properties and corrosion behaviour of aluminium hybrid composites reinforced with silicon carbide and bamboo leaf ash. *Tribology in Industry,* **2013**, *35*(1), 25-35.

[61] Tiruvenkadam, N.; Thyla, P.R.; Senthilkumar, M.; Bharathiraja, M. Development of optimum friction new nano hybrid composite liner for biodiesel fuel engine. *Transp. Res. Part D Transp. Environ.,* **2016**, *47*, 22-43. [http://dx.doi.org/10.1016/j.trd.2016.03.020]

[62] Radha, A.; Vijayakumar, K.R. An investigation of mechanical and wear properties of AA6061 reinforced with silicon carbide and graphene nano particles-Particulate composites. *Mater. Today Proc.,* **2016**, *3*(6), 2247-2253. [http://dx.doi.org/10.1016/j.matpr.2016.04.133]

[63] Krishna, M.V.; Xavior, A.M. An investigation on the mechanical properties of hybrid metal matrix composites. *Procedia Eng.,* **2014**, *97*, 918-924.

[http://dx.doi.org/10.1016/j.proeng.2014.12.367]

[64] Suresh, P.; Marimuthu, K.; Ranganathan, S.; Rajmohan, T. Optimization of machining parameters in turning of Al-SiC-Gr hybrid metal matrix composites using grey-fuzzy algorithm. *Trans. Nonferrous Met. Soc. China,* **2014**, *24*(9), 2805-2814.
[http://dx.doi.org/10.1016/S1003-6326(14)63412-9]

[65] Rajmohan, T.; Palanikumar, K. Modeling and analysis of performances in drilling hybrid metal matrix composites using D-optimal design. *Int. J. Adv. Manuf. Technol.,* **2013**, *64*(9-12), 1249-1261.
[http://dx.doi.org/10.1007/s00170-012-4083-6]

[66] Fatile, O.B.; Akinruli, J.I.; Amori, A.A. Microstructure and mechanical behaviour of stir-cast Al-M-
-SI alloy matrix hybrid composite reinforced with corn cob ash and silicon carbide. *Int. J. Eng. Technol. Innovat.,* **2014**, *4*(4), 251.

[67] Sangeetha, M.; Prakash, S.; Ramya, D. Wear properties estimation and characterization of coated ceramics with multiwall carbon nano tubes (MWCNT) reinforced in aluminium matrix composites. *Mater. Today Proc.,* **2016**, *3*(6), 2537-2546.
[http://dx.doi.org/10.1016/j.matpr.2016.04.173]

[68] Johnson, W.B.; Sonuparlak, B. Diamond/Al metal matrix composites formed by the pressureless metal infiltration process. *J. Mater. Res.,* **1993**, *8*(5), 1169-1173.
[http://dx.doi.org/10.1557/JMR.1993.1169]

[69] Yang, J.; Chung, D.D.L. Casting particulate and fibrous metal-matrix composites by vacuum infiltration of a liquid metal under an inert gas pressure. *J. Mater. Sci.,* **1989**, *24*(10), 3605-3612.
[http://dx.doi.org/10.1007/BF02385746]

[70] Kaptay, G.; BÁrczy, T. On the asymmetrical dependence of the threshold pressure of infiltration on the wettability of the porous solid by the infiltrating liquid. *J. Mater. Sci.,* **2005**, *40*(9-10), 2531-2535.
[http://dx.doi.org/10.1007/s10853-005-1987-7]

[71] Morgan, P. *Carbon fibers and their composites*; CRC press, **2005**.
[http://dx.doi.org/10.1201/9781420028744]

[72] Baker, A.A. Carbon fibre reinforced metals — a review of the current technology. *Mater. Sci. Eng.,* **1975**, *17*(2), 177-208.
[http://dx.doi.org/10.1016/0025-5416(75)90231-1]

[73] Aghajanian, M.K.; Burke, J.T.; White, D.R.; Nagelberg, A.S. A new infiltration process for the fabrication of metal matrix composites *SAMPE quarterly,* **1989**, *20*, 43-46.

[74] Yan, C.; Lifeng, W.; Jianyue, R. Multi-functional SiC/Al composites for aerospace applications. *Chin. J. Aeronauti.,* **2008**, *21*(6), 578-584.
[http://dx.doi.org/10.1016/S1000-9361(08)60177-6]

[75] Xu, F.; WU Lawrence, C.; Han, G.; Tan, Y. Compression creep behavior of high volume fraction of SiC particles reinforced Al composite fabricated by pressureless infiltration. *Chin. J. Aeronauti.,* **2007**, *20*(2), 115-119.
[http://dx.doi.org/10.1016/S1000-9361(07)60016-8]

[76] Lee, K.B.; Sim, H.S.; Kim, S.H.; Han, K.H.; Kwon, H. Fabrication and characteristics of AA6061/SiCp composites by pressureless infiltration technique. *J. Mater. Sci.,* **2001**, *36*(13), 3179-3188.
[http://dx.doi.org/10.1023/A:1017978117837]

[77] Yang, L.; Zhang, M. Fabrication of SiC$_p$/Cu–Al electronic packaging material by pressureless infiltration method. *Mater. Sci. Technol.,* **2013**, *29*(3), 326-331.
[http://dx.doi.org/10.1179/1743284712Y.0000000152]

[78] Hemambar, C.; Rao, B.S.; Jayaram, V. *Al–SiC electronic packages with controlled thermal expansion coefficient by a new method of pressureless infiltration.,* **2001**.
[http://dx.doi.org/10.1081/AMP-100108698]

[79]   Zhang, Q.; Ma, X.; Wu, G. Interfacial microstructure of SiCp/Al composite produced by the pressureless infiltration technique. *Ceram. Int.,* **2013**, *39*(5), 4893-4897. [http://dx.doi.org/10.1016/j.ceramint.2012.11.082]

[80]   Chen, Y.; Chung, D.D.L. Aluminium-matrix silicon carbide whisker composites fabricated by pressureless infiltration. *J. Mater. Sci.,* **1996**, *31*(2), 407-412. [http://dx.doi.org/10.1007/BF01139158]

[81]   Cui, Y.; Jin, T.; Cao, L.; Liu, F. Aging behavior of high volume fraction SiCp/Al composites fabricated by pressureless infiltration. *J. Alloys Compd.,* **2016**, *681*, 233-239. [http://dx.doi.org/10.1016/j.jallcom.2016.04.127]

[82]   Mohd Salim, N.N.; Zuhailawati, H.; Mohamad, H.; Anasyida, A.S. Sic-reinforced aluminum-silicon composite *via* pressureless infiltration using polystyrene as external binder for electronic assemblies. *Mater. Sci. Forum,* **2015**, *819*, 220-225. Trans Tech Publications Ltd. [http://dx.doi.org/10.4028/www.scientific.net/MSF.819.220]

[83]   Yang, Y.; Li, X. Ultrasonic cavitation-based nanomanufacturing of bulk aluminum matrix nanocomposites **2007**, *129*, 497-508.

[84]   Poovazhagan, L.; Kalaichelvan, K.; Rajadurai, A.; Senthilvelan, V. Characterization of hybrid silicon carbide and boron carbide nanoparticles-reinforced aluminum alloy composites. *Procedia Eng.,* **2013**, *64*, 681-689. [http://dx.doi.org/10.1016/j.proeng.2013.09.143]

[85]   Rana, R.S.; Purohit, R.; Sharma, A.; Rana, S. Optimization of wear performance of Aa 5083/10 Wt.% Sicp composites using Taguchi method. *Procedia. Mater. Sci.,* **2014**, *6*, 503-511. [http://dx.doi.org/10.1016/j.mspro.2014.07.064]

[86]   Gopalakannan, V.; Viswanathan, N. Synthesis of magnetic alginate hybrid beads for efficient chromium (VI) removal. *Int. J. Biol. Macromol.,* **2015**, *72*, 862-867. [http://dx.doi.org/10.1016/j.ijbiomac.2014.09.024] [PMID: 25256552]

[87]   Murthy, N.V.; Reddy, A.P.; Selvaraj, N.; Rao, C.S.P. Preparation of SiC based Aluminium metal matrix nano composites by high intensity ultrasonic cavitation process and evaluation of mechanical and tribological properties. *IOP Conf. Series Mater. Sci. Eng.,* **2016**, *149*(1), 012106. IOP Publishing. [http://dx.doi.org/10.1088/1757-899X/149/1/012106]

[88]   Towata, S.I.; Sen, I.Y.; Ohwaki, T. Strength and interfacial reaction of high modulus carbon fiber-reinforced aluminum alloys. *Trans. Jpn. Inst. Met.,* **1985**, *26*(8), 563-570. [http://dx.doi.org/10.2320/matertrans1960.26.563]

[89]   Dhanashekar, M.; Kumar, V.S. Squeeze casting of aluminium metal matrix composites-an overview. *Procedia Eng.,* **2014**, *97*, 412-420. [http://dx.doi.org/10.1016/j.proeng.2014.12.265]

[90]   Vijayaram, T.R.; Sulaiman, S.; Hamouda, A.M.S.; Ahmad, M.H.M. Fabrication of fiber reinforced metal matrix composites by squeeze casting technology. *J. Mater. Process. Technol.,* **2006**, *178*(1-3), 34-38. [http://dx.doi.org/10.1016/j.jmatprotec.2005.09.026]

[91]   Yang, L.J. The effect of casting temperature on the properties of squeeze cast aluminium and zinc alloys. *J. Mater. Process. Technol.,* **2003**, *140*(1-3), 391-396. [http://dx.doi.org/10.1016/S0924-0136(03)00763-5]

[92]   Mosisa, G.E.; Bazhin, V.Y. Properties of aluminum metal matrix composites reinforced by particles of silicon carbide using powder metallurgy *Metall. Min. Ind.,* **2017**, *2*

[93]   Seyed Reihani, S.M. Processing of squeeze cast Al6061–30vol% SiC composites and their characterization. *Mater. Des.,* **2006**, *27*(3), 216-222. [http://dx.doi.org/10.1016/j.matdes.2004.10.016]

[94] Singh, S.; Gangwar, S.; Yadav, S. A review on mechanical and tribological properties of micro/nano filled metal alloy composites. *Mater. Today Proc.,* **2017**, *4*(4), 5583-5592.
[http://dx.doi.org/10.1016/j.matpr.2017.06.015]

[95] Taufik, M.; Widyastuti, M.T.; Sulaiman, A.; Murdiyarso, D.; Santikayasa, I.P.; Minasny, B. An improved drought-fire assessment for managing fire risks in tropical peatlands. *Agric. For. Meteorol.,* **2022**, *312*, 108738.
[http://dx.doi.org/10.1016/j.agrformet.2021.108738]

[96] Kalkanlı, A.; Yılmaz, S. Synthesis and characterization of aluminum alloy 7075 reinforced with silicon carbide particulates. *Mater. Des.,* **2008**, *29*(4), 775-780.
[http://dx.doi.org/10.1016/j.matdes.2007.01.007]

[97] Karnezis, P.A.; Durrant, G.; Cantor, B. Microstructure and tensile properties of squeeze cast SiC particulate reinforced AI-7Si alloy. *Mater. Sci. Technol.,* **1998**, *14*(2), 97-107.
[http://dx.doi.org/10.1179/mst.1998.14.2.97]

[98] Raju, K.; Ojha, S.N.; Harsha, A.P. Spray forming of aluminum alloys and its composites: An overview. *J. Mater. Sci.,* **2008**, *43*(8), 2509-2521.
[http://dx.doi.org/10.1007/s10853-008-2464-x]

[99] Gui, M.; Kang, S.B.; Euh, K. Microstructure characteristics of SiC particle-reinforced aluminum matrix composites by plasma spraying. *J. Therm. Spray Technol.,* **2004**, *13*(4), 537-543.
[http://dx.doi.org/10.1361/10599630421451]

[100] Srivastava, V.C.; Ojha, S.N. Microstructure and electrical conductivity of Al-SiCp composites produced by spray forming process. *Bull. Mater. Sci.,* **2005**, *28*(2), 125-130.
[http://dx.doi.org/10.1007/BF02704231]

[101] Sharifitabar, M.; Sarani, A.; Khorshahian, S.; Shafiee Afarani, M. Fabrication of 5052Al/Al$_2$O$_3$ nanoceramic particle reinforced composite *via* friction stir processing route. *Mater. Des.,* **2011**, *32*(8-9), 4164-4172.
[http://dx.doi.org/10.1016/j.matdes.2011.04.048]

[102] Dhayalan, R.; Kalaiselvan, K.; Sathiskumar, R. Characterization of AA6063/SiC-Gr surface composites produced by FSP technique. *Procedia Eng.,* **2014**, *97*, 625-631.
[http://dx.doi.org/10.1016/j.proeng.2014.12.291]

[103] Mishra, R.S.; Ma, Z.Y.; Charit, I. Friction stir processing: A novel technique for fabrication of surface composite. *Mater. Sci. Eng. A,* **2003**, *341*(1-2), 307-310.
[http://dx.doi.org/10.1016/S0921-5093(02)00199-5]

[104] Kurt, A.; Uygur, I.; Cete, E. Surface modification of aluminium by friction stir processing. *J. Mater. Process. Technol.,* **2011**, *211*(3), 313-317.
[http://dx.doi.org/10.1016/j.jmatprotec.2010.09.020]

[105] Mahmoud, E.R.I.; Takahashi, M.; Shibayanagi, T.; Ikeuchi, K. Wear characteristics of surface-hybri--MMCs layer fabricated on aluminum plate by friction stir processing. *Wear,* **2010**, *268*(9-10), 1111-1121.
[http://dx.doi.org/10.1016/j.wear.2010.01.005]

[106] Devaraju, A.; Kumar, A.; Kotiveerachari, B. Influence of addition of Grp/Al$_2$O$_3$p with SiCp on wear properties of aluminum alloy 6061-T6 hybrid composites *via* friction stir processing. *Trans. Nonferrous Met. Soc. China,* **2013**, *23*(5), 1275-1280.
[http://dx.doi.org/10.1016/S1003-6326(13)62593-5]

[107] Puviyarasan, M.; Pushkaran, S.; Senthil, T.S.; Karthikeyan, K.R. Effect of process parameters on material strength (AA6061-T6) during Friction Stir Processing: Simulation and experimental validation *Materials Today: Proceedings,* **2022**.

[108] Sahraeinejad, S.; Izadi, H.; Haghshenas, M.; Gerlich, A.P. Fabrication of metal matrix composites by friction stir processing with different Particles and processing parameters. *Mater. Sci. Eng. A,* **2015**,

626, 505-513.
[http://dx.doi.org/10.1016/j.msea.2014.12.077]

[109] Puhan, D.; Mahapatra, S.S.; Sahu, J.; Das, L. A hybrid approach for multi-response optimization of non-conventional machining on AlSiCp MMC. *Measurement,* **2013**, *46*(9), 3581-3592.
[http://dx.doi.org/10.1016/j.measurement.2013.06.007]

[110] Ogawa, F.; Masuda, C. Microstructure evolution during fabrication and microstructure–property relationships in vapour-grown carbon nanofibre-reinforced aluminium matrix composites fabricated *via* powder metallurgy. *Compos., Part A Appl. Sci. Manuf.,* **2015**, *71*, 84-94.
[http://dx.doi.org/10.1016/j.compositesa.2015.01.005]

[111] Ropars, L.; Dehmas, M.; Gourdet, S.; Delfosse, J.; Tricker, D.; Aeby-Gautier, E. Structure evolutions in a Ti–6Al–4V matrix composite reinforced with TiB, characterised using high energy X-ray diffraction. *J. Alloys Compd.,* **2015**, *624*, 179-188.
[http://dx.doi.org/10.1016/j.jallcom.2014.10.203]

[112] Jayakumar, K.; Mathew, J.; Joseph, M.A.; Kumar, R.S.; Shukla, A.K.; Samuel, M.G. Synthesis and characterization of A356-SiCp composite produced through vacuum hot pressing. *Mater. Manuf. Process.,* **2013**, *28*(9), 991-998.

[113] Alves, S.J.F.; de Sousa, M.M.S.; de Araújo, E.R.; Filho, F.A.; dos Santos, M.J.; de Araújo Filho, O.O. Processing of metal matrix AA2124 aluminium alloy composites reinforced by alumina and silicon carbide by powder metallurgy techniques. *Mater. Sci. Forum,* **2014**, *802*, 84-89. Trans Tech Publications Ltd.
[http://dx.doi.org/10.4028/www.scientific.net/MSF.802.84]

[114] Abhik, R.; Umasankar, V.; Xavior, M.A. Evaluation of properties for Al-SiC reinforced metal matrix composite for brake pads. *Procedia Eng.,* **2014**, *97*, 941-950.
[http://dx.doi.org/10.1016/j.proeng.2014.12.370]

[115] Lee, H.S.; Yeo, J.S.; Hong, S.H.; Yoon, D.J.; Na, K.H. The fabrication process and mechanical properties of SiCp/Al–Si metal matrix composites for automobile air-conditioner compressor pistons. *J. Mater. Process. Technol.,* **2001**, *113*(1-3), 202-208.
[http://dx.doi.org/10.1016/S0924-0136(01)00680-X]

[116] Kwok, J.K.M.; Lim, S.C. High-speed tribological properties of some Al/SiCp composites: I. Frictional and wear-rate characteristics. *Compos. Sci. Technol.,* **1999**, *59*(1), 55-63.
[http://dx.doi.org/10.1016/S0266-3538(98)00055-4]

[117] Bodukuri, A.K.; Eswaraiah, K.; Rajendar, K.; Sampath, V. Fabrication of Al–SiC–B4C metal matrix composite by powder metallurgy technique and evaluating mechanical properties. *Perspect. Sci.,* **2016**, *8*, 428-431.
[http://dx.doi.org/10.1016/j.pisc.2016.04.096]

[118] Karabulut, Ş.; Gökmen, U.; Çinici, H. Study on the mechanical and drilling properties of AA7039 composites reinforced with Al₂O₃/B4C/SiC particles. *Compos., Part B Eng.,* **2016**, *93*, 43-55.
[http://dx.doi.org/10.1016/j.compositesb.2016.02.054]

[119] Kumar, C.A.V.; Rajadurai, J.S. Influence of rutile (TiO2) content on wear and microhardness characteristics of aluminium-based hybrid composites synthesized by powder metallurgy. *Trans. Nonferrous Met. Soc. China,* **2016**, *26*(1), 63-73.
[http://dx.doi.org/10.1016/S1003-6326(16)64089-X]

[120] Li, S.; Su, Y.; Ouyang, Q.; Zhang, D. *In-situ* carbon nanotube-covered silicon carbide particle reinforced aluminum matrix composites fabricated by powder metallurgy. *Mater. Lett.,* **2016**, *167*, 118-121.
[http://dx.doi.org/10.1016/j.matlet.2015.12.155]

[121] Ghasali, E.; Yazdani-rad, R.; Asadian, K.; Ebadzadeh, T. Production of Al-SiC-TiC hybrid composites using pure and 1056 aluminum powders prepared through microwave and conventional heating methods. *J. Alloys Compd.,* **2017**, *690*, 512-518.

[http://dx.doi.org/10.1016/j.jallcom.2016.08.145]

[122] Ghasali, E.; Pakseresht, A.; Rahbari, A.; Eslami-shahed, H.; Alizadeh, M.; Ebadzadeh, T. Mechanical properties and microstructure characterization of spark plasma and conventional sintering of Al–SiC–TiC composites. *J. Alloys Compd.,* **2016,** *666,* 366-371.
[http://dx.doi.org/10.1016/j.jallcom.2016.01.118]

[123] Hu, C.; Yan, H.; Chen, J.; Su, B. Microstructures and mechanical properties of 2024Al/Gr/SiC hybrid composites fabricated by vacuum hot pressing. *Trans. Nonferrous Met. Soc. China,* **2016,** *26*(5), 1259-1268.
[http://dx.doi.org/10.1016/S1003-6326(16)64226-7]

[124] Madeira, S.; Carvalho, O.; Carneiro, V.H.; Soares, D.; Silva, F.S.; Miranda, G. Damping capacity and dynamic modulus of hot pressed AlSi composites reinforced with different SiC particle sized. *Compos., Part B Eng.,* **2016,** *90,* 399-405.
[http://dx.doi.org/10.1016/j.compositesb.2016.01.008]

[125] Montalba, C.; Ramam, K.; Eskin, D.G.; Ruiz-Navas, E.M.; Prat, O. Fabrication of a novel hybrid AlMg5/SiC/PLZT metal matrix composite produced by hot extrusion. *Mater. Des.,* **2015,** *69,* 213-218.
[http://dx.doi.org/10.1016/j.matdes.2014.12.061]

[126] Carvalho, O.; Buciumeanu, M.; Madeira, S.; Soares, D.; Silva, F.S.; Miranda, G. Mechanisms governing the mechanical behavior of an AlSi–CNTs–SiCp hybrid composite. *Compos., Part B Eng.,* **2016,** *90,* 443-449.
[http://dx.doi.org/10.1016/j.compositesb.2016.01.032]

[127] Saito, Y.; Utsunomiya, H.; Tsuji, N.; Sakai, T. Novel ultra-high straining process for bulk materials—development of the accumulative roll-bonding (ARB) process. *Acta Mater.,* **1999,** *47*(2), 579-583.
[http://dx.doi.org/10.1016/S1359-6454(98)00365-6]

[128] Naseri, M.; Hassani, A.; Tajally, M. An alternative method for manufacturing Al/B4C/SiC hybrid composite strips by cross accumulative roll bonding (CARB) process. *Ceram. Int.,* **2015,** *41*(10), 13461-13469.
[http://dx.doi.org/10.1016/j.ceramint.2015.07.137]

[129] Fattah-alhosseini, A.; Naseri, M.; Alemi, M.H. Corrosion behavior assessment of finely dispersed and highly uniform Al/B4C/SiC hybrid composite fabricated *via* accumulative roll bonding process. *J. Manuf. Process.,* **2016,** *22,* 120-126.
[http://dx.doi.org/10.1016/j.jmapro.2016.03.006]

[130] Baazamat, S.; Tajally, M.; Borhani, E. Fabrication and characteristic of Al-based hybrid nanocomposite reinforced with WO3 and SiC by accumulative roll bonding process. *J. Alloys Compd.,* **2015,** *653,* 39-46.
[http://dx.doi.org/10.1016/j.jallcom.2015.08.267]

[131] Ardakani, M.R.K.; Amirkhanlou, S.; Khorsand, S. Cross accumulative roll bonding—A novel mechanical technique for significant improvement of stir-cast Al/Al$_2$O$_3$ nanocomposite properties. *Mater. Sci. Eng. A,* **2014,** *591,* 144-149.
[http://dx.doi.org/10.1016/j.msea.2013.10.073]

[132] Hosseini, S.M.; Habibolahzadeh, A.; Králík, V.; Němeček, J. Significant improvement in structural features, mechanical and physical properties of a novel CAR processed Al foam by nano-SiCp addition. *Mater. Sci. Eng. A,* **2016,** *670,* 342-350.
[http://dx.doi.org/10.1016/j.msea.2016.06.035]

[133] Darmiani, E.; Danaee, I.; Golozar, M.A.; Toroghinejad, M.R. Corrosion investigation of Al–SiC nano-composite fabricated by accumulative roll bonding (ARB) process. *J. Alloys Compd.,* **2013,** *552,* 31-39.
[http://dx.doi.org/10.1016/j.jallcom.2012.10.069]

[134] Alizadeh, M.; Paydar, M.H. Fabrication of nanostructure Al/SiCP composite by accumulative roll-

bonding (ARB) process. *J. Alloys Compd.,* **2010**, *492*(1-2), 231-235.
[http://dx.doi.org/10.1016/j.jallcom.2009.12.026]

[135] Ana, S.V.A.; Reihanian, M.; Lotfi, B. Accumulative roll bonding (ARB) of the composite coated strips to fabricate multi-component Al-based metal matrix composites. *Mater. Sci. Eng. A,* **2015**, *647*, 303-312.
[http://dx.doi.org/10.1016/j.msea.2015.09.006]

[136] Reihanian, M.; Jalili Shahmansouri, M.; Khorasanian, M. High strength Al with uniformly distributed Al_2O_3 fragments fabricated by accumulative roll bonding and plasma electrolytic oxidation. *Mater. Sci. Eng. A,* **2015**, *640*, 195-199.
[http://dx.doi.org/10.1016/j.msea.2015.05.104]

[137] Amirkhanlou, S.; Jamaati, R.; Niroumand, B.; Toroghinejad, M.R. Fabrication and characterization of Al/SiCp composites by CAR process. *Mater. Sci. Eng. A,* **2011**, *528*(13-14), 4462-4467.
[http://dx.doi.org/10.1016/j.msea.2011.02.037]

[138] Alizadeh, M.; Paydar, M.H.; Terada, D.; Tsuji, N. Effect of SiC particles on the microstructure evolution and mechanical properties of aluminum during ARB process. *Mater. Sci. Eng. A,* **2012**, *540*, 13-23.
[http://dx.doi.org/10.1016/j.msea.2011.12.026]

[139] Hosseini, S.M.; Habibolahzadeh, A.; Králík, V.; Petráňová, V.; Němeček, J. Nano-SiCp effects on the production, microstructural evolution and compressive properties of highly porous Al/CaCO3 foam fabricated *via* continual annealing and roll-bonding process. *Mater. Sci. Eng. A,* **2017**, *680*, 157-167.
[http://dx.doi.org/10.1016/j.msea.2016.10.091]

[140] Chayong, S.; Atkinson, H.V.; Kapranos, P. Thixoforming 7075 aluminium alloys. *Mater. Sci. Eng. A,* **2005**, *390*(1-2), 3-12.
[http://dx.doi.org/10.1016/j.msea.2004.05.004]

[141] Hirt, G.; Kopp, R., Eds. *Thixoforming: Semi-solid metal processing*; John Wiley & Sons, **2009**.
[http://dx.doi.org/10.1002/9783527623969]

[142] Lowe, A.; Ridgway, K.; Atkinson, H. The pros and cons of semi-solid processing. *Mater. World,* **1999**, *7*(9), 541-543.

[143] Zhang, X.Z.; Chen, T.J.; Qin, Y.H. Effects of solution treatment on tensile properties and strengthening mechanisms of SiCp/6061Al composites fabricated by powder thixoforming. *Mater. Des.,* **2016**, *99*, 182-192.
[http://dx.doi.org/10.1016/j.matdes.2016.03.068]

[144] Özyürek, D.; Kalyon, A.; Yıldırım, M.; Tuncay, T.; Çiftçi, İ. Experimental investigation and prediction of wear properties of Al/SiC metal matrix composites produced by thixomoulding method using artificial neural networks. *Mater. Des.,* **2014**, *63*, 270-277.
[http://dx.doi.org/10.1016/j.matdes.2014.06.005]

[145] Kandemir, S.; Atkinson, H.V.; Weston, D.P.; Hainsworth, S.V. Thixoforming of A356/SiC and A356/TiB2 nanocomposites fabricated by a combination of green compact nanoparticle incorporation and ultrasonic treatment of the melted compact. *Metall. Mater. Trans., A Phys. Metall. Mater. Sci.,* **2014**, *45*(12), 5782-5798.
[http://dx.doi.org/10.1007/s11661-014-2501-0]

[146] Guo, M.; Liu, J.; Jia, C.; Jia, Q.; Guo, S. Microstructure and properties of electronic packaging shell with high silicon carbide aluminum-base composites by semi-solid thixoforming. *J. Cent. South Univ.,* **2014**, *21*(11), 4053-4058.
[http://dx.doi.org/10.1007/s11771-014-2396-3]

[147] Pai, B.C.; Pillai, R.M.; Kelukutty, V.S.; Srinivasa Rao, H.; Soman, T.; Pillai, S.G.K.; Sukumaran, K.; Satyanarayana, K.G.; Ravikumar, K.K.; Gupta, A.K.; Sikand, R. Semi-solid slurry process for making short carbon fibre dispersed aluminium alloy matrix composites. *J. Mater. Sci. Lett.,* **1994**, *13*(17), 1278-1280.

[http://dx.doi.org/10.1007/BF00270960]

[148] Nath, H. A review on <i>in situ</i> synthesis of Al/TiC and Al/SiC-composites. *Key Eng. Mater.,* **2016**, *684*, 287-292.
[http://dx.doi.org/10.4028/www.scientific.net/KEM.684.287]

[149] Wu, B.; Reddy, R.G. *In-situ* formation of SiC-reinforced Al-Si alloy composites using methane gas mixtures. *Metall. Mater. Trans., B, Process Metall. Mater. Proc. Sci.,* **2002**, *33*(4), 543-550.
[http://dx.doi.org/10.1007/s11663-002-0033-2]

[150] Aikin, R.M. The mechanical properties of *in-situ* composites. *J. Miner. Met. Mater. Soc.,* **1997**, *49*(8), 35-39.
[http://dx.doi.org/10.1007/BF02914400]

[151] Shivalingappa, D. In-situ magnesium based composites-development and tribological behavior *(Doctoral dissertation, Ph. D. Thesis, Metallurgical Engineering Department, IIT Roorkee, Roorkee). ,* **2007**.

[152] Tong, X.C.; Fang, H.S. Al-TiC composites in situ-processed by ingot metallurgy and rapid solidification technology: Part i. microstructural evolution. *Metall. Mater. Trans., A Phys. Metall. Mater. Sci.,* **1998**, *29*(3), 875-891.
[http://dx.doi.org/10.1007/s11661-998-0278-8]

[153] Tjong, S.; Ma, Z.Y. Microstructural and mechanical characteristics of in situ metal matrix composites. *Mater. Sci. Eng. Rep.,* **2000**, *29*(3-4), 49-113.
[http://dx.doi.org/10.1016/S0927-796X(00)00024-3]

[154] Moazami Goudarzi, M.; Akhlaghi, F. Fabrication of Al/SiC nanocomposite powders *via in-situ* powder metallurgy method. *Adv. Mat. Res.,* **2011**, *295-297*, 1347-1352. Trans Tech Publications Ltd.
[http://dx.doi.org/10.4028/www.scientific.net/AMR.295-297.1347]

[155] Clarke, D.R.; Adar, F.; Rossington, D.R.; Condrate, R.A.; Snyder, R.L. Advances in Materials Characterization. *Material Science Research,* **1983**, *15*, 449-464.

[156] Hashim, J.; Looney, L.; Hashmi, M.S.J. The wettability of SiC particles by molten aluminium alloy. *J. Mater. Process. Technol.,* **2001**, *119*(1-3), 324-328.
[http://dx.doi.org/10.1016/S0924-0136(01)00975-X]

[157] Ray, S. Casting of composite components. *Proceeding of the 1995 Conference on Inorganic Matrix Composites,* Bangalore, India **1996**, pp. 69-89.

[158] Zhou, W.; Xu, Z.M. Casting of SiC reinforced metal matrix composites. *J. Mater. Process. Technol.,* **1997**, *63*(1-3), 358-363.
[http://dx.doi.org/10.1016/S0924-0136(96)02647-7]

[159] Andersson, C-H.; Warren, R. Silicon carbide fibres and their potential for use in composite materials. Part 1. *Composites,* **1984**, *15*(1), 16-24.
[http://dx.doi.org/10.1016/0010-4361(84)90956-X]

[160] Metcalf, A.G. Interfaces in Metal Matrix Composites. *Composite Materia*; 1st Edition - February 28, **1974**, *1*.

[161] Hekner, B.; Myalski, J.; Pawlik, T.; Sopicka-Lizer, M. Effect of carbon in fabrication Al-SiC nanocomposites for tribological application. *Materials,* **2017**, *10*(6), 679.
[http://dx.doi.org/10.3390/ma10060679] [PMID: 28773039]

[162] Bartuli, C.; Carassiti, F.; Valente, T. Interfacial reactions in Al/SiC composites produced by low pressure plasma spray. *Adv. Perform. Mater.,* **1994**, *1*(3), 231-242.
[http://dx.doi.org/10.1007/BF00711205]

[163] Kawai, C. Effect of interfacial reaction on the thermal conductivity of Al–SiC composites with SiC dispersions. *J. Am. Ceram. Soc.,* **2001**, *84*(4), 896-898.
[http://dx.doi.org/10.1111/j.1151-2916.2001.tb00764.x]

[164] Krizik, P.; Balog, M.; Matko, I.; Svec, P., Sr; Cavojsky, M.; Simancik, F. The effect of a particle–matrix interface on the Young's modulus of Al–SiC composites. *J. Compos. Mater.,* **2016,** *50*(1), 99-108.
[http://dx.doi.org/10.1177/0021998315571028]

[165] Valente, T.; Bartuli, C. A plasma spray process for the manufacture of long-fiber reinforced Ti-6Al-4V composite monotapes. *J. Therm. Spray Technol.,* **1994,** *3*(1), 63-68.
[http://dx.doi.org/10.1007/BF02649001]

[166] Schwabe, U.; Wolff, L.R.; van Loo, F.J.J.; Ziegler, G. Corrosion of technical ceramics by molten aluminium. *J. Eur. Ceram. Soc.,* **1992,** *9*(6), 407-415.
[http://dx.doi.org/10.1016/0955-2219(92)90101-I]

[167] Viala, J.C.; Fortier, P.; Bouix, J. Stable and metastable phase equilibria in the chemical interaction between aluminium and silicon carbide. *J. Mater. Sci.,* **1990,** *25*(3), 1842-1850.
[http://dx.doi.org/10.1007/BF01045395]

[168] Li, S.; Arsenault, R.J.; Jena, P. Quantum chemical study of adhesion at the SiC/Al interface. *J. Appl. Phys.,* **1988,** *64*(11), 6246-6253.
[http://dx.doi.org/10.1063/1.342082]

[169] Pech-Canul, M.I.; Katz, R.N.; Makhlouf, M.M. Optimum parameters for wetting silicon carbide by aluminum alloys. *Metall. Mater. Trans., A Phys. Metall. Mater. Sci.,* **2000,** *31*(2), 565-573.
[http://dx.doi.org/10.1007/s11661-000-0291-z]

[170] Kimoto, Y.; Nagaoka, T.; Mizuuchi, K.; Fukusumi, M.; Morisada, Y.; Fujii, H. Thermal conductivity of Al/SiC particulate composites produced by friction powder sintering. *J. Jpn. Soc. Powder. Powder. Metall.,* **2016,** *63*(7), 563-567.
[http://dx.doi.org/10.2497/jjspm.63.563]

[171] Shi, Z.; Ochiai, S.; Hojo, M.; Lee, J.; Gu, M.; Lee, H.; Wu, R. The oxidation of SiC particles and its interfacial characteristics in Al-matrix composite. *J. Mater. Sci.,* **2001,** *36*(10), 2441-2449.
[http://dx.doi.org/10.1023/A:1017977931250]

[172] Sozhamannan, G.G.; Balasivanandha Prabu, S. Evaluation of interface bonding strength of aluminum/silicon carbide. *Int. J. Adv. Manuf. Technol.,* **2009,** *44*(3-4), 385-388.
[http://dx.doi.org/10.1007/s00170-008-1871-0]

[173] Shi, Z.; Yang, J.M.; Lee, J.C.; Zhang, D.; Lee, H.I.; Wu, R. The interfacial characterization of oxidized SiC(p)/2014 Al composites. *Mater. Sci. Eng. A,* **2001,** *303*(1-2), 46-53.
[http://dx.doi.org/10.1016/S0921-5093(00)01943-2]

[174] Gu, M.; Jin, Y.; Mei, Z.; Wu, Z.; Wu, R. Effects of reinforcement oxidation on the mechanical properties of SiC particulate reinforced aluminum composites. *Mater. Sci. Eng. A,* **1998,** *252*(2), 188-198.
[http://dx.doi.org/10.1016/S0921-5093(98)00674-1]

[175] Rahmani Fard, R.; Akhlaghi, F. Effect of extrusion temperature on the microstructure and porosity of A356-SiCp composites. *J. Mater. Process. Technol.,* **2007,** *187-188*, 433-436.
[http://dx.doi.org/10.1016/j.jmatprotec.2006.11.077]

[176] Kandpal, B.C.; Kumar, J.; Singh, H. Optimization of electrical discharge machining AA6061/ 10 %Al$_2$O$_3$ composite using Taguchi optimization technique. *Mater. Today Proc.,* **2018,** *5*(9), 18946-18955.
[http://dx.doi.org/10.1016/j.matpr.2018.06.245]

[177] Khanna, V.; Singh, K.; Kumar, S.; Bansal, S.A.; Channegowda, M.; Kong, I.; Khalid, M.; Chaudhary, V. Engineering electrical and thermal attributes of two-dimensional graphene reinforced copper/aluminium metal matrix composites for smart electronics. *ECS J. Solid State Sci. Technol.,* **2022,** *11*(12), 127001.
[http://dx.doi.org/10.1149/2162-8777/aca933]

[178] Kumar, K.; Kumar, S.; Gill, H.S. Role of surface modification techniques to prevent failure of components subjected to the fireside of boilers. *J. Fail. Anal. Prev.,* **2022.**
[http://dx.doi.org/10.1007/s11668-022-01556-w]

[179] Kumar, S. Influence of processing conditions on the mechanical, tribological and fatigue performance of cold spray coating: a review. *Surf. Eng.,* **2022,** *38*(4), 324-365.
[http://dx.doi.org/10.1080/02670844.2022.2073424]

[180] Mohan, R.; Saxena, N.V.; Kumar, S. Performance optimization and numerical analysis of boiler at husk fuel based thermal power plant. *E3S Web of Conferences 405,* **2023,** *02010,* 1-12.
[http://dx.doi.org/10.1051/e3sconf/202340502010]

[181] Kumar, S.; Singh, H.; Kumar, R.; Singh Chohan, J. Parametric optimization and wear analysis of AISI D2 steel components. *Mater. Today Proc.,* **2023.**
[http://dx.doi.org/10.1016/j.matpr.2023.01.247]

[182] Kumar, R.; Kumar, M.; Singh Chohan, J.; Kumar, S. Effect of process parameters on surface roughness of 316L stainless steel coated 3D printed PLA parts. *Mater. Today Proc.,* **2022,** *68*(4), 734-741.
[http://dx.doi.org/10.1016/j.matpr.2022.06.004]

[183] Chauhan, A.; Kumar, M.; Kumar, S. Fabrication of polymer hybrid composites for automobile leaf spring application. *Mater. Today Proc.,* **2022,** *48*(5), 1371-1377.
[http://dx.doi.org/10.1016/j.matpr.2021.09.114]

[184] Singh, H.; Sarabjit, N.J.; Tyagi, A.K. An overview of metal matrix composite: Processing and SiC based mechanical properties. *Journal of Engineering Research and Studies,* **2011,** *2*(4), 72-78.

[185] Hashim, J.; Looney, L.; Hashmi, M.S.J. Metal matrix composites: Production by the stir casting method. *J. Mater. Process. Technol.,* **1999,** *92-93,* 1-7.
[http://dx.doi.org/10.1016/S0924-0136(99)00118-1]

[186] Dahiya, M.; Khanna, V.; Anil Bansal, S. Effect of graphene size variation on mechanical properties of aluminium graphene nanocomposites: A modeling analysis. *Mater. Today Proc.,* **2022,** (Jul)
[http://dx.doi.org/10.1016/j.matpr.2022.07.259]

[187] Gupta, P.; Ahamad, N.; Kumar, D.; Gupta, N.; Chaudhary, V.; Gupta, S.; Khanna, V.; Chaudhary, V. Synergetic effect of CeO_2 doping on structural and tribological behavior of $Fe-Al_2O_3$ metal matrix nanocomposites. *ECS J. Solid State Sci. Technol.,* **2022,** *11*(11), 117001.
[http://dx.doi.org/10.1149/2162-8777/ac9c92]

[188] Dahiya, M.; Khanna, V.; Anil Bansal, S. Aluminium-graphene metal matrix nanocomposites: Modelling, analysis, and simulation approach to estimate mechanical properties. *Mater. Today Proc.,* **2022,** (Nov)
[http://dx.doi.org/10.1016/j.matpr.2022.10.181]

[189] Singh, K.; Bansal, S.A.; Khanna, V.; Singh, S. Effects of performance measures of non-conventional joining processes on mechanical properties of metal matrix composites. *Metal Matrix Composites,* **2022,** (Aug), 135-165.
[http://dx.doi.org/10.1201/9781003194897-7]

[190] Khanna, V.; Kumar, V.; Bansal, S.A.; Prakash, C.; Ubaidullah, M.; Shaikh, S.F.; Pramanik, A.; Basak, A.; Shankar, S. Fabrication of efficient aluminium/graphene nanosheets (Al-GNP) composite by powder metallurgy for strength applications. *J. Mater. Res. Technol.,* **2023,** *22,* 3402-3412.
[http://dx.doi.org/10.1016/j.jmrt.2022.12.161]

CHAPTER 4

Synthesis Approaches and Traits of Carbon Fibers-Reinforced Metal Matrix-Based Composites

Himanshi[1], Rohit Jasrotia[1,*], Suman[2], Ankit Verma[3], Sachin Kumar Godara[4], Abhishek Kandwal[1], Pawan Kumar[1], Jahangeer Ahmed[5] and Susheel Kalia[6]

[1] *School of Physics and Materials Science, Shoolini University, Bajhol, Solan, H.P., India*

[2] *Department of Mathematics, School of Basic and Applied Sciences, Maharaja Agrasen University, Baddi, H.P., India*

[3] *Faculty of Science and Technology, ICFAI University, Baddi, Himachal Pradesh, India*

[4] *Department of Chemistry, Guru Nanak Dev University, Punjab, Amritsar, India*

[5] *Department of Chemistry, College of Science, King Saud University, Riyadh, Saudi Arabia*

[6] *Department of Chemistry, ACC Wing (Academic Block), Indian Military Academy, Dehradun (Uttarakhand) India*

Abstract: In this chapter, an overview of the advancement and research efforts that have been undertaken on CFR-MMC (carbon-fiber reinforced metal matrix-based composites) during the last several decades is presented. Carbon fiber is widely implemented in the construction sector for rehabilitation and structural repair projects. Although, studies show that carbon fiber-reinforced metal-matrix (CFR-MMC) has a bright future, the use of carbon fibre as a reinforcement in metal matrix is still in its development. The uses, and traits of carbon fiber are discussed in general terms in this study. The various traits such as mechanical, and structural properties of the resultant CFR-MMC, are significantly influenced by the structure and content of the carbon fibre as well as its bonding to the MM (Metal matrix). The effect on the various traits of MMCs by CFs (Carbon fibers) was investigated. In addition, a detailed study on the various synthesis approaches for the preparation of CFR-MMC has been taken into practice in this book chapter.

Keywords: Composite, Carbon fibers, Cellicious precursor, Diffusion bonding, Fiber reinforcement, Ion platting, Mechanical properties, Melt stirring, Metal matrix composite precursors, Powder metallurgy, PAN precursor, Pitch precursor, Plasma spraying, Reinforcement, Synthesis approaches, Squeeze casting, Structural properties, Scanning electron microscopy, Tow.

---

* **Corresponding author Rohit Jasrotia:** School of Physics and Materials Science, Shoolini University, Bajhol, Solan, H.P., India; E-mail: rohitsinghjasrotia4444@gmail.com

Virat Khanna, Prianka Sharma & Santosh Kumar (Eds.)

INTRODUCTION

The progress of technology has led to the increasing demand for innovative materials to handle day-to-day difficulties for a variety of uses. MMCs (Metal Matrix Composites), which is the combination of two or more elements to create a new composite, have emerged as one of the most significant material systems in recent decades. The demand for lightweight structural materials for aerospace and automotive applications is quite high. Conventional materials, including titanium and aluminium, are unable to handle most of the modern-day obstacles [1]. The properties of these materials decrease rapidly at extreme temperatures, limiting their use in essential components. Combining inherent matrix toughness and ductility with high specific strength and high stiffness materials, such as CFs (carbon fibers) or ceramic filaments, allows for the production of materials that can be used in advanced applications and overcome performance difficulties. The introduction of such reinforcements into metal matrix enhances elastic modulus, tensile strength, hardness, as well as additional mechanical characteristics substantially [2]. Other characteristics of metal matrix composites, such as corrosion resistance, CTE, wear resistance, coefficient of friction, and thermal conductivity (TC) can be tuned to meet application necessities. At the same time, the material's CTE must be small so that it creates a lesser amount of thermal stress during the heat process. Copper and aluminium have comparatively high values of coefficient of thermal expansion, resulting in significant thermal stresses [3]. According to research, SiC (Silicon carbide) reinforced metal matrix composite with low CTE can give maximum heat absorption rates as well as minimum thermal stresses. As an alternative to bulk metal heat sinks, diamond particle-reinforced copper metal matrix composite has also been researched. Even though, it has been demonstrated that these MMCs have improved thermo-mechanical properties, up to 60% more volume of reinforcement may still be necessary. In an effort to reconcile, thermo-mechanical and machinability characteristics of CFR-MMCs have been studied for heat sink applications. The aeroplane, aerospace, vehicle, and electronics sectors have all expressed interest in CF-MMCs [4]. Simply said, conventional metallic alloys and pure metals are usually unable to bring effective characteristics in challenging applications. This resulted in the creation of composites with certain traits that are highly suitable for a specific purpose. MMCs are already being used in a variety of real-world applications. Multi-chip modules, power electronics modules, disc brakes, drive shafts, and, automobile engine cylinders, are a few prominent applications of MMCs.

CARBON FIBER

Carbon fibers (CFs) are the fibers that have at least 92 wt% of carbon and construct a fibre shape. The early background of the carbon fibers is shown in Fig. (**1**). Carbon fibers are fibers with a diameter of 5 to 10 μm and are mostly made of carbon atoms. Commercially, bundles of carbon fibers are described as tow, which basically means a bundle of thousands of numbers of carbon fibers (1 tow stands for 1000 fibers in a bundle). CFs are appropriate for applications requiring high young's modulus, strong tensile strength, high chemical resistance, low density, moderate thermal expansion, and outstanding electrical and thermal conductivity. Commercial production and usage of carbon fibers as a reinforcing material occurred more than a half-century later [5]. A CF is composed of layers of carbon atoms (graphene sheets) organised in a regular hexagonal pattern and with an atomic structure similar to graphite [6]. CF layer planes can either have a hybrid structure, a turbostratic structure, or a graphitic structure depending on the precursors used and the production methods. Layer planes are consistently arranged parallel to one another in graphitic crystalline regions. The atoms on a plane are covalently joined by sp^2 bonding, while the sheets are bonded by van der waals forces, which are rather weak [1]. The d-spacing between two the graphene layers in a single graphitic crystal is approximately 0.335 nm. In mostly carbon fibers, the basic units of structures are stacked in turbostratic layers. The parallel graphene sheets in a turbostratic structure are folded, slanted, or divided randomly or haphazardly. Uneven stacking and the presence of sp^3 bonding have been seen to increase d-spacing [7, 8]. The microstructure of carbon fibre is defined by the precursors and the processing conditions [9]. Carbon fibers were synthesized by pyrolyzing a precursor material in an inert atmosphere at high temperatures. Different precursors (PAN, Pitch, and cellulosic-based precursors) can create carbon fiber using various processing methods, but the basic thermal conversion mechanisms are the same for all methods. The initial stage in the production of carbon fibre is the stabilisation of precursor fibres between (200 and 400 °C) in air. Stabilization increases the thermal stability and reduces fiber collapse at higher temperatures. The stabilised precursor fibers are next carbonization, which involves placing oxidised fibres in non-reactive environment at high temperatures (1900 °C) to eliminate the non-carbonaceous components like hydrogen, nitrogen and oxygen. Carbon fibers are generated at this stage, but to raise carbon concentration and young's modulus, graphitization is used, in which the fibres are heated (at maximum temperature).

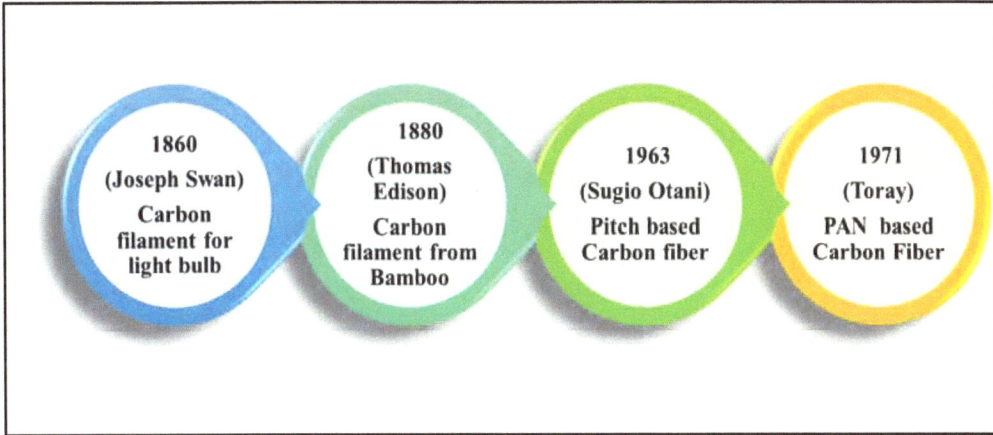

Fig. (1). A schematic that summarises the background of CF research.

(PAN, PITCH, AND CELLICIOUS) PRECURSOR FOR THE FABRICATION OF CFS

Pan Precursor

PAN, often known as polyacrylonitrile, is a synthetic thermoplastic polymer. Polyacrylonitrile (PAN), which contains 68% of the carbon, is presently the most extensively utilized precursor for the manufacture of CFs. It is a semi-crystalline polymer having melting and glass transition temperatures of 623 K and 353 K, respectively. PAN precursor is used in a wide range of applications, including air filtration, textiles, protective garments, supercapacitors, and water purification. The most common use of PAN precursor is the production of fiber for the synthesis of carbon fibres [2]. The PAN-based carbon fiber production method includes fiber spinning, thermal stabilisation, and carbonization steps. Determining the precursor fiber's state, stability, tensile, and flexural properties of carbon fiber depends heavily on the spinning processes [10]. Without a high-quality precursor fiber, it is difficult to fabricate excellent CF by the stabilisation approach. Firstly, acrylic fiber is created using the dry-wet-jet, wet, or dry spinning processes. Inorganic and/or organic solvents such as sodium thiocyanate, DMF (dimethyl formamide), and DMAc (dimethyl acetamide) are utilized in wet spinning to create polyacrylonitrile polymer dope. The polyacrylonitrile solution is filtered before being put into a spinneret submerged in an aqueous sodium thiocyanate-containing solvent bath. As a result, the fiber precipitates and solidifies as it leaves the spinneret hole [11]. The leftover solvent remained in the filaments is eliminated in the next steps by rinsing. Whenever the precursor fiber diameter was smaller, mass and energy (heat) dispersion would be easy and aids in the homogeneous thermal stabilisation (the next step of CF manufacture).

Following the final washing process, the fibers are passed through a succession of heat rollers to dry [12]. In case of dry-jet-wet spinning, an air gap in the middle of the spinneret and the coagulation bath will be maintained ranging from 10 to 200 μm. This method facilitates filament stretching and enhances molecular orientation [13]. Additionally, because the dope filaments and the coagulation bath are at different temperatures, it is possible to reduce high diffusion rates in the coagulation bath. Some processing parameters, such as pH, extrusion rate and temperature, dope concentration, coagulation bath temperature, stretching, and washing, are studied [14]. There are several recognised procedures for the production of CF, including plasma oxidation, in addition to the standard thermal treatment. In the beginning, PAN is thermally stabilised (at a temperature between 500 to 600 °C in the air), resulting in the development of cross-linked ladder structures which allow the polyacrylonitrile fibers to withstand high temperatures *via* carbonization [2]. After that, the carbonization process takes place in which the stabilized fibers heat up (1800 °C in non-reactive atmosphere) and assist in making the turbostatic graphitic structure. Last optional step: Fibers are again treated at elevated temperatures, often in the range of 2000 °C, in a non-reactive environment, to improve the fibre stiffness and the alignment of the basal planes. The stabilisation step is critical in the production of high-performance CFs [15, 16]. Fig. (**2**) depicts the traits of a PAN-based precursor.

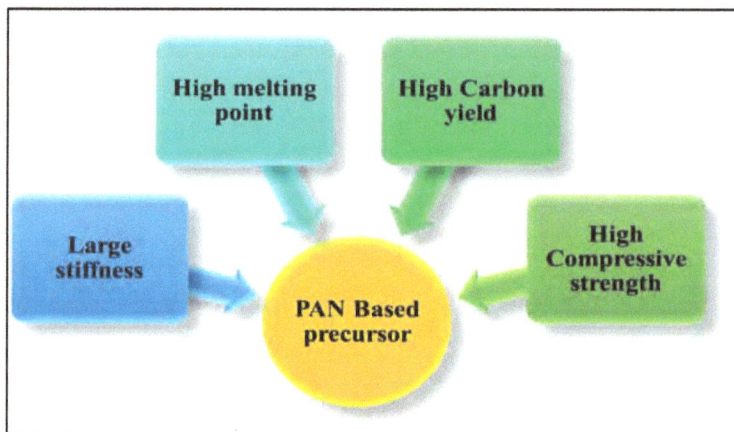

Fig. (2). Different Traits of PAN Based precursor.

Pitch Precursor

Carbon fibers based on pitch are generally classified into 2 types: Anisotropic mesophase pitch and isotropic pitch. The differences between these two types of carbon fibers include their microstructure, optical texture, and physical characteristics in addition to their carbonaceous sources. Mesophase pitch has been considered a suitable precursor for the production of a wide range of advanced engineering and industrial carbon products, particularly high-performance pitch-based carbon fibers, due to its characteristic nematic liquid crystalline structure and ease of graphitization [17]. In an effort to find extremely inexpensive and high-carbon (>80% carbon) raw materials, pitch-based carbon fibres have been developed. Both coal tar and petroleum by-product are used to make pitch. The distinction between these two pitches (coal and petroleum) is that the former is high in aliphatic molecules, and the latter is high in aromatic substances. Both pitches can be utilised as carbon fibre precursors. Because it is more challenging to eliminate natural inclusions and defect-sensitive particles from coal pitch, petroleum pitch is recommended [18]. Pitch precursors have a lower molecular weight than PNA precursors. Pitch precursor is cleaned first in order to remove the volatile compounds in a vacuum. Raw materials pre-treatment, melt-spinning, pitch preparation stabilisation, carbonization, and graphitization are essentially identical fabrication procedures. The pitch is then heated at 200 to 300 °C in the non-reactive atmosphere. The condensation of pitch oligomers creates a mesophase at this temperature treatment. Controlling this procedure is quite challenging. However, the benefit of the resulting mesophase was that it was able to generate fibers utilising a melt-spinning approach rather than a solution-based procedure [19]. The precursor fiber receives a very high orientation of the aromatic layers during the melt-spinning process, and then, quickly transforms into the resultant carbon fiber. Stabilization occurs after the melt-spinning process. Stabilization helps to transform pitch fibers from the thermoplastic to the thermoset phase through an oxidative process, providing pitch fiber structure with a stable shape. Carbonization removes non-carbonaceous atoms and side chains from the aromatic groups, improving carbon aromaticity [20]. After carbonization, carbon fiber with certain mechanical, physical, and chemical characteristics is created. Optional graphitization is used to produce graphitic structure, boost carbon aromaticity, and improve the organised orientation of aromatic groups in carbon fibers to improve mechanical qualities. The fundamental benefit of utilising pitch-based precursor is that it produces ultra-high modulus CFs with significantly larger graphite crystallite size. However, purifying and refining pitch precursor in order to make the CF with constant high mechanical properties is difficult. Fig. (**3**) displays the pitch-based precursor's characteristics [21].

Cellicious Precursor

Thomas Alva Edison originally employed cellulose-based precursor material for manufacturing the carbon fibers during his research to develop an appropriate filament substance for electric light. Soon after that, Edison created a technique for dispersing the natural cellulose or cotton in $ZnCl_2$ solvent to create a dope that could be injected into a spin bath and carbonised in an oxygen-free environment. It studied different techniques to enhance the mechanical traits. When carbonised synthetic rayon was used to make the CFs for high-temperature radar applications in the late 1950s, interest in CFs was renewed. The importance of cellulose-based precursors was reduced after the invention of the polyacrylonitrile procedure in the 1960s due to its advantages of improved mechanical qualities, reduced environmental issues, and high carbon yield [22]. But as the demand for carbon fibers with reduced thermal conductivity, low cost of raw materials, high purity, and high flexibility increases for use in medical and aerospace, cellulose-based carbon fibers gain interest once again. However, it has significant drawbacks, such as weaker mechanical characteristics, a lesser yield, and a superior processing expense to produce the starting material. Natural cellulose fibers cannot be used as precursors for high-performance carbon fibers due to their discontinuous character [23]. They also have impurities, a porous structure, and a low degree of orientation. Thus, potential materials include textile-grade rayon and redeveloped cellulose fiber precursors like viscose and cuprammonium rayon [24]. At this time, lyocell fibers, are a modern type of cellulose fiber that can become a solution with the ability to harness the mechanical capabilities of cellulose-based CFs [25]. Lyocell fibers feature a more environmentally friendly manufacturing technique and superior fiber characteristics. In the dry state, they show greater tenacity, modulus, and shrinkage, but in the wet state, they show decreased tenacity and modulus. The molecular chains are closely packed along the fiber axis and their surface is smooth as well as spherical.

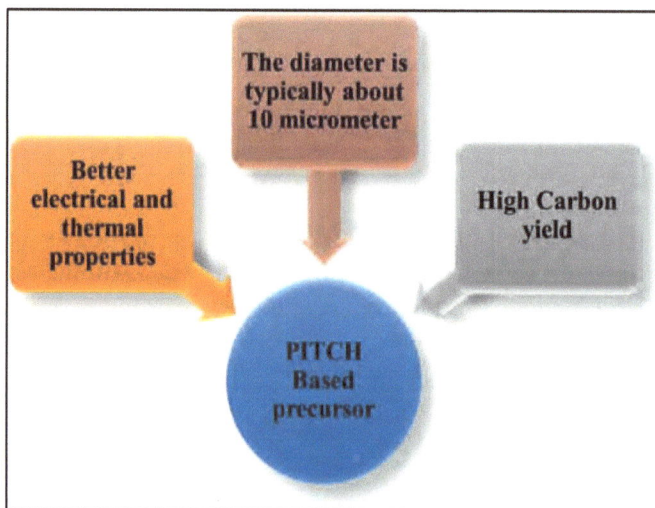

Fig. (3). Properties of PITCH Based precursor.

METAL MATRIX-BASED COMPOSITE (MMCs)

In an MMC, at least one alloy or metal component creates at least one percolating network. Reinforcement is the name of the additional component that is embedded into the matrix material. To create MMC material, a high strength substance called reinforcement is scattered on the surface of matrix material. To avoid the chemical interaction of the reinforcement with the matrix, the surface of the reinforcement is coated [26]. MMCs are classified into two types based on their reinforcing form: (a) fiber reinforced MMCs with either continuous or discontinuous fiber reinforcements. Basically, metal matrix composites are made up of a continuous matrix of metal or alloy with reinforcement that could be continuous fiber, short fiber or whisker [27]. Due to their lower price compared to continuous fiber, reinforced composites, and higher isotropic properties, particle or discontinuously reinforced MMCs have gained popularity (28). The fibers act as reinforcing material in the FR-MMC material, which holds all of the reinforcement constituents in an appropriate form and transfers stress between them. MMCs with fiber reinforcement have high heat conductivity, high strength, large specific stiffness, and a low coefficient of thermal expansion [28]. These characteristics have attracted great attention in CF-MMC. Metal matrix composites with fiber reinforcement are utilised in space shuttles, electronic substrates, and a variety of other specific applications. A metal matrix composite's ultimate performance is determined by three major factors: the reinforcement, the matrix, and the reinforcement/matrix contact [3, 4]. Based on their excellent mechanical performance, simplicity, and cost, many metallic matrices have been employed to fabricate carbon fiber-reinforced metal matrix composites. Different metals and alloys, including copper, titanium, magnesium, and aluminium alloys, are used as matrix materials [28].

Aluminium alloys have essential uses in the aeronautical field due to their outstanding toughness, strength, corrosion resistance and low density. Al-lithium alloys are among the most significant precipitation-hard aluminium alloys [29, 30]. Lithium has the remarkable property of increment in the elastic modulus while lower the density of the alloy when used as a major alloying element in aluminium. Naturally, the aircraft industry has been the main focus of this advancement [28]. Titanium is an important material in the aerospace industry. Its young's Young's modulus is 115 GPa, and its density is 4.5 g cm^{-3}. While the modulus of titanium alloys can be anywhere between 80 and 130 GPa, but the density can vary between 4.3 and 5.1 g cm^{-3}. It is critical to have a high strength/weight and modulus/weight ratio. Titanium has a relatively high retain strength and melting point (1672 °C) at high temperatures while being resistant to oxidation and corrosion. It is a great material for aeronautical applications because of all of these characteristics [13]. In Jet engines (compressor blades and turbine), fuselage components, and other applications use titanium alloys [4]. Shalu *et al.* (2009) used the squeeze casting infiltration technique to create an Al MMC reinforced with carbon preform. To investigate the improvement in characteristics compared to the matrix, tests of impact, hardness, and tension were performed on the squeeze cast matrix and the composite. The alloy seems to have lower dendritic sizes and also, the carbon fibers are equally dispersed throughout the matrix [31]. Lv *et al.*, (2021) used TC4 powder and carbon fibers as raw materials to synthesised a discontinuous 1%CFs/TC4 composite using powder metallurgy route, which includes hot isostatic pressing sintering and high-energy ball-milling. Heat treatment changed the interfacial structure (HT). To create various interfacial structures, heat treatment up to 1000 °C was performed to powder-metallurgy Ti-6Al-4V MMC supplemented with discontinuous CFs. The results of mechanical tests conducted at 500 °C and room temperature revealed that the optimum mechanical qualities were reached at a temperature of 900 °C. Interfacial structure, phase structure and grain morphology were studied to better understand the fundamental process. It was found that the interfacial thickness increased gradually as the heat treatment temperature was raised, and there was obvious grain growth at 1000 °C. The production of high performance CFs/TMCs depends on having an appropriate interfacial reaction and fine grain size [13].

MECHANICAL AND STRUCTURAL PROPERTIES OF CFR-MMC

Mechanical properties

MMCs particle reinforced metal matrix composites have the ability to provide ultra-high mechanical characteristics, such as particular stiffness and specific strength, in civil and defence applications, as well as the automotive and aerospace industries. In general, interfacial bonding of the reinforcement and matrix, which has a significant impact on the final composite features, also influences the mechanical performance of composites. To enhance the mechanical traits of composite, carbon fibers are intended to be employed primarily in a matrix and to assist with the transmission of stress away from the matrix. Most non-carbonaceous components are eliminated during the carbonization process, leaving only carbon in the fibers. Carbon atoms contribute to the formation of a strong and chemically stable fiber. All known matrices have a weak interaction with the carbon atoms in graphite. Delamination or debonding, a major fault in composite structures, is caused by a decrease in ILSS (Interlaminar shear strength), which is the result of a poor connection between the fibers and matrix. Despite the type of precursor used, a surface treatment is usually necessary to enhance the bonding between the fiber and matrix. By using a hot coining technique, Eid *et al.* (2021) fabricated a CF/Aluminium composite from 0, 5, 10, 15, and 20 wt% carbon fiber. One of them is made from uncoated 12 weight percent copper (Cu) coated CF. The uncoated CF/Al has much better microhardness and wear characteristics up to 10 wt% CF. Whereas the compressive strength for the 20 wt % uncoated CF sample dropped from 320.8 MPa to 179.8 MPa. Comparing coated and uncoated CF/Al composites, the coated composites show better mechanical traits. The wear rate test showed that there was a 68.5% reduction in wear rate between coated and uncoated composites. In microhardness and compression tests, the greatest rising percentages were 31.6% and 23.44%, respectively. The outcomes suggested that the carbon fiber surface modification might enhance CF/Aluminum composites mechanical characteristics [32]. Gatea *et al.* (2018) utilized an aluminum alloy incorporated with a 17.5% volume fraction of silicon carbide (SiC) particles. These sheets were manufactured through the powder metallurgical method. The composite of Al/SiCp was tested under two distinct heat treatment conditions and subjected to tensile testing at a strain rate of 8×10^{-5} s^{-1}. In Fig. (**4a**), the fracture surface of the sample subjected to the T6 condition is depicted. This surface displays shallow dimples along with numerous fractured SiC particles. Additionally, there is evident debonding occurring at the interface between the SiC particles and the aluminum matrix. This combination of broken particles and debonding plays a pivotal role in causing fracture and contributes to the reduced ductility observed in the Al/SiCp composite under the T6-condition. On the other hand, Fig. (**4b**) illustrates the fracture surface resulting from the O-condition annealing. Notably, a significant proportion of deep dimples are visible on this fracture surface when compared to the sample under the T6-condition. Furthermore, there is a limited presence of SiC particles that are surrounded by the matrix [33] Table **1**.

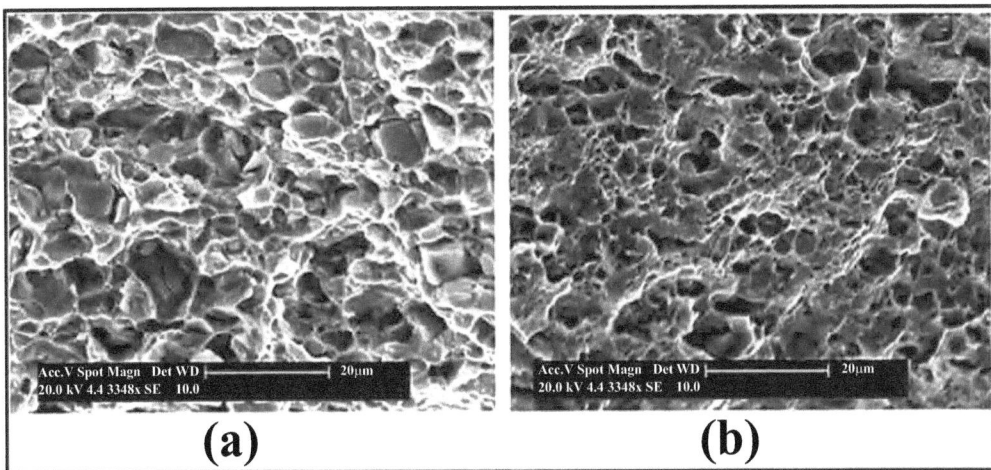

Fig. (4). Demonstrates the impact of heat treatment on the fracture surface at a strain rate (**a**) T6-treatment, (**b**) depicts the fracture surface resulting from O-condition annealing.

Table 1. Table shows the different mechanical properties of metal-matrix based composite.

Synthesis	Matrix	Reinforcement	Mechanical Traits	References
Liquid Metallurgy	A356	ZrSiO$_4$	Hardness= 79 HB	[34]
Liquid Metallurgy	TiB2	A356	Modulus=78.31GPa Yield Strength=244 MPa	[30]
Liquid Metallurgy	CF	Al-17%Si	Hardness = 125 BHN	[35]
Solid-Liquid mixed casting	Al$_2$O$_3$	Al 2024	Yield Strength = 150 MPa	[29]
Powder metallurgy	MWCNT	Al-1.0 wt% Ni	Hardness=0.65GPa Modulus= 95.4GPa	[36]
Stir casting	TiC	AA6061	Ultimate tensile Strength=235 MPa	[37]
Hot press sintering	Al$_2$O$_3$	Cu	Hardness=79.4 HB	[38]
Powder metallurgy	Al$_2$O$_3$	Cu	Hardness=125 Hv	[39]
Stir casting	Al	SiC	Ultimate tensile strength=293	[40]
Powder metallurgy	Graphene	Al	Ultimate tensile strength= 249 MPa	[41]

Structural Properties

Alten *et al.*, (2019) fabricated nickel coated CFs matrix composites reinforced Al-6063 alloy through the squeeze method. CFs were coated with Ni using an electroless approach to solve the wetting issue between the aluminium matrix and reinforcing agent. Hypophosphite (HP), coating times, pH levels, and bath temperatures were all studied. To characterise the layers, SEM, EDS, and XRD techniques were used. The thickness of the Ni coatings on the carbon fibers has been observed to increase with the increasing bath pH, coating duration, bath temperature, and hypophosphite quantity. The properties of the coating depend on its structure and chemical composition. At different pH, the microstructure of nickel- phosphorous coated CFs performed at 80 °C for 1 hour. Increment in the pH value of the solution, the thickness of the coating increases. In their research, they discovered that the structural properties of the composite altered depending on the temperature. As the temperature increases, the coating thickness also increased. The A1 plating bath was used to conduct research on the impact of bath temperature at Ni-P coating at 70 and 90 °C at a pH value of 8 for one hour. Additionally, the result showed that when the deposition temperature is raised, the phosphorous concentration is reduced dramatically. It was discovered that the coating thickness was significantly influenced by the HP concentration. When the concentration was raised from 10 g/l to 30 g/l for 1 hour at a pH between 8 and 9, the thickness rise from 4 mm to 12 mm [42]. In order to increase the wettability and decrease the interfacial reaction between molten Al alloy and carbon fibers,

Zhu *et al*. (2020) synthesised alumina (Al_2O_3) coated carbon fiber. The Al_2O_3-coated CF was then utilized to synthesize CF/Al composites using the pressure infiltration route. Effects of heat treatment, immersion time, and immersion-drying cycle on the surface morphologies of CF coated with Al_2O_3 were examined. As the temperature increases, the phase change was nearly complete and the XRD peaks became sharper. The surface morphology of heat-treated Al_2O_3-coated carbon fiber at 500 °C is fairly smooth. The majority of CFs were coated with Al_2O_3 when the temperature reached 600 °C, along with a small amount of material on the surface that may have resulted from the Al_2O_3 coating breaking. The large-scale loss of Al_2O_3 coating on the surface of CFs after the heat treatment at 700, 800, and 900 °C [43].

DIFFERENT SYNTHESIS TECHNIQUES

The potential of carbon fibre as a reinforcing material for MMC has been thoroughly researched. The development of a material with high strength, lightweight materials is the major goal of (CF-MMCs). Improved thermal characteristics *i.e.*, decreased thermal expansion coefficient and improved thermal conductivity enhanced mechanical characteristics specifically, elastic modulus, specific strength, *etc.* The fabrication methods used to create CF-MMCs may be broadly categorised into three groups: processing in the following three states: solid state process, liquid state process, and deposition process [44]. The fabrication of carbon fibre (CFs) reinforced MMCs has been widely studied in the past, and the classification of fabrication techniques is shown in Fig. (5).

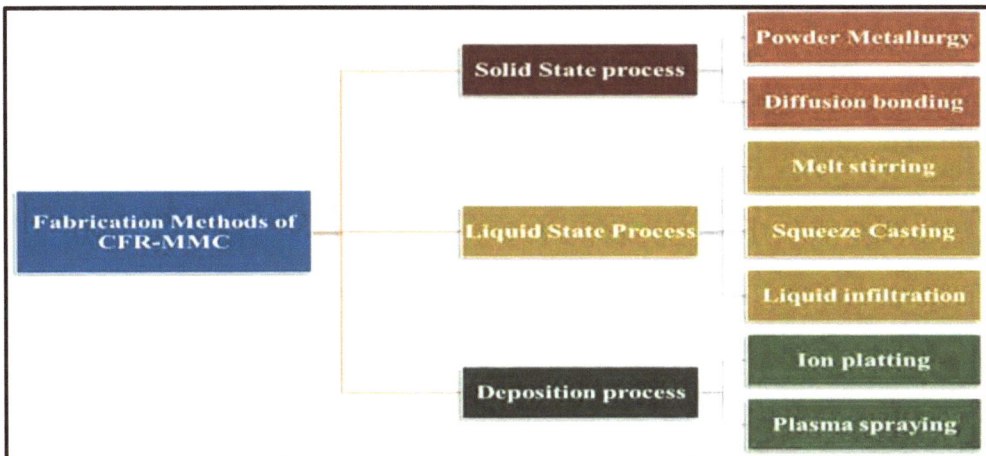

Fig. (5). Fabrication methods for the CF-metal matrix composite.

Solid State Method

The most extensively utilised method for fabricating CFR-MMC is a solid-state method. The starting processes take place at room temperature without undergoing any phase transitions. The bonding of the metallic matrix and carbon fiber is a result of the mutual diffusion that occurs between MM and CFs in this method at high pressure/temperature during the solid-state manufacture of CF-MMCs technology.

Powder Metallurgy Approach

The powder metallurgy synthesis is a cost-effective as well as simple approach for producing carbon fiber metal matrix composite using the solid-state approaches. Ball milling is commonly used to produce homogeneous reinforcement dispersion in MMC. Ball to powder ratio, milling conditions, milling time, milling speeds, and so on are the important factors of ball milling to identify the ultimate morphology of the powder mixture of composites [45]. High-density composites are created by pressing the powder mixtures that have been ball milled in a graphite die and then, heating them. A diagram of the powder metallurgy process is depicted in Fig. (**6**). Particle reinforced MMCs are mostly created *via* powder metallurgy in the industrial sectors and academic. It has been observed that using a high intensity ball milling process, is particularly successful in combining powders in stoichiometric ratios. During the deformation processing of consolidated composites, the CNFs are further aligned in the metal matrices, such as cold and hot rolling, hot extrusion, and so on which enhances the mechanical characteristics of CF-Metal matrix composite. These consolidation strategies have size constraints for the metal matrix composite. The distinctive aspect ratio of CNFs may be reduced during the blending and consolidation phases, so it is important to choose the crucial aspects of ball milling and sintering carefully. Some of the main challenges in pursuing the powder metallurgy approach: Carbon fiber dispersion in the metal matrix is uniform and homogeneous, keeping carbon fibres' distinctive high aspect ratio throughout the composite production phases [46]. Dhanashekar *et al*., 2020 used powder metallurgy to synthesize the AA6061/Silicon Carbide (SiC) composites. In this process, SiC with an average particle size of 10 μm was reinforcement. The base matrix was dried in a furnace at 120 °C during the production process to eliminate the moisture content. Toluene was used to avoid the oxidation during powder mixing process, which was carried out in a planetary tumbler mixer at a ball to powder weight ratio of 10:1 at 55 rpm. The matrix material was then mixed with varied weight percentages of SiC particles (0, 2.5, 5, and 7.5 wt%) in the tumbler mixer. The green compact was made from the mixed powder using an 850 MPa uniaxial press. The green compact composite was sintered for 60 minutes in a closed furnace in an argon environment at 525 °C [47].

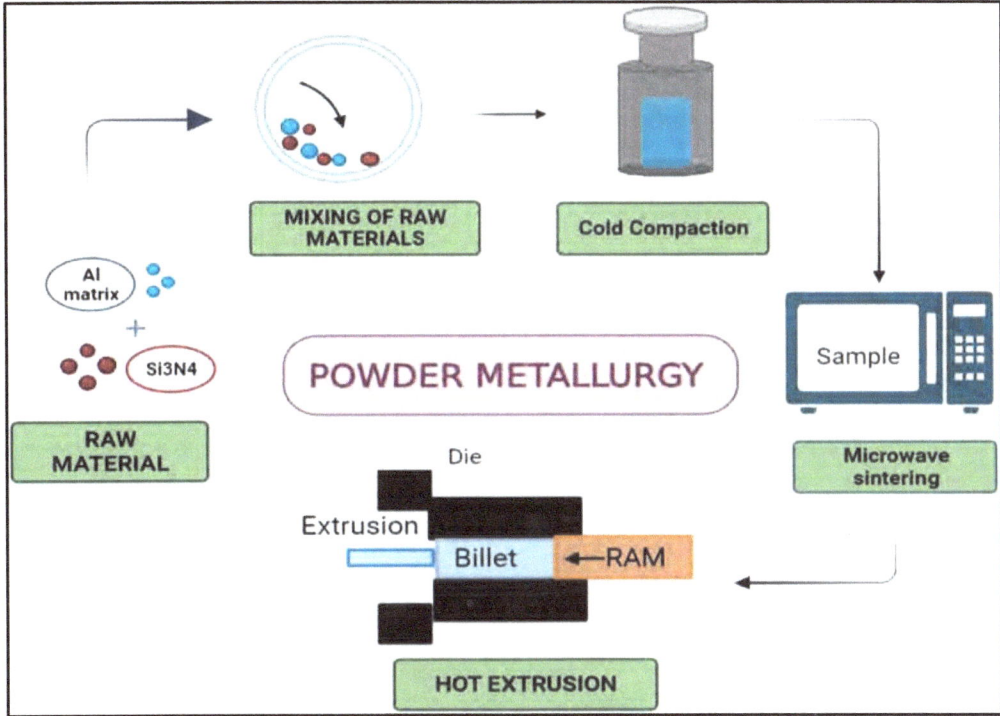

Fig. (6). Different processes in the powder metallurgy manufacturing of nanocomposites.

Diffusion Bonding

Diffusion bonding is a low temperature process to fabricate the CFs metal-matrix composite with the help of equipment like dies and rams. A preform carbon fiber is commonly generated in these manufacturing approaches by infiltrating carbon fiber into the polymer binder. Then, a metal sheet is stacked on top of the preform by stacking the multiple layers of preform with the metal sheet. To avoid oxidation, the stacked layers are pressed at a low temperature of 15-30 MPa under a high vacuum. The approach of a diffusion bonding setup for the manufacture of CFR-MMC is shown in Fig. (**7**). A clean surface on both the carbon fiber and the metal preforms is necessary to produce a high strength interface between the metal matrices and carbon fiber. Carbon fibers are often treated with acidic or alkaline solutions to establish the appropriate interfacial bonding [48]. Coatings on CFs are applied using several coating processes, including chemical coating, plasma spraying, chemical vapour deposition (CVD), electrochemical plating, and PVD. Surface modification of metallic sheets improves the bonding strength by removing thin oxide layers from the surfaces. By layering an array of fibers on top of the metal matrix during this conventional manufacturing process, the possibility of intermediate strength composites was activated [49]. This technology has gained tremendous interest to be applied for monofilament fibers such as SiC, SiC/W, and Al_2O_3.

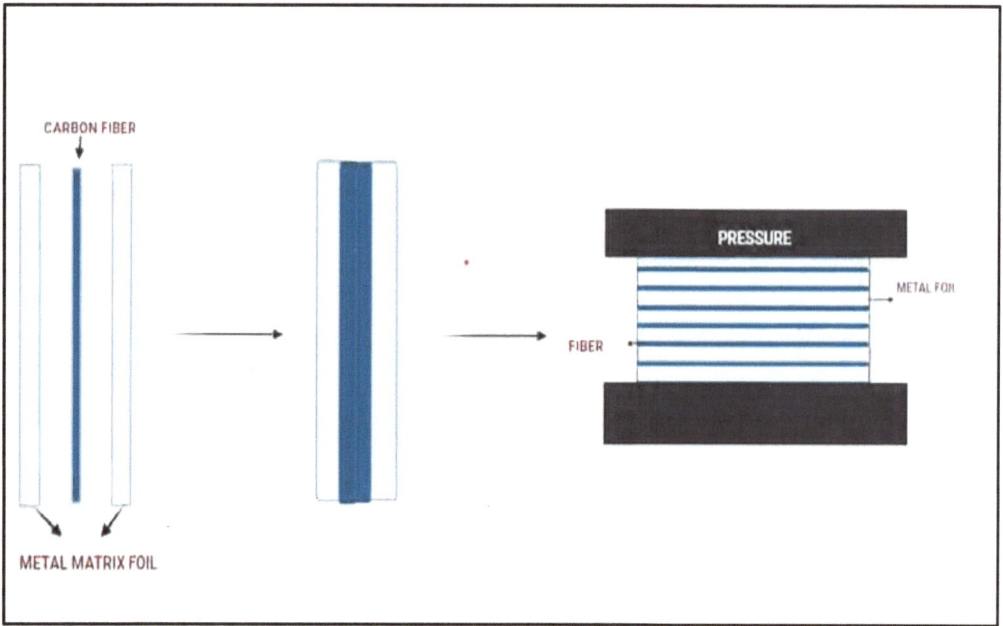

Fig. (7). Schematic representation of Diffusion Bonding method.

Liquid-Based Approach

MMC may be manufactured in a variety of ways depending on cost efficiency, weight/complexity, and simplicity of the final product. The liquid-based approach includes vacuum infiltration and casting. Infiltration squeeze casting, and stir casting, are the most often used processes for the processing of coated CFR-MMCs. This technology has several advantages, including low cost, quick processing periods, and high fiber content. However, the necessity to use high temperatures for composite production can result in increased fiber-matrix interfacial interaction and wetting difficulties. Weak interactions between carbon and other metal elements may occur during the production of the carbon fibers and metal matrix composites due to the weak wetting of carbon fiber and liquid metal. The finished parameters of the CF-MMCs are determined by interfacial wetting and reaction control which is difficult to find out [50]. Protective coating and external pressure on the surface of CFs are frequently used to overcome these issues. Another barrier to implementing the liquid state approach is the great difference in densities between the low weight CFs and metal matrices. Because of considerable density variations, carbon fibers are expected to split and float on the surface. In general, liquid-based synthesis of carbon fibers metal matrix composite is ideal for the creation of complicated geometries. By modifying the carbon fiber surface to make it more wettable by molten metal and optimising the CF-MMCs manufacturing parameters, homogeneous dispersion of carbon fibers can be achieved [51].

Melt Stirring Approach

Due to cost-effectiveness and simplicity when compared to alternative processes, the manufacturing of MMCs utilising the melt stirring methodology has attracted the interest of industry and academics. Carbon fibers are consistently mixed during the melt stirring process into a molten matrix material, which is subsequently cast using the standard casting technique to create the final product. In most cases, the reinforcing material is mixed into the molten matrix at a particular molar ratio [52]. After that, mechanical stirring at high temperatures is used to combine the components for 30-60 minutes. The melt stirring setup is seen schematically in Fig. (**8**). The high temperatures and potential for interfacial interactions between the fiber and reinforcement limit, the use of the well-established and relatively simple melt stirring method in the creation of carbon fiber metal matrix composites [53]. Singh *et al.*, (2021) used stir casting to create an AL/TIB2 metal matrix composite. The matrix material is aluminium 7075 (Al7075), and the hard artistic material is titanium di-boride (TiB$_2$). First, the dispersed phase of the ceramic particles was mechanically mixed with the liquid framework. Mix casting is the simplest and most effective approach for creating fluid states. The fluid composite material is subsequently cast using the standard projection techniques and can also be created using the typical metal framing methods. In this cycle, particles are typically broken up in general structure and agglomerates by excessive mixing. The mixing setup consists of a computer-controlled suppress heater and a graphite stirrer connected to an electric engine with a speed range of 22-840 rpm. To increase their wettability with liquid aluminium, TiB particles were deceptively oxidised at 1000 °C in the air. This treatment promotes particle consolidation while reducing undesirable interfacial reactions. Bundles of network amalgam were dissolved in a 1.5 kilogram mud reinforced graphite pot with a small stifling heater. The temperature of the mixture was initially elevated to roughly 800 °C before being blended at 540 rpm using a graphite impeller operated by a variable air conditioning motor [54].

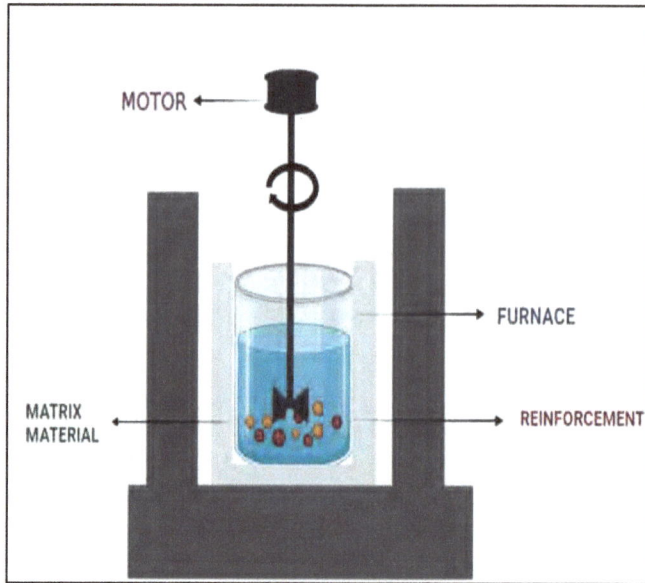

Fig. (8). Diagram illustrating the Melt stirring apparatus.

Semi-Solid Slurry Casting Approach

Semi-solid slurry casting is frequently performed at lower metallic matrix freezing temperatures. Slurry casting is the process of casting composites using the rheo-slurry at a temperature between the solid and liquid. After melting the alloy above liquid, it is gradually cooled to the semi-solid region. Carbon fibers are introduced to the semi-solid slurry during the stirring process. A low temperature method has the advantage of reducing or completely eliminating the possibility of interfacial reactions and interface degradation. As it did with earlier liquid processing methods like melt stirring, the high viscosity of matrix prevents CF separation. Mechanical stirring is used to disperse Carbon fibers in the slurry casting, which is then followed by gravity casting at lower pressures Fig. (**9**) depicts the slurry casting process.

Squeeze Casting Process

The technique consists of preheating the mould, melting the matrix, inserting it into the mould, and compressing the mould. In lower fixed mould half, a reinforcement phase precursor (particles, fibres) is inserted. The preform is typically inserted into the appropriate die, and a certain amount of molten matrix is poured into it. The die is then subjected to pressures between 70 and 100 MPa while under the high vacuum. This technology may produce highly densified yet smaller CF-MMCs components. Squeeze casting allows for the reduction of porosity and shrinkage, as well as increased mechanical parameters of the objects [55, 56]. High pressure is typically used during the squeeze casting to obtain maximum infiltration thickness, which might damage preforms (Fig. **10**). Natrayan *et al.* (2020) used squeeze casting to manufacture an AA6061/Al$_2$O$_3$/SiC metal matrix composite (MMC). Al 6061 alloy has been used as the basic material, with reinforcing materials of 5% Al$_2$O$_3$ and 5% SiC. Reinforcement particles typically have a diameter of 10 m. In order to enhance the wettability, it has been ensured that the reinforcements are equally distributed throughout the matrix, and Mg particles (1 wt%) are added to the melt. The reinforcement particles were warmed at 250 °Celsius to remove any moisture content before being added and melted in the calculated amounts. Pressure holding time (15 s), Squeeze pressure (100 MPa), die temperature (150 °C), and pouring time (15 s) are constant squeeze casting process parameters. The alloy was stirred for 10 minutes at 250 rpm [57].

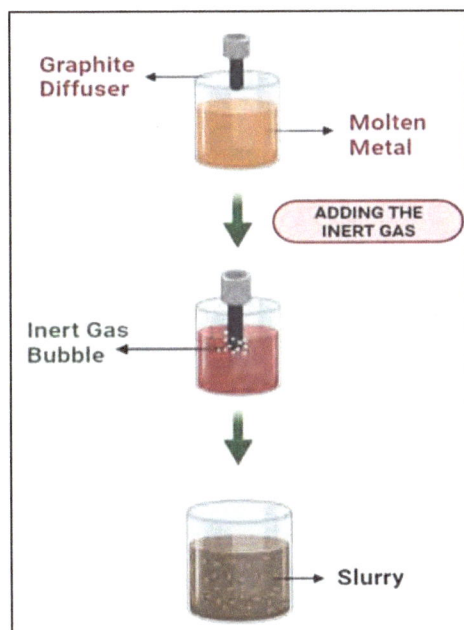

Fig. (9). Schematic representation of slurry casting process.

Deposition Processing

Plasma spraying and Ion plating are two popular deposition processes employed in the fabrication of CF-MMCs. These approaches have been employed in various research to improve the CF wettability and chemical bonding in the carbon fiber metal matrix composite.

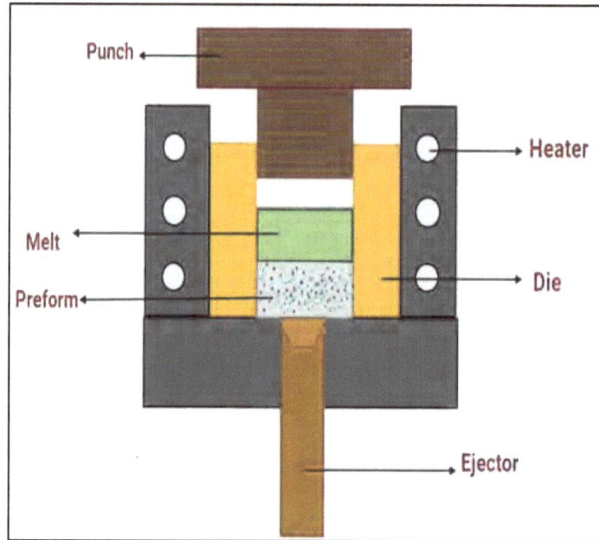

Fig. (10). Diagram showing how the squeeze infiltration equipment is designed.

Ion Plating

Ion plating is a (PVD) physical vapor deposition technique. Carbon fibers are distributed with an air knife in a vacuum chamber with argon serving as the inert gas during the ion plating process. The Argon plasma is then formed by applying a high potential. Metal is subsequently injected into the system to evaporate on the moving fibers. Hot pressing and diffusion bonding are used as the final consolidation procedure for the metal coated fibers. The ion plating procedure for the CF-MMC manufacturing is identical to the spraying process. The cathode is a spool of carbon fibers that is partially immersed in the electrolytic solution [58].

Plasma Spraying Approach

Plasma spraying is a process used to apply the matrix on the surface of continuous fibers. A continuous fiber spool can be wound into a carbon fiber metal matrix composite and coated with matrix material wires or matrix powders utilising arc wire spraying or plasma spraying. This procedure involves winding a carbon fiber spool, spraying the matrix material with arc wire or plasma, laminating, and consolidating [44]. Cold or hot isostatic pressing methods can be used to consolidate the plasma sprayed carbon fibers. The production of metal-coated CF can result in a carbon fiber metal matrix composite in which the fibers are aligned in any predetermined orientation and volume ratio. The high plasma temperature is one of the most significant benefits of plasma spraying, making this approach appropriate for treating the high melting point materials. Xiong *et al.*, 2018 prepared CF/Al specimens by plasma spraying method. In this method, aluminium powder is used as a raw material. A 15 g/min powder feed rate, a 23kW gun, a 160 mm spray distance, and a 30 slpm Ar flow rate were all used in the experiment. The relative pass speed between the torch and substrate was also around 110 mm/min. CF/Aluminium composites were created by layer-by-layer spraying aluminium powder over the carbon fibers. The process began with spraying aluminium powder onto an aluminium base plate to create an aluminium layer with a thickness of about 2 mm. A layer of unidirectional CFB (carbon fibre bundle), consisting of two S-twisted carbon threads with a diameter of 8 m and 300–700 CFs on each, was then applied to cover the plate. The space between each of the adjacent CFB was approximately 0.5 mm. After that, the second layer of Al coating was created by spraying the Al powder onto the CFB- and first-coated Al layer. This process was carried out and repeated to produce CF/Al composites with one to three CFB layers grown on one side of the aluminium base plate. By machining, the extra carbon fibers and the aluminium base plate were eliminated. The CF/Al composite was heated in vacuum at 350 °C for 1 hour to eliminate the micro-void and raise the density of the aluminium matrix [44].

CONCLUSION

The aerospace, automotive, and petrochemical industries have a great deal of potential for replacing current unreinforced metals and alloys with CFR-MMC. CFR-MMCs have excellent strength and mechanical performance. They are simple to manufacture and have outstanding thermal and electrical traits, improved corrosion resistance/wear, and a decreased coefficient of friction, making them suitable for various technical applications. The properties of the composite will deteriorate due to the poorly dispersed carbon fillers in the metal matrix. Therefore, more investigation is needed to create the best dispersion methods and processes in order to take advantage of the superior qualities that carbon fibres have. One of the most intriguing research areas that have been emerged in the recent years is the surface modification of carbon fibre. The final characteristic of CF-MMC can change with or without post-processing treatments, so this is a potential area for research. The next generation of carbon fiber, which comprises carbon fibers surface-modified with various carbon nanotubes (CNT), graphene, or polymers, has improved thermal, mechanical, and physical conductivity in comparison to pure carbon fibre. Future studies on the possibility of using this type of carbon fibers as an alternative to pure CFs may be fruitful. The mechanical performance of Metal Matrix Composites can be predicted using mathematical modelling techniques like neural or fuzzy processing networks. This is an area of research for MMCs that has yet to be completely explored.

LIST OF ABBREVIATIONS

CFR- MMC Carbon Fiber Metal Matrix Composite

MM Metal Matrix

CTE Coefficient of Thermal Expansion

PAN Polyacrylonitrile

DMF Dimethyl Formamide

DAMc Dimethyl Acetamide

ILSS Interlaminar Shear Strength

SEM Scanning Electron Microscopy

CNT Carbon Nanotube

REFERENCES

[1]   Shirvanimoghaddam, K.; Hamim, S.U.; Karbalaei Akbari, M.; Fakhrhoseini, S.M.; Khayyam, H.; Pakseresht, A.H.; Ghasali, E.; Zabet, M.; Munir, K.S.; Jia, S.; Davim, J.P.; Naebe, M. Carbon fiber reinforced metal matrix composites: Fabrication processes and properties. *Compos., Part A Appl. Sci. Manuf.,* **2017**, *92*, 70-96.
[http://dx.doi.org/10.1016/j.compositesa.2016.10.032]

[2]   Yusof, N.; Ismail, A.F. Post spinning and pyrolysis processes of polyacrylonitrile (PAN)-based carbon fiber and activated carbon fiber: A review. *J. Anal. Appl. Pyrolysis,* **2012**, *93*, 1-13.
[http://dx.doi.org/10.1016/j.jaap.2011.10.001]

[3]   Casati, R.; Vedani, M. Metal matrix composites reinforced by nano-particles—a review. *Metals,* **2014**, *4*(1), 65-83.
[http://dx.doi.org/10.3390/met4010065]

[4]   Miracle, D. Metal matrix composites – From science to technological significance. *Compos. Sci. Technol.,* **2005**, *65*(15-16), 2526-2540.
[http://dx.doi.org/10.1016/j.compscitech.2005.05.027]

[5]   Huang, X. Fabrication and properties of carbon fibers. *Materials,* **2009**, *2*(4), 2369-2403.
[http://dx.doi.org/10.3390/ma2042369]

[6]   Soutis, C. Carbon fiber reinforced plastics in aircraft construction. *Mater. Sci. Eng. A,* **2005**, *412*(1-2), 171-176.
[http://dx.doi.org/10.1016/j.msea.2005.08.064]

[7]   Cho, T.; Lee, Y.S.; Rao, R.; Rao, A.M.; Edie, D.D.; Ogale, A.A. Structure of carbon fiber obtained from nanotube-reinforced mesophase pitch. *Carbon,* **2003**, *41*(7), 1419-1424.
[http://dx.doi.org/10.1016/S0008-6223(03)00086-1]

[8]   MInus, M.; Kumar, S. The processing, properties, and structure of carbon fibers. *J. Miner. Met. Mater. Soc.,* **2005**, *57*(2), 52-58.
[http://dx.doi.org/10.1007/s11837-005-0217-8]

[9]   Newcomb, B.A. Processing, structure, and properties of carbon fibers. *Compos., Part A Appl. Sci. Manuf.,* **2016**, *91*, 262-282.
[http://dx.doi.org/10.1016/j.compositesa.2016.10.018]

[10]   Dunham, M.G.; Edie, D.D. Model of stabilization for pan-based carbon fiber precursor bundles. *Carbon,* **1992**, *30*(3), 435-450.
[http://dx.doi.org/10.1016/0008-6223(92)90042-U]

[11]   Le, N.D.; Trogen, M.; Ma, Y.; Varley, R.J.; Hummel, M.; Byrne, N. Cellulose-lignin composite fibers

as precursors for carbon fibers: Part 2 – The impact of precursor properties on carbon fibers. *Carbohydr. Polym.,* **2020**, *250*, 116918.
[http://dx.doi.org/10.1016/j.carbpol.2020.116918] [PMID: 33049890]

[12]   Khayyam, H.; Jazar, R.N.; Nunna, S.; Golkarnarenji, G.; Badii, K.; Fakhrhoseini, S.M.; Kumar, S.; Naebe, M. PAN precursor fabrication, applications and thermal stabilization process in carbon fiber production: Experimental and mathematical modelling. *Prog. Mater. Sci.,* **2020**, *107*, 100575.
[http://dx.doi.org/10.1016/j.pmatsci.2019.100575]

[13]   Lv, S.; Li, J.S.; Li, S.F.; Kang, N.; Chen, B. Effects of heat treatment on interfacial characteristics and mechanical properties of titanium matrix composites reinforced with discontinuous carbon fibers. *J. Alloys Compd.,* **2021**, *877*, 160313.
[http://dx.doi.org/10.1016/j.jallcom.2021.160313]

[14]   Frank, E.; Hermanutz, F.; Buchmeiser, M.R. Carbon fibers: Precursors, manufacturing, and properties. *Macromol. Mater. Eng.,* **2012**, *297*(6), 493-501.
[http://dx.doi.org/10.1002/mame.201100406]

[15]   Damodaran, S.; Desai, P.; Abhiraman, A.S. Chemical and physical aspects of the formation of carbon fibres from PAN-based precursors. *J. Textil. Inst.,* **1990**, *81*(4), 384-420.
[http://dx.doi.org/10.1080/00405009008658719]

[16]   Zhang, H.; Guo, L.; Shao, H.; Hu, X. Nano-carbon black filled Lyocell fiber as a precursor for carbon fiber. *J. Appl. Polym. Sci.,* **2006**, *99*(1), 65-74.
[http://dx.doi.org/10.1002/app.22184]

[17]   Jones, S.P.; Fain, C.C.; Edie, D.D. Structural development in mesophase pitch based carbon fibers produced from naphthalene. *Carbon,* **1997**, *35*(10-11), 1533-1543.
[http://dx.doi.org/10.1016/S0008-6223(97)00106-1]

[18]   Kim, B.J.; Kotegawa, T.; Eom, Y.; An, J.; Hong, I-P.; Kato, O.; Nakabayashi, K.; Miyawaki, J.; Kim, B.C.; Mochida, I.; Yoon, S-H. Enhancing the tensile strength of isotropic pitch-based carbon fibers by improving the stabilization and carbonization properties of precursor pitch. *Carbon,* **2016**, *99*, 649-657.
[http://dx.doi.org/10.1016/j.carbon.2015.12.082]

[19]   Liu, J.; Chen, X.; Liang, D.; Xie, Q. *Development of pitch-based carbon fibers: A review*; Energy Sources Part Recovery Util. Environ. Eff, **2020**, pp. 1-21.

[20]   Wazir, A.H.; Kakakhel, L. Preparation and characterization of pitch-based carbon fibers. *N. Carbon Mater.,* **2009**, *24*(1), 83-88.
[http://dx.doi.org/10.1016/S1872-5805(08)60039-6]

[21]   Mora, E.; Blanco, C.; Prada, V.; Santamaría, R.; Granda, M.; Menéndez, R. A study of pitch-based precursors for general purpose carbon fibres. *Carbon,* **2002**, *40*(14), 2719-2725.
[http://dx.doi.org/10.1016/S0008-6223(02)00185-9]

[22]   Choi, D.; Kil, H.S.; Lee, S. Fabrication of low-cost carbon fibers using economical precursors and advanced processing technologies. *Carbon,* **2019**, *142*, 610-649.
[http://dx.doi.org/10.1016/j.carbon.2018.10.028]

[23]   Frank, E.; Steudle, L.M.; Ingildeev, D.; Spörl, J.M.; Buchmeiser, M.R. Carbon fibers: precursor systems, processing, structure, and properties. *Angew. Chem. Int. Ed.,* **2014**, *53*(21), 5262-5298.
[http://dx.doi.org/10.1002/anie.201306129] [PMID: 24668878]

[24]   Bengtsson, A.; Bengtsson, J.; Jedvert, K.; Kakkonen, M.; Tanhuanpää, O.; Brännvall, E.; Sedin, M. Continuous stabilization and carbonization of a lignin–cellulose precursor to carbon fiber. *ACS Omega,* **2022**, *7*(19), 16793-16802.
[http://dx.doi.org/10.1021/acsomega.2c01806] [PMID: 35601329]

[25]   Peng, S.; Shao, H.; Hu, X. Lyocell fibers as the precursor of carbon fibers. *J. Appl. Polym. Sci.,* **2003**, *90*(7), 1941-1947.

[http://dx.doi.org/10.1002/app.12879]

[26]   Bahl, S. Fiber reinforced metal matrix composites - A review. *Mater. Today Proc.,* **2021**, *39*, 317-323.
       [http://dx.doi.org/10.1016/j.matpr.2020.07.423]

[27]   Clyne, T.W.; Withers, P.J. *An introduction to metal matrix composites*; Cambridge university press,
       1995.

[28]   Samal, P.; Vundavilli, P.R.; Meher, A.; Mahapatra, M.M. Recent progress in aluminum metal matrix
       composites: A review on processing, mechanical and wear properties. *J. Manuf. Process.,* **2020**, *59*,
       131-152.
       [http://dx.doi.org/10.1016/j.jmapro.2020.09.010]

[29]   Su, H.; Gao, W.; Feng, Z.; Lu, Z. Processing, microstructure and tensile properties of nano-sized
       Al2O3 particle reinforced aluminum matrix composites In: *Mater. Des*; , **2012**; 36, pp. 590-596.
       [http://dx.doi.org/10.1016/j.matdes.2011.11.064]

[30]   Akbari, M.K.; Baharvandi, H.R.; Shirvanimoghaddam, K. Tensile and fracture behavior of nano/micro
       TiB2 particle reinforced casting A356 aluminum alloy composites In: *Mater. Des*; , **2015**; 66, pp. 150-
       161.

[31]   Shalu, T.; Abhilash, E.; Joseph, M.A. Development and characterization of liquid carbon fibre
       reinforced aluminium matrix composite. *J. Mater. Process. Technol.,* **2009**, *209*(10), 4809-4813.
       [http://dx.doi.org/10.1016/j.jmatprotec.2008.12.012]

[32]   Eid, M.; Kaytbay, S.; Elkady, O.; El-Assal, A. Microstructure and mechanical properties of CF/Al
       composites fabricated by hot coining technique. *Ceram. Int.,* **2021**, *47*(15), 21890-21904.
       [http://dx.doi.org/10.1016/j.ceramint.2021.04.207]

[33]   Gatea, S.; Ou, H.; McCartney, G. Deformation and fracture characteristics of Al6092/SiC/17.5p metal
       matrix composite sheets due to heat treatments. *Mater. Charact.,* **2018**, *142*, 365-376.
       [http://dx.doi.org/10.1016/j.matchar.2018.05.050]

[34]   Shirvanimoghaddam, K.; Akbari, M.K.; Abdizadeh, H.; Pakseresht, A.; Abdi, F.; Shahbazkhan, A.
       Investigation of discontinuously reinforced aluminium metal matrix composite fabricated by two
       different micron ceramic reinforcements (ZrSiO 4, B 4 C): Comparative study. *Kov. Mater.,* **2015**,
       53(3), 139-146.

[35]   Ramesh, C. S.; Prasad, T. B. Friction and wear behavior of graphite-carbon short fiber reinforced
       Al–17% Si alloy hybrid composites, **2009**.
       [http://dx.doi.org/10.1115/1.2991124]

[36]   He, C.N.; Zhao, N.Q.; Shi, C.S.; Song, S.Z. Mechanical properties and microstructures of carbon
       nanotube-reinforced Al matrix composite fabricated by *in situ* chemical vapor deposition. *J. Alloys
       Compd.,* **2009**, *487*(1-2), 258-262.
       [http://dx.doi.org/10.1016/j.jallcom.2009.07.099]

[37]   Gopalakrishnan, S.; Murugan, N. Production and wear characterisation of AA 6061 matrix titanium
       carbide particulate reinforced composite by enhanced stir casting method. *Compos., Part B Eng.,*
       2012, *43*(2), 302-308.
       [http://dx.doi.org/10.1016/j.compositesb.2011.08.049]

[38]   Fathy, A.; Shehata, F.; Abdelhameed, M.; Elmahdy, M. Compressive and wear resistance of
       nanometric alumina reinforced copper matrix composites *Mater. Des.,* **2012**, *36*, 100-107.
       [http://dx.doi.org/10.1016/j.matdes.2011.10.021]

[39]   Dash, K.; Ray, B.C.; Chaira, D. Synthesis and characterization of copper–alumina metal matrix
       composite by conventional and spark plasma sintering. *J. Alloys Compd.,* **2012**, *516*, 78-84.
       [http://dx.doi.org/10.1016/j.jallcom.2011.11.136]

[40]   Boopathi, M.M.; Arulshri, K.P.; Iyandurai, N. Evaluation of mechanical properties of aluminium alloy
       2024 reinforced with silicon carbide and fly ash hybrid metal matrix composites. *Am. J. Appl. Sci.,*
       2013, *10*(3), 219-229.

[http://dx.doi.org/10.3844/ajassp.2013.219.229]

[41]   Wang, J.; Li, Z.; Fan, G.; Pan, H.; Chen, Z.; Zhang, D. Reinforcement with graphene nanosheets in aluminum matrix composites. *Scr. Mater.,* **2012**, *66*(8), 594-597.
[http://dx.doi.org/10.1016/j.scriptamat.2012.01.012]

[42]   Alten, A.; Erzi, E.; Gürsoy, Ö.; Hapçı Ağaoğlu, G.; Dispinar, D.; Orhan, G. Production and mechanical characterization of Ni-coated carbon fibers reinforced Al-6063 alloy matrix composites. *J. Alloys Compd.,* **2019**, *787*, 543-550.
[http://dx.doi.org/10.1016/j.jallcom.2019.02.043]

[43]   Zhu, C.; Su, Y.; Zhang, D.; Ouyang, Q. Effect of Al2O3 coating thickness on microstructural characterization and mechanical properties of continuous carbon fiber reinforced aluminum matrix composites. *Mater. Sci. Eng. A,* **2020**, *793*, 139839.
[http://dx.doi.org/10.1016/j.msea.2020.139839]

[44]   Xiong, J.; Zhang, H.; Peng, Y.; Li, J.; Zhang, F. Fabrication and characterization of plasma-sprayed carbon-fiber-reinforced aluminum composites. *J. Therm. Spray Technol.,* **2018**, *27*(4), 727-735.
[http://dx.doi.org/10.1007/s11666-018-0696-0]

[45]   Manohar, G.; Dey, A.; Pandey, K.M.; Maity, S.R. Fabrication of metal matrix composites by powder metallurgy: A review In: *AIP conference proceedings*; AIP Publishing LLC, **2018**; p. 020041.
[http://dx.doi.org/10.1063/1.5032003]

[46]   Samal, P.; Newkirk, J. *Powder metallurgy methods and applications*; ASM Handb. Powder Metall, **2015**, *7*.
[http://dx.doi.org/10.31399/asm.hb.v07.9781627081757]

[47]   Dhanashekar, M.; Loganathan, P.; Ayyanar, S.; Mohan, S.R.; Sathish, T. Mechanical and wear behaviour of AA6061/SiC composites fabricated by powder metallurgy method. *Mater. Today Proc.,* **2020**, *21*, 1008-1012.
[http://dx.doi.org/10.1016/j.matpr.2019.10.052]

[48]   Lee, H-S. Diffusion bonding of metal alloys in aerospace and other applications. In: *Welding and Joining of Aerospace Materials*; Elsevier, **2021**; pp. 305-327.
[http://dx.doi.org/10.1016/B978-0-12-819140-8.00010-9]

[49]   Morgan, P. Carbon fibers and their composites. *CRC Press.,* (1st ed..), Available from: https://scholar.google.com/scholar?q=+morgan+P.+CARBON+FIBERS+and+their+composites.+CRC +Press%3B+2005.&hl=en&as_sdt=0,5 (accessed Jan. 31, **2023**).

[50]   Kumar, A.; Vichare, O.; Debnath, K.; Paswan, M. Fabrication methods of metal matrix composites (MMCs). *Mater. Today Proc.,* **2021**, *46*, 6840-6846.
[http://dx.doi.org/10.1016/j.matpr.2021.04.432]

[51]   Bains, P.S.; Sidhu, S.S.; Payal, H.S. Fabrication and machining of metal matrix composites: A review. *Mater. Manuf. Process.,* **2016**, *31*(5), 553-573.
[http://dx.doi.org/10.1080/10426914.2015.1025976]

[52]   Chandra Kandpal, B.; Kumar, J.; Singh, H. Manufacturing and technological challenges in Stir casting of metal matrix composites– A review. *Mater. Today Proc.,* **2018**, *5*(1), 5-10.
[http://dx.doi.org/10.1016/j.matpr.2017.11.046]

[53]   Annigeri Veeresh Kumar, U.K.G.B.; Kumar, U. K. A. V. Method of stir casting of Aluminum metal matrix composites: A review. *Mater. Today Proc.,* **2017**, *4*(2), 1140-1146.
[http://dx.doi.org/10.1016/j.matpr.2017.01.130]

[54]   Kumar Singh, P.; Kumar Singh, P.; Sharma, K. Manufacturing and categorization of AL/TIB2 metal matrix compound by means of stir casting method. *Mater. Today Proc.,* **2021**, *45*, 3568-3573.
[http://dx.doi.org/10.1016/j.matpr.2020.12.1091]

[55]   Vijayaram, T.R.; Sulaiman, S.; Hamouda, A.M.S.; Ahmad, M.H.M. Fabrication of fiber reinforced metal matrix composites by squeeze casting technology. *J. Mater. Process. Technol.,* **2006**, *178*(1-3),

34-38.
[http://dx.doi.org/10.1016/j.jmatprotec.2005.09.026]

[56]   Dhanashekar, M.; Kumar, V.S. Squeeze casting of aluminium metal matrix composites-an overview. *Procedia Eng.,* **2014**, *97*, 412-420.
[http://dx.doi.org/10.1016/j.proeng.2014.12.265]

[57]   Natrayan, L.; Senthil Kumar, M. Optimization of wear behaviour on AA6061/Al2O3/SiC metal matrix composite using squeeze casting technique – Statistical analysis. *Mater. Today Proc.,* **2020**, *27*, 306-310.
[http://dx.doi.org/10.1016/j.matpr.2019.11.038]

[58]   Shang, S.M.; Zeng, W. *4 - Conductive nanofibres and nanocoatings for smart textiles*; T. Kirstein, Ed., in Woodhead publishing series in textiles. Woodhead publishing, **2013**, pp. 92-128.
[http://dx.doi.org/10.1533/9780857093530.1.92]

CHAPTER 5

Fabrication and Interfacial Bonding of CNT-reinforced Metal Matrix Composites

Prianka Sharma[1], **Vidushi Karol**[2], **Sarabjeet Kaur**[1,2] and **Manish Taunk**[3,*]

[1] *Department of Physics, School of Basic & Applied Sciences, Maharaja Agrasen University, Solan, H.P., India*

[2] *Department of Applied Science, Chandigarh Engineering College, Landran, Mohali, Panjab, India*

[3] *Department of Physics, Chandigarh University, Mohali, Panjab, India*

Abstract: Recent advances in various engineering applications demand new materials that have multi-functionality along with suitable structural properties. Metal matrix composites are the class of materials that satisfy this purpose due to their lightweight, increased strength, and other improved mechanical properties. These composite materials can be prepared by various conventional techniques which aim reducing the cost of production and meeting the demand of the industries efficiently. The properties and functionality of these materials are greatly influenced by the type of reinforced particulates and their composition in the metal matrix. Many reinforcement particles or fibers can be used in MMC depending upon the applications. Commonly used reinforced materials are graphene, polymers, carbon fibers, ceramic materials, *etc.* Among the carbon family, carbon nanotubes (CNT) exhibit enhanced performance as an ideal reinforcement material for MMCs. With outstanding intrinsic physical properties, CNTs are considered a promising candidate for reinforcement. CNT owes its properties due to its small diameter, high tensile strength, stiffness, high Young's modulus, and good chemical stability. They exhibit thermal stability even at high temperatures and exhibit good electrical conductivity. They also show improved fatigue resistance and plasticity and thus broaden the performance of the MMC. In this chapter, various fabrication techniques along with blending and processing methods of CNT-reinforced MMC have been discussed. The main methods have been explained with their schematic representations. The advantages and limitations of these methods have also been discussed. A strong interfacial bonding between the reinforced particulate and the metal matrix affects the performance of the material. This chapter also deals with a deep understanding of the various interfacial bonds that can exist between CNT and the metal matrix.

Keywords: Bonding interface, Carbon nanotubes, Mechanical properties, Metal matrix composite, Reinforcement effect, Solid state processing.

* **Corresponding author Manish Taunk:** Department of Physics, Chandigarh University, Mohali, Panjab, India; E-mail: manishphy@gmail.com

Virat Khanna, Prianka Sharma & Santosh Kumar (Eds.)
All rights reserved-© 2024 Bentham Science Publishers

INTRODUCTION

Rapid industrialization demands the development of high-performance materials for varied engineering applications. With technological advancements, automobile, aircraft, and aerospace industries especially expect the fabrication of next-generation materials that are high in strength and hardness besides being ultra-light, environmentally friendly [1], and comparatively economical. In this context, the spotlight is focused on composite materials. These materials maintain the original characteristics of the host material and can also be tailored for additional properties by doping or reinforcement with other mono-lithic compounds. The reinforcement particle is added to attain improved properties like strength, toughness, stiffness, electrical and thermal conductivity, electromagnetic shielding, coefficient of thermal expansion, damping and wear resistance [2 - 4]. These technologically improved materials then exhibit enhanced and unique properties that might not be exhibited by the base material and hence show multifunctional applications. Composite materials can be categorized into different types: polymer, metal, and ceramic [5, 6] depending upon the host matrix and the reinforcement particle. The reinforcement particle can be incorporated in the form of homogenous or discontinuous dispersion in the host matrix at the microstructural level. Another way of reinforcement of particles can be in the form of a continuous layer surrounding or within the matrix. The reinforced particles can be oxides, nitrides, and carbides of metals and metalloids. Depending upon the large volume fraction of the host matrix, the composite material is termed a Polymer matrix composite (PMC), Metal matrix composite (MMC), and Ceramic matrix composite (CMC). For lightweight and low-temperature applications, polymer matrix composites are used. On the other hand, for high-temperature, inert atmosphere applications ceramic matrix composites are used. These CMCs exhibit high mechanical properties. The metal matrix composites are lightweight, work under high temperatures, and exhibit outstanding mechanical properties as well. MMCs can be easily tailored for microstructure and other physical properties' modifications. Since the host matrix is a metal or an alloy of metal, it can undergo a wide range of thermo-mechanical processing. Technological interest in MMCs arose due to their profound stiffness even at low density MMCs, which offer the advantage of high strength to weight ratio, high elastic modulus, good thermal and electrical conductivity, lower coefficient of thermal expansion, superior elevated temperature properties like rupture strength and improved creep resistance, better fatigue performance and ability to resist moisture and radiations. Due to these advantages, MMCs find wide applications in the automotive and aerospace industries [7 - 11]. Fig. (**1**) shows the various classification of matrix & reinforcement types in MMC.

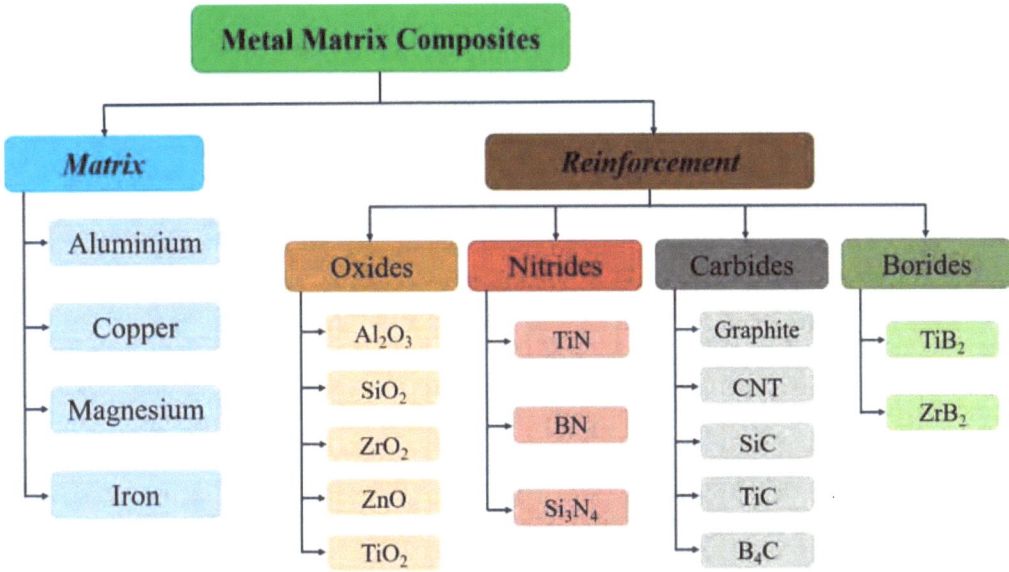

Fig. (1). Classification of matrix & reinforcement types in MMC.

Various reinforcement particles or fibers can be used in MMC depending upon the applications. Depending on the type of reinforcement material used, MMCs can be particle-reinforced composites or whisker-type short-fiber composites or sheets of continuous fiber composite. In particle-reinforced MMCs, ceramics, metals, amorphous material or glasses can be used as reinforcement material. These types of MMCs have high modulus than the host matrix but have low permeability and ductility. Thus they can sustain high tensile strength and shear and compressive stress. The fiber-reinforced MMCs also have high modulus and are bonded with the matrix along the fiber length with strong covalent bonds. The fibers are relatively oriented and thereby make a strong impact on the mechanical properties of the MMC. Commonly reinforced materials used are graphene, polymers, carbon fibers, ceramic materials [12], *etc*. Though ceramic-reinforced MMCs exhibit good mechanical properties but they exhibit poor electrical and thermal properties [13]. Carbon fiber-reinforced MMCs (CF-MMC) exhibit a good balance between thermo-mechanical properties and machinability and can be effectively utilized for heat sink applications [14, 15]. Due to their capability of high wear resistance, CF-MMCs can be used as bearings and wear parts with improved friction coefficients. Owing to their self-lubricating effect and high-temperature strength, CF-MMCs have superior, strength, modulus, and electrical conductivity. Hence they find important applications in aerospace, automobile, and electronic industries [16 - 19]. In the recent era, carbon materials like graphene find wide applications as reinforced materials in MMC for the synthesis of high-performance composite material for improved material performance. The two-dimensional nanometric structure of graphene endows superior mechanical, electrical, and thermal properties to graphene-reinforced MMC (GMMC).

CNT REINFORCED METAL MATRIX COMPOSITE

Another member of the carbon family, carbon nanotubes (CNTs) [20] exhibit enhanced performance as an ideal reinforcement material for MMCs. Outstanding intrinsic physical properties of CNTs might overcome the drawbacks of ceramic and polymer-reinforced MMC. CNTs are considered a promising candidate for reinforcement due to their lightweight and high composite strength. CNT owes these properties due to its small diameter, high tensile strength greater than 100 GPa, stiffness up to 1000GPa, high Young's modulus 1 TPa, and good chemical stability. They exhibit thermal stability even at high temperatures (3000^{0}C) and exhibit good electrical conductivity. They also show improved fatigue resistance and plasticity [21] and thus broaden the performance of the MMC.

CNT is formed from the sp^2 planar sheet of graphene. In a graphene sheet, atoms of carbon are tightly bonded and densely packed with carbon atoms at the corners of the hexagon having sp^2 hybridization (Fig. **2**). These graphene sheets are rolled up in the form of a cylinder to form CNT. When a graphene sheet is rolled up along the axis that cannot be superimposed along with its mirror image, also known as the chiral axis denoted by C_h, then the circumference of the resultant nanotube is equal to the length of the chiral axis. The chiral axis is expressed as $C_h = na_1 + ma_2$, where 'n' and 'm' are integers and a_1 and a_2 are lattice translation vectors. Metallic nanotubes are formed in the case when $n=m$ also known as 'armchair' nanotubes. However, when $n/m = 0$, semiconducting 'zigzag' nanotubes are formed [1, 22] as shown in Fig. (**3**).

Fig. (2). Basic structure of graphene.

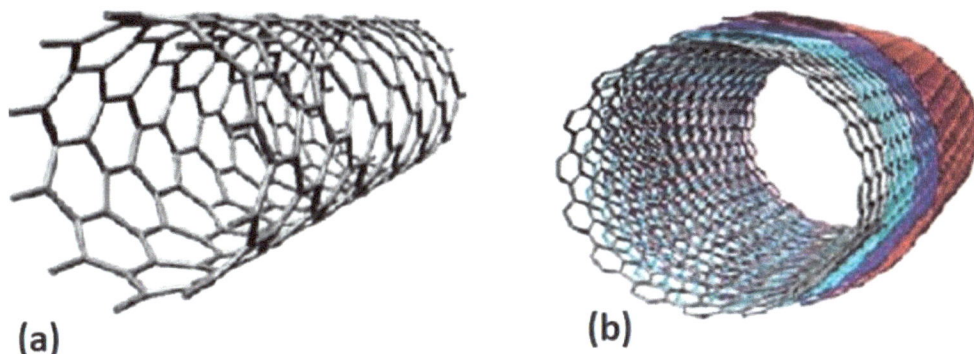

Fig. (3). Types of CNTs (**a**) Single walled carbon nanotube (SWCNT) (**b**) Multi walled carbon nanotube (MWCNT).

CNTs' length can extend up to 18 centimeters, however, the diameter is very small (a few nanometers). There are two types of CNTs–single-walled carbon nanotube (SWCNT) and multi-walled carbon nanotube (MWCNT) as represented in Fig. (**3**). In SWCNT, only one graphene sheet is rolled up to form the nanotube. In MWCNT, the layer of graphene sheets rolls up to cover the inner nanotube. There are about 2-50 concentric tubes with an interlayer distance between concentric cylinders of approximately 0.33 nm so that the inner tube is protected from any chemical interaction or contamination. Though SWCNT and MWCNT exhibit similar properties, but the tensile strength of MWCNT is comparatively high.

Though CNTs prove to be an effective reinforcement material for enhancing the thermal and mechanical properties of MMCs, yet some difficulties pave in the incorporation of CNTs in the host matrix of MMC. CNTs are entangled with each other due to the strong Van der Waal forces, hence they are prone to agglomeration, create defects, and form clusters due to high surface area (approx. $200 \, m^2$). CNT clusters reduce the strength and increase the porosity of the MMC which become the detrimental factors. Thus, inhomogeneity in the structure of the host matrix leads to a decrease in the mechanical properties of the CNT-MMC. Also, the weak interfacial bond strength between the CNT and the host matrix limits the stress transfer capability between the host matrix and the CNT. Thus, for the successful incorporation of CNT as a reinforcement phase, these drawbacks must be minimized for efficient fabrication of CNT-reinforced MMC.

CNT REINFORCED MMC FABRICATION

Incorporating CNT in the host matrix improves the wear resistance, elastic modulus, strength, stiffness, and various other properties of the MMC [23, 24]. However, proper alignment and uniform dispersion of CNTs into the MMC matrix have always remained a challenging task. Commercially available or synthesized CNTs are mostly in the agglomerated state. Firstly they need to be treated with acid or their surface functionalization is done with a surfactant to modify the surface energy of the specific CNT. This helps to improve the adhesion or wetting characteristics and thereby reduces the tendency of agglomeration in the CNT. Some processing techniques like powder metallurgy [25], mechanical milling [26, 27], spark plasma sintering [28, 29] electroless plating [30, 31] and plasma spray [32] methods are used for the fabrication of CNT-reinforced MMC. These fabrication methods are classified into two basic groups depending upon whether the host matrix is in solid or liquid or any other form as represented in Fig. (**4**).

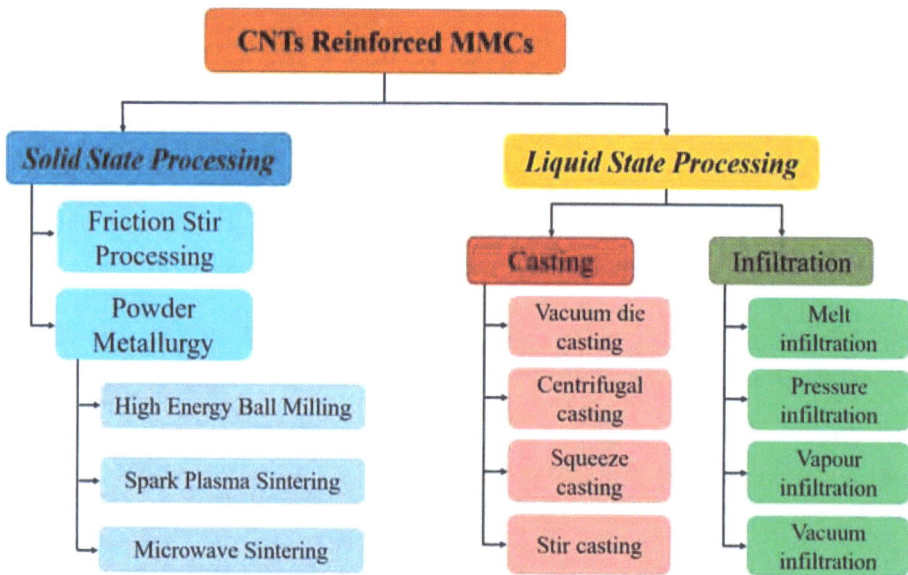

Fig. (4). Classification of different fabrication techniques.

Solid State Processing

Mostly, powder metallurgy along with sintering is an attractive method used for solid-state processing of CNT-reinforced MMCs. In this method, the processing is carried out at a low temperature which helps to control the kinetics of the interface in a better way. This method helps to employ proper mixing of the reinforced particle and the metal matrix and modify the microstructural refinements. Powder metallurgy is a two-step process. Firstly, the CNT and the host metal matrix are blended and mixed by different techniques. Secondly, the mixture is pressed to form a compacted disc and sintered in a vacuum or in an inert atmosphere to get the desired product [33].

Blending of CNT with Metal Matrix

Since CNTs are usually in the agglomerated form, different techniques are adopted to disaggregate them and blend them with the host metal matrix material. Different blending or dispersing processes are ball milling [34 - 41], colloidal mixing [42 - 46], molecular level mixing [47 - 51], magnetic stirring [52] roller mixing [41, 53], dipping [40], layer stacking [54], friction stir processing [55], particle composite system mixing [56, 57], nanoscale dispersion processing [58, 59], *etc.* Among all three processing techniques, ball milling, colloidal mixing, and molecular-level mixing are commonly adopted.

Ball Milling Blending

A traditional ball mill setup is used for blending the metal powder with the reinforced material. Different apparatus like planetary mill, attritor, or a horizontal ball mill are used for carrying out the ball milling process. Though, the principle operations of all the techniques are similar, however, planetary ball mill is frequently utilized for the mechanical ball milling process. The advantage of this technique is the requirement of a small amount of powder which makes it quite suitable for research. This ball mill apparatus consists of one turntable disc with two or four clamps for holding bowls or jars. The rotation of the disc is in one direction and the bowls or jars rotate in the opposite direction. The rotation of the jar along with the rotation of the disc about its axis creates the centrifugal forces applied for the mixing of the powder with milling balls in the jar. The powdered mixture gets fractured andundergoes crushing due to the high energy impact. The important parameters to be considered for the ball milling process are: (i) ball to powder weight ratio (BPR), (ii) milling balls size, (iii) medium type, (iv) rotational speed per minute (rpm)of the planetary ball milling and (v) the time of milling. Two methods namely, Factorial and Taguchi method can be utilized for statistical modelling and optimization of the parameters. However, Taguchi method is considered to be more efficient and optimal for the determination of the combination of controlled parameters. Generally, milling time of 0.5 hrs to 7 hrs and even higher can be considered with a rotational speed of 100-250 rpm. Another important parameter to be considered is the BPR which can be taken as 5:1, 10;1, 20:1. However, BPR of 10:1-20:1 and 2-4 hrs milling with 200-250 rpm is usually adopted in most of the cases. Though different sizes of balls can be used for the milling process but usually the balls of 10-12 mm diameter give optimum results. All these factors influence the milling process and the reduction in particle size strongly. The reinforced material (CNT) and the metal powders are taken in a mixing jar along with zirconia balls. Mixing is done by rotating the jar continuously for long hours with a specific rotational speed (Fig. **5a**). While rotating, the balls and the powder material are continuously hit and crushed Fig.

(**5b**) leading to a reduction in size and incorporation of CNTs into the host matrix powder material Fig. (**5c-d**) due to the impact energy. Balls of different sizes and materials, rotational speed, any gas atmosphere, and even ball-to-powder ratio can influence the mixing of CNT with the host metal. Generally, ethanol is added to the mixing jar as a process control agent to prevent particle welding of the metal matrix material. This ball-milling blending method gives a homogenous distribution of the CNT-reinforced phase into the MMC since it inhibits agglomeration and leads to proper dispersion of partially welded CNTs with the host metal matrix material [60]. However, this method gives rise to defects in CNTs due to the high contact pressure of 30 GPa at the time of mixing. These defects in CNTs can be reduced by using low-energy ball milling but still few defects will still exist [61].

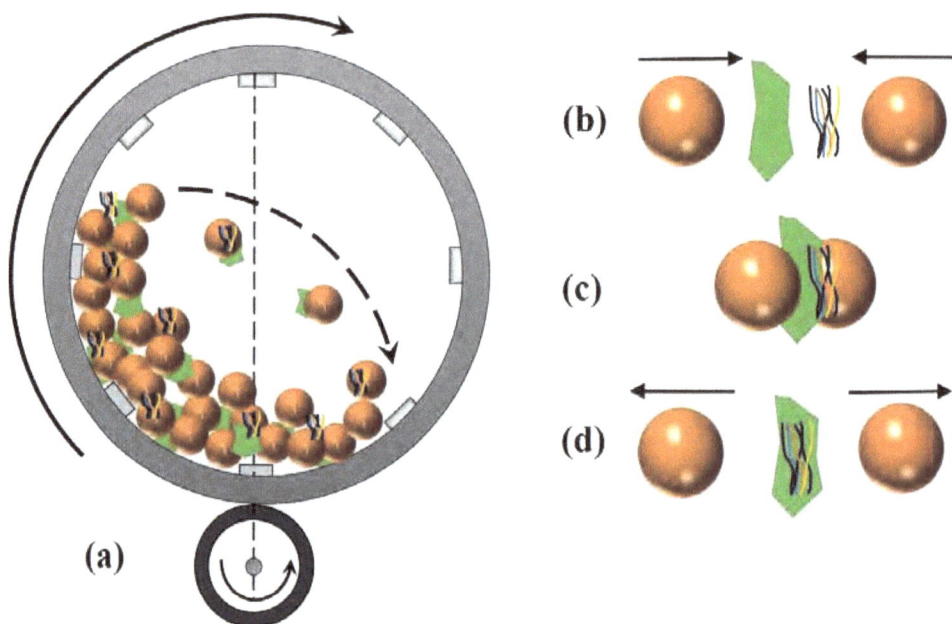

Fig. (5). Schematic representation of ball milling.

Colloidal Mixing

In this method, CNTs are dispersed in a solvent like ethylene glycol or DMF using an ultrasonic bath or magnetic stirrer. The stable dispersion of the CNT depends upon the solvent as well as the surface functionalization of the CNT used. Pristine CNTs can also be used to retain the actual properties of CNT. However, this colloidal mixing by ultrasonic agitation also leads to an increase in defect density. After proper dispersion of CNTs, the metal powder is added and finally mixed with the dispersed CNT by ultrasonic agitation or stirring again. Finally, the solvent is evaporated by heating and a dry mixture in powdered form is obtained. Fig. (**6**) shows the schematic representation of the colloidal mixing method.

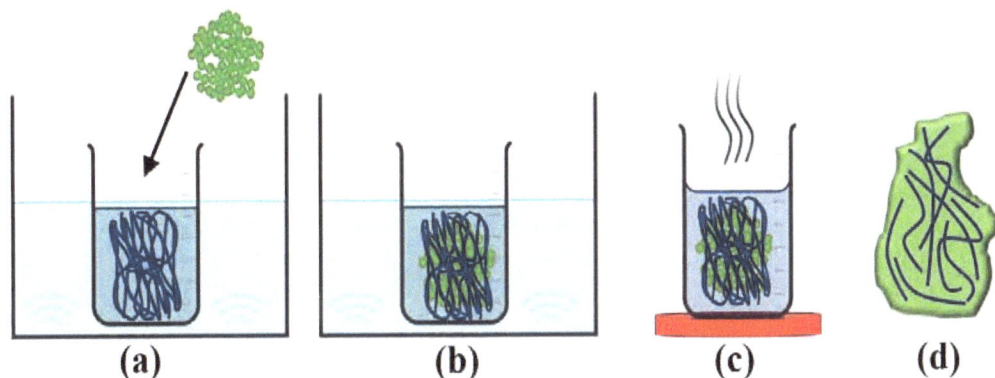

Fig. (6). Schematic representation of the colloidal mixing method.

Molecular Level Mixing

In the molecular mixing method, the important step is the surface functionalization of the CNT by acid treatment or any other method. After surface functionalization, CNTs are dispersed in any solvent (for eg. ethylene glycol or DMF) by ultrasonication to get a stable suspension. After that, metal salt of the host matrix material is added and undergoes a reduction process by any reducing agent to form a CNT metal oxide suspension. The CNTs here act as nucleation centers for the formation of the metal oxide. Finally, the CNT-reinforced MMC slurry is washed off many times to remove any chemicals and the powder undergoes calcination under a hydrogen atmosphere to reduce into CNT-reinforced MMC powder. This method enjoys the advantage that the host metal matrix coats the CNT particles or in another way, the CNT particles are embedded into the metal matrix. In other blending methods, the reinforced particle is situated only at the grain boundaries, thereby offering weak interfacial bonding between the host matrix material and the reinforcement phase which reduces the homogeneity of distribution. However, in the molecular-level blending process, the reinforced phase embeds deep into the host matrix network giving the homogenous distribution needed for thermal and electrical applications. Despite homogenous distribution of CNT into the host metal matrix, this method has a drawback. The required functionalization of the CNT breaks the covalent of the carbon network to accept the functional groups into the CNT surface which diminishes the excellent properties of CNTs for which it has been used as a reinforced phase.

Processing Techniques

After mixing or blending CNT and metal composite powder, different techniques are adopted for the densification of the final product. A few of these methods are spark plasma sintering (SPS) [62 - 69], hot uniaxial pressing (HUP) [70 - 76], cold pressed sintering (CPS) [77 - 80], friction stir processing [81], microwave sintering [82], *etc.* Commonly used processing techniques are discussed below.

Cold Pressed Sintering (CPS)

In this processing technique, the CNT-MMC blended powder undergoes densification by putting pressure by a uniaxial press to form a pre-compact disc of the desired shape. The disc is then sintered in an inert atmosphere or vacuum to form a densified sample without any further pressure (Fig. **7**). Though the heating-cooling cycle may take much time, but samples of different shapes can be sintered at the same time. This makes the process of sintering quite a time-efficient.

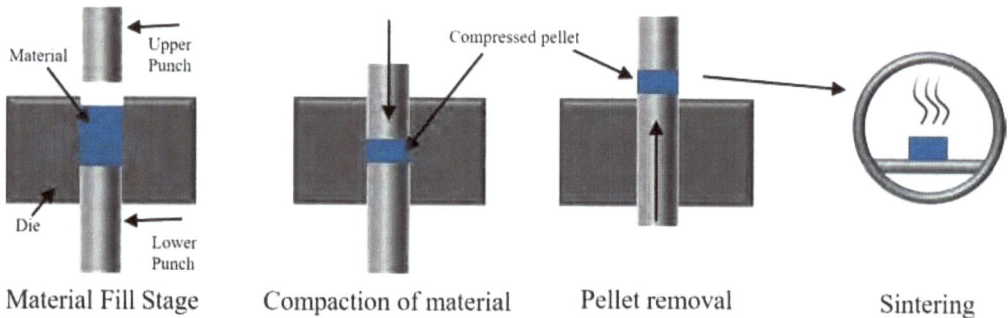

Fig. (7). Schematic representation of cold-pressed sintering.

However, this technique has a disadvantage in that the resultant composite material has poor density since the densification of composite material depends on the proper diffusion of the grain and lattice boundaries [83] and as no further pressure is applied at the time of sintering, thus, the material remains porous [84].

Hot Uniaxial Pressing (HUP)

In CPS, due to pressureless sintering along with grain and lattice boundary diffusion, plastic deformation and creep also contribute to the sintering mechanism. As porosity is more in pressure-less sintering, thus pressure-assisted sintering becomes much more effective. When external pressure is applied, compact densification of the sample takes place but the grain size is large. To reduce the grain size, sintering time, as well as temperature, is reduced with external pressure [72]. In HUP, the CNT-MMC blended powder pre-compacted into pellets with a uniaxial press is inserted into a die usually a steel die, and uniaxial pressure is exerted onto the sample by two alumina punches (Fig. **8**). Heating of the sample is performed with limited heating rate by the induction process which makes the HUP sintering process quite time-consuming. In this process, sintering can undergo high pressure of several hundred MPa and the highly densified samples can be obtained by exhibiting good mechanical properties.

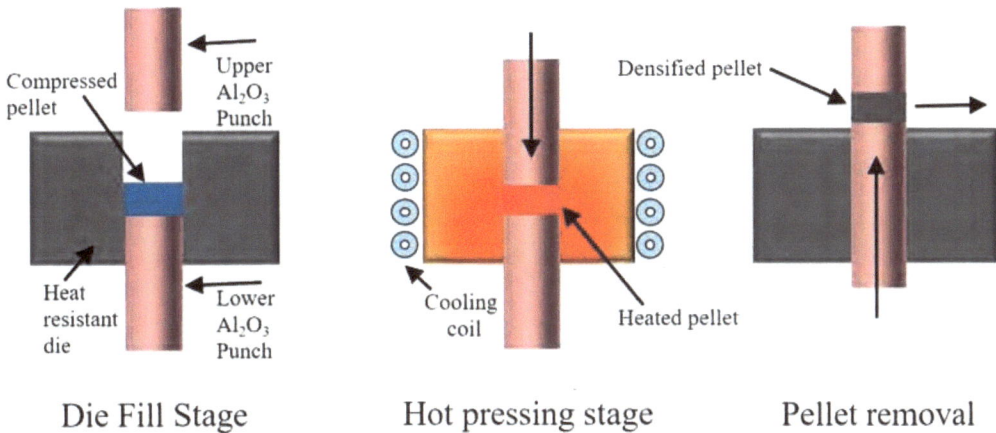

Fig. (8). Schematic representation of hot uniaxial pressed sintering.

Spark Plasma Sintering (SPS)

In SPS, the sintering mechanism is quite similar to that of HUP but the densified composite powders are achieved quickly Fig. (**9**) shows the Schematic representation of spark plasma sintering. Like in HUP, the blended CNT-MMC powder is compacted by a uniaxial press, and pellets are obtained. These pellets are then inserted into graphite die instead of steel die like in HUP. The graphite punches induce pressure on the sample and at the same time, a conducting electric DC pulse is sent through the sample. The sample gets heated due to the arisen electric resistance. By using the current pulse in a controlled manner, high heating rate of around several hundred °C/min can be adjusted for more efficiency. However, compared to HUP, low pressure (about 50 Mpa) can be applied during the sintering process since graphite punches are used. Full densification of the sample can be achieved by this process. Overall, this process is a time-efficient and an effective way to obtain CNT-reinforced MMC powders.

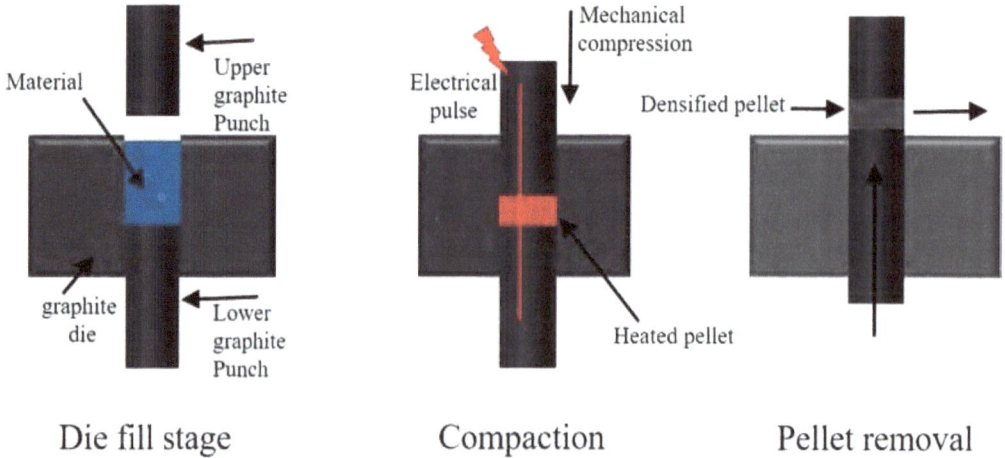

Fig. (9). Schematic representation of spark plasma sintering.

Liquid State Processing

In the liquid state processing, the reinforced particulates are incorporated into the molten matrix material. They are classified broadly into two groups: liquid infiltration and casting methods.

Liquid Infiltration Process

In this process, the molten matrix material is put into a preform of reinforcing material made similar to a die to fill all the porosity. The molten material flows into the reinforced particulate by adopting different means like capillarity force or any external mechanical force that might overcome the resistive force arising due to dragging by the capillary action [85]. The pressure required for combining the reinforced material and the molten metal depends upon frictional forces arising due to the viscous drag of the molten metal into the CNT preform. Wetting of the CNT preform depends upon various factors like temperature, CNT preform, alloy material composition, surface morphology, *etc* [86, 87].

Casting Methods

Different casting techniques like centrifugal die casting, gravity pouring, squeeze casting, and vacuum die casting are adopted for obtaining the desired casts of the reinforcement particulate and the metal matrix above the solidification temperature of the matrix material. The commonly used casting method is the conventional stir mixing with gravity casting as the most economical way of producing large net-shaped reinforced MMCs. Mostly stir mixing is done by stirrers designed in a way that helps to get improved vortex formation of the molten metal for obtaining homogenous dispersion of the reinforced particulate into the metal matrix. Efficient dispersion of the reinforcement particulate and the properties of the MMC are strongly dependent on the temperature of the molten metal and the mold, stirring speed and time, crucible size, rate of inflow of the mixture, *etc*. However, some drawbacks arise since reinforcement materials do not wet the molten metal matrix, and most reinforcing particulates either float or sink due to the relative density of the molten metal. Hence, heterogeneous dispersion of the reinforced particulate may happen [88] which defies the basic purpose of reinforcement.

In-Situ Deposition Process

Another fabrication process of MMC is the *in-situ* physical vapor deposition (PVD). In this process, a high-power electron beam falls on a solid bar of reinforcing material as feedstock and vapors are produced. The metal matrix to be coated is passed through the vapors of the reinforcing material. The metal is then coated with the vapors of the reinforcing material that condenses on the metal matrix. The coated metal matrix is assembled sequentially in arrays and compressed in heat to obtain the final reinforced MMC [89, 90]. Spray forming is another deposition technique for getting shape MMCs. In this process, there are two steps. Firstly the molten liquid is atomized into fine droplets by passing through argon an inert atomizing agent. Next, the atomized particles are deposited on the metallic substrate. In this deposition process different parameters like deposition rate, atomizing pressure, angle of spraying, atomizing gas, and nozzle to substrate distance [91]. The advantages of the deposition process are that fine grain size microstructure can be obtained and the segregation of grains is low. This process has some disadvantages also. It can be utilized with discontinuous reinforcements, the cost of the product is high and it is difficult to get intricate shape products. However, the composites obtained by the spray deposition process are inexpensive in comparison to other fabrication processes like powder metallurgy or stir casting [92 - 98].

THE BONDING INTERFACE OF CNT-REINFORCED MMC

Till now, we have discussed various processes for the fabrication of CNT-reinforced MMC. These fabrication processes are chosen in a way so that there is an increase in dislocation density and a uniform distribution of reinforcement in the matrix can be obtained [81 - 84]. A uniform interface between the host matrix and the reinforcement particulate can be obtained [85 - 87], which directly affects the performance of the reinforced MMC. A strong interfacial bonding between the reinforcement particulate CNT and the host metal matrix leads to enhancement in the mechanical properties of the CNT-reinforced MMC. There are four types of bonding interfaces: i) free diffusion interface, ii) mechanical bonding interface, iii) reaction bonding interface, and iv) hybrid bonding interface. These bonding interfaces have different strengths in the order of mechanical bonding interface < free diffusion interface < hybrid bonding interface < and reaction bonding interface. Different morphology of the interface and the bonding mechanism can be obtained by using different fabrication techniques.

Free Diffusion Interface

In the free diffusion interface, there is random diffusion of the reinforcement particulate into the metal matrix. The intermolecular interaction force affects the bonding strength of the free diffusion interface. Since there is a vast difference in the properties of CNT and any host metal matrix, hence the reinforcement can occur under certain conditions only. For this type of interface, it is necessary to promote wettability and fluidity between the reinforcement particulate and the host metal matrix. In such conditions, a chemical reaction does not happen at the interface surface rather the reinforcement particulate is uniformly distributed at the interface and then at the failure fracture point, a pull-out mechanism occurs [99]. This interface is obtained under the application of external factors so that the particles can effectively move towards the junction between the host and the reinforcement material and then there is mutual penetration. This type of interface occurs in the case of casting and spraying fabrication techniques.

In CNT-reinforced MMC, free diffusion of CNT into the liquid by casting method leads to the formation of the interface. Typical casting fabrication techniques for free diffusion interface are vacuum-assisted [100], induction melting [101], and molten state stirring casting technique [90]. By vacuum-assisted casting technique, uniform dispersion of the CNT reinforcement particulate into the host metal matrix is obtained at high temperatures. CNTs are strongly bonded to the broken portion of the MMC which forms a bridge and shifts the stress to the MMC. Hence, the strength of the interface increases. CNTs are pre-coated in the induction melting method which enhances the wettability and dispersibility of the CNT and the host metal matrix. By both techniques, a strong free diffusion interface is formed and the reinforced MMC exhibits enhanced hardness and compressive strength. Another casting technique is mechanical stirring casting. This method gives MMCs with different mechanical properties. In this method, different CNT concentrations can be used and the addition of adequate CNT content reduces agglomeration and enhances the mechanical properties.

Smooth and flat microstructures can be obtained with a good interface bonding state. The spraying technique also gives a free diffusion interface. Random dispersion of CNT over the solid happens under this fabrication method. Two types of spray fabrication processes are adopted: plasma spraying and thermal spraying [99]. However, the spraying process leads to weak interface bonding. When reinforcement particulate is sprayed on the matrix solid, due to weak van der Waals force between the CNT and the host metal matrix, bonds break down and the interface curls up near the CNTs. Further, stress is concentrated in a particular area which makes the CNTs prone to easily peel off from the host metal matrix surface due to weak interface bonding. During the spraying process, chemical reaction at the defect points on the CNTs releases carbon monoxide which generates bubbles and leads to the weakening of the interface bonding. Also, the thermal activation reaction reduces the strength of the bonding. Hence, these chemical reactions hinder the formation of the CNT and the metal matrix interface, thereby reducing the strength of the interface bonding [101]. Fig. (**10**) shows the various Schematic representation of MWCNT fabrication.

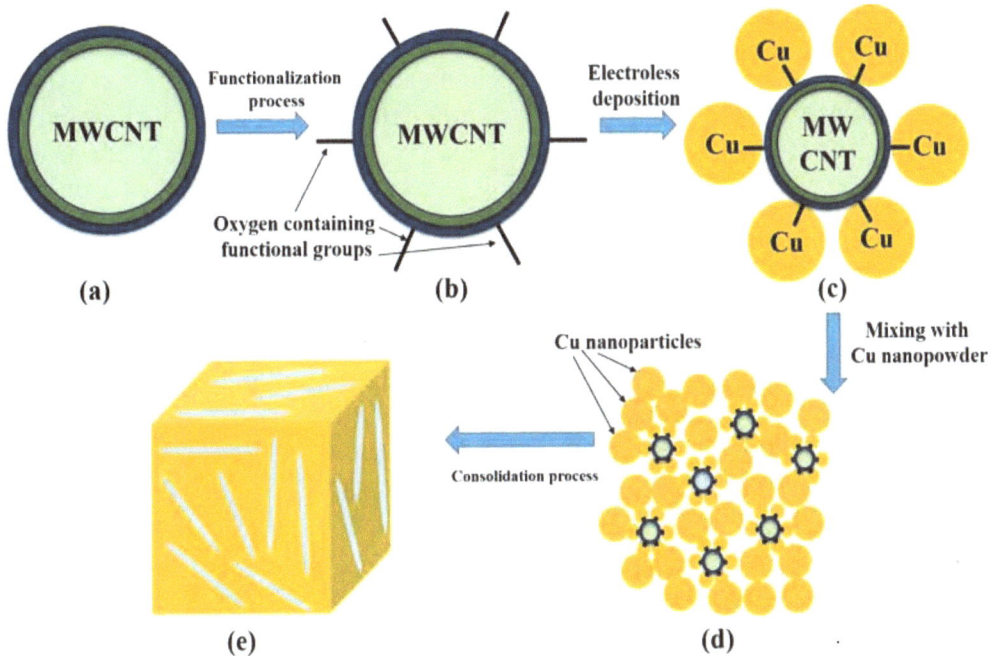

Fig. (10). Schematic representation of MWCNT fabrication; a) pure MWCNT; b) functionalization of MWCNT; c) metal encapsulated CNT; d) mixing of encapsulated MWCNT with pure metal nanoparticles; and e) fabricated CNT reinforced MMC.

Mechanical Bonding Interface

In this type of interface bonding, the reinforcement particulate is combined with the host metal matrix under an external force and no chemical reaction should occur. The bonding strength of the mechanical bonding interface is less than the free diffusion interface. Since the reinforcement particulate and the host metal matrix are forcibly bonded, when a large external force is applied, the interface bonds break down since the force cannot be transmitted to the reinforcement particulate continuously. Hence, post-fabrication treatments like solution treatment and aging treatment are done to erase the internal stress and enhance the strength of the reinforced MMC. Since, during the fabrication process, the CNT in the interface is subjected to external mechanical force, the arrangement in the interface becomes flat, regular, and ordered in the direction of applied force [102]. Due to relatively weak bonding strength, the fabrication process must include heat treatment so that the reinforcement particulate is easily dispersed and embedded into the host matrix for strengthening the interface bonding. Most fabrication techniques for mechanical bonding interfaces are the friction stir process, the hot rolling composite method [91], and the sandwich technique [103].

In sandwich technology, distinct layers of the CNTs and the host matrix are rolled up to form the composite. A regular and flat interface is formed with no damage or agglomeration. However several pores occur between the interlayers [92]. Another approach to obtaining a mechanical bonding interface is using a combination of friction stir technology and the hot rolling method for mixing and stirring the CNTs with the host metal matrix [104]. The CNTs are orderly arranged in the host metal matrix along the hot rolling direction to obtain a smooth, relatively flat, and tightly bonded interface. This enhances the tensile strength and leads to the elongation of the composite material.

Reaction Bonding Interface

An *in situ* fabrication process of growing CNT onto the host metal matrix composite leads to the formation of a reaction bonding interface. Chemical reaction with additional products under certain conditions occurs in this fabrication process. In this process, CNTs are grown on the host metal matrix by chemical vapor deposition (CVD) process, dried and ball milled for a short time. Then the resultant product of CNT-MMC powder is compacted sintered and undergoes hot extrusion. Parameters like the temperature of the heat treatment during the fabrication process and ball milling time influence the formation of the reaction bonding interface. Since the CNT-metal matrix interface is directly grown on the host matrix substrate, thus reinforcement particulates of CNTs are uniformly distributed. No amorphous carbon grows on the matrix and no agglomeration of CNT is observed. The CNTs are tightly bonded with the host matrix and fused to form the largest interface bonding surface area thereby, the bonding force is maximum and gives the best dynamics enhancement effect [105]. This enhances the strength of the material and the elongation of the CNT-MMC is highly improved.

Hybrid Bonding Interface

A hybrid interface is formed by the combination of the above-mentioned interface types. This bonding interface is commonly visible in composite materials. This bonding is complicated and formed due to high temperature, high extrusion force, and other fabrication means adopted for the synthesis of the composite material. There are many preparation techniques for obtaining a hybrid interface. This includes powder metallurgy used along with friction stir technology and spark plasma sintering (SPS). The formation of the interface is influenced by different fabrication technologies adopted and the processing parameters. In the SPS process, high sintering temperature and high-pressure lead to the formation of a reaction bonding interface, and during the hot extrusion process, a mechanical bonding interface is formed [106]. The CNT-reinforced MMC prepared by powder metallurgy along with the friction stir process leads to the formation of different interfaces [107]. A mechanical bonding interface is formed at the time of ball milling and friction stir processing. When later high-temperature heat treatment is given, this gives rise to a reaction bonding interface. At the same time, a free diffusion interface is also formed. Hence, a hybrid bonding interface comes into existence which enhances the strength of the material.

FACTORS INFLUENCING THE BONDING INTERFACE OF CNT REINFORCED MMC

The interfacial bonding state between CNT and the host metal matrix is influenced by various fabrication processes and parameters.

Effect of the Fabrication Process on the Bonding Interface

Different interfacial bonding arises from different fabrication processes which will influence the properties of the reinforced MMC. For example, as discussed earlier, a mechanical bonding interface can be obtained by combining the friction stir method with the hot rolling process. Vacuum-assisted fabrication method leads to a free diffusion interface. *In-situ* fabrication techniques can be utilized to get a reaction bonding interface. However, a mixed bonding interface known as the hybrid interface can be obtained from powder metallurgy. Varied fabrication parameters also influence the interfacial bonding states. In the case of SiCw + CNTs/6061Al hybrid composites, the material is treated under three processing conditions: cast state, hot extrusion, and post-hot extrusion heat treatment. A strong interfacial bonding state is obtained when the material undergoes hot extrusion and reaction heat inhibits the formation of an inadequate interface [108]. However, interface formation between host metal and CNT can be directly influenced by keeping control over the sintering temperature in powder metallurgy fabrication [109 - 112]. Strong interfacial bonding can be obtained at a higher annealing temperature than the melting point of the host metal. Hence, the interface phase can be precisely controlled by tuning the processing temperature. Jiang *et al.* [95] altered the conventional powder metallurgy fabrication process with sheet metallurgy. Double the size nanocomposites were obtained and CNT changed from 1-dimension to an array of 2-dimension thereby increasing the contact area between the host metal matrix and the CNT. As a result, more stable interfacial bonds were formed increasing the ductility and tensile strength of the MMC. Similarly, other parameters like mechanical impact by ball milling can also bring significant effects. Short CNT fibers can be obtained by a small duration of ball milling which can increase the dispersibility of the CNT and influence the interfacial bonding [99 - 101]. Longer ball milling time can damage the CNTs [113 - 116]. Some process control agents or smelting aids can also be added during ball milling as dispersants to avoid the formation of oxide films on the host metal and damage to the CNT [103 - 105]

Effect of Material Composition on the Bonding Interface

The bonding interface is directly affected by the composition wettability and some other properties of CNTs.

Wettability

Properties of CNTs like surface tension, density, modulus, coefficient of thermal expansion, *etc.* differ from the properties of the host metal. This gives rise to poor wettability conditions between the CNT and the host metal. The wettability condition can be increased by generating functional bonds like hydrogen bonds and covalent bonds [117 - 119], reducing the contact angle, and applying mechanical rolling processes to borrow transition layers and alloy elements of high affinity. Coating of CNT with metals like copper or nickel reduces the contact angle between the CNT and the host metal and leads to the formation of covalent bonds [107, 108]. The coating of metal dissolves in the matrix to form a stable compound [120]. Thus, metal-coated CNT-MMC exhibits a smooth and compact interface without much porosity. Another strategy to enhance wettability between the CNT and the host metal matrix is by creating a mismatch between these two different materials and forming interfacial dislocations. In this case, CNTs have to be heated before metal plating [121]. Another way is rolling CNT over the host metal for a forced fusion [32]. However, this method does not prove to be as efficient as the wetting method.

Dispersion

The degree of dispersion of CNT on the host metal surface has an effective impact on the interfacial bonding of the CNT and the host metal. Since CNT has high surface energy, the Van der Waals bonds between pair of CNTs is strong. For the low degree of dispersion, CNTs segregate and result in uneven interfacial bonding. This lowers the strength of CNTs which is a major limitation for the fabrication of CNT-reinforced MMC. The ball milling vibration is the most commonly adopted dispersion technique for uniform dispersion of CNT in the metal matrix. This method effectively prohibits the CNT from agglomeration at the interface [122]. However, ball milling has certain limitations like thick metal flakes cannot be broken and hinder the dispersion of CNT [107]. Other methods like *in situ* synthesis can also be adapted to enhance the dispersibility of CNTs in the host metal matrix [123]. Additionally, high pressure and high temperature can enhance dispersion but will cause irreversible damage to CNTs which affects their practical applications [124 - 126].

Form and Content

Interfacial bonding is also affected by the morphology of the CNT and the host metal matrix. Particularly, the content of CNT, size of host metal particles, shape, integrity, and aspect ratio significantly affect the bonding at the interface. Instead of CNTs' traditional tubular structure, more defects are present in the CNTs having chevron shape which reacts weakly with the host metal forming an intermediate thin layer. This prevents the formation of the interface [74]. Small aspect ratio CNT has many advantages. Such CNTs show a large effective area of contact with the host metal matrix and easily form interfacial bonding [114, 115]. Also, small-diameter CNTs exhibiting strong capillary action produce non-covalent bonds and thereby form smooth and bubble-free interfacial bonding [114]. The addition of an appropriate amount of CNT to the host metal matrix creates many nucleation sites and finer grains that promote interfacial bonding between the CNT and the host metal. However, the addition of a higher content of CNT causes agglomeration and results in improper interfacial bonding with degraded mechanical properties [127]. Besides CNTs, the size of the host metal particles also affects the interfacial bonding. During ball milling, the metal particle undergoes fracture, cold welding, and deformations which lead to finer metal particles. Dislocation annihilation occurs for grain diameters less than 70 nm leading to weak interfacial bonding [128].

CHALLENGES & FUTURE PROSPECTS OF CNT REINFORCED MMC

Till now, we have discussed the fabrication processes of adding CNT as reinforcement particulate into any host metal matrix and the interfacial bonding between them. However, certain challenges are associated with the successful incorporation of CNT as a reinforcement phase. These challenges include interfacial bonding, dispersion, defects, alignment, *etc*. The first challenge is to prohibit the agglomeration of CNTs into clusters and their effective dispersion into the host metal matrix. Formation of clusters or agglomeration of CNTs for more than 1 wt% leads to a reduction in mechanical and other properties of the CNT-reinforced MMCs.

Taylton *et al*. [129] studied the influence of CNT agglomeration on the conductivity of the matrix and found that the concentration of CNTs is near the percolation threshold, uniform distribution of CNTs can be obtained which enhances the conductivity. Hence, effective fabrication processes have been extensively investigated along with high-energy ball-milling, functionalization with surfactants, and metal encapsulation of CNTs with composite precursors, *etc* [130, 131]. Despite good progress, optimization of parameters for uniform dispersion without degradation of CNTs has remained a big challenge.

Secondly, maintaining an intact CNT structure in the fabricated MMC is also a major challenge. In this case, CNTs can be pre-treated for example acidic, mechanical, thermal, and chemical methods before the fabrication processes for composite formation [120]. However, these pre-treatments may damage the CNTs significantly, thereby reducing their unique properties [132]. Additionally, heat treatment like sintering may cause the degradation of CNTs by the formation of carbides [133]. Thus, optimization of the parameters of pre-treatment must be done to avoid damage to the actual structure of CNT.

The third challenge is the formation of an interface between the CNT and the host metal matrix. For effective mechanical, thermal, and electrical performances, interfacial bonding of atomic scale is required [134, 135]. With weak interfacial bonding, the bonds may break, hence, the photon-electron scattering will be subsequently increased which may deteriorate the properties of the composite. Strong interfacial bonding can be achieved by pre-coating of CNTs with metal particles. Another challenge is the proper alignment of CNTs in the metal matrix composite [119, 136, 137, 138, 139, 140]. Alignment of CNTs along the axis yields improved properties of the composite than those aligned along the radial direction. CNTs can be aligned along the axis by deformation processes like rolling, extrusion, *etc.* to achieve improved properties. Hence, significant advancement in successful fabrication techniques must be done to promote uniform alignment and dispersion of CNTs in the host metal matrix, without damaging the basic feature of CNTs. This will help to obtain CNT-reinforced MMC with excellent physical and mechanical properties.

REFERENCES

[1]   Jindal, H.; Kumar, S.; Kumar, R. The Asian review of civil engineering environmental pollution and its impact on public health. *Crit. Rev.,* **2020,** *9*(1), 11-18.

[2]   Agarwal, A.; Lahiri, D.; Bakshi, S. R. *Carbon Nanotubes: Reinforced Metal Matrix Composites*; , **2018.**

[3]   Singh, N.; Taunk, M. *In-situ* chemical synthesis, microstructural, morphological and charge transport studies of polypyrrole-CuS hybrid nanocomposites. *J. Inorg. Organomet. Polym. Mater.,* **2021,** *31*(1), 437-445.
[http://dx.doi.org/10.1007/s10904-020-01747-8]

[4]   Taunk, M. Charge transport studies in flexible and rollable Polypyrrole-PVDF composite films. *Mater. Today Proc.,* **2023.**
[http://dx.doi.org/10.1016/j.matpr.2023.01.168]

[5]   Karol, V.; Prakash, C.; Sharma, A. Observation of high dielectric properties of Mg-substituted BST ceramic synthesized by conventional solid-state route. *Journal of Materials Science: Materials in Electronics,* **2021,** (14), 19478-19486.
[http://dx.doi.org/10.1007/s10854-021-06465-6]

[6]   Singh, N.; Chand, S.; Taunk, M. Facile *in-situ* synthesis, microstructural, morphological and electrical transport properties of polypyrrole-cuprous iodide hybrid nanocomposites. *J. Solid State Chem.,* **2021,** *303*, 122501.
[http://dx.doi.org/10.1016/j.jssc.2021.122501]

[7]   Senapati, A.K.; Panda, S.S.; Dutta, B.K.; Mishra, S. Effect of stirring speed during casting on mechanical properties of Al–Si based MMCs. *Smart Innovation, Systems and Technologies,* **2020,** *169*, 703-710.
[http://dx.doi.org/10.1007/978-981-15-1616-0_68]

[8]   Krishna, A.R.; Arun, A.; Unnikrishnan, D.; Shankar, K.V. An investigation on the mechanical and tribological properties of alloy A356 on the addition Of WC. *Mater. Today Proc.,* **2018,** *5*(5), 12349-12355.
[http://dx.doi.org/10.1016/j.matpr.2018.02.213]

[9]   Joseph, J.; Pillai, B.S.; Jayanandan, J.; Jayagopan, J.; Nivedh, S.; Balaji, U.S.S.; Shankar, K.V. Mechanical behaviour of age hardened A356/TiC metal matrix composite. *Mater. Today Proc.,* **2021,** *38*, 2127-2132.
[http://dx.doi.org/10.1016/j.matpr.2020.05.013]

[10]   Kumar, V.A. An of B4C/Al2O3 on the wear behavior of Al-6.6Si-0.4Mg alloy using response surface methodology *IJSEIMS,* **2020,** *8*(2), 14.
[http://dx.doi.org/10.4018/IJSEIMS.2020070105]

[11]   Anilkumar, V.; Shankar, K.V.; Balachandran, M.; Joseph, J.; Nived, S.; Jayanandan, J.; Jayagopan, J.; Surya Balaji, U.S. Impact of heat treatment analysis on the wear behaviour of al-14.2si-0.3mg-tic composite using response surface methodology. *Tribology in Industry,* **2021,** *43*(4), 590-602.
[http://dx.doi.org/10.24874/ti.988.10.20.04]

[12]   Khanna, V.; Singh, K.; Kumar, S.; Bansal, S.A.; Channegowda, M.; Kong, I.; Khalid, M.; Chaudhary, V. Engineering electrical and thermal attributes of two-dimensional graphene reinforced copper/aluminium metal matrix composites for smart electronics. *ECS J. Solid State Sci. Technol.,* **2022,** *11*(12), 127001.
[http://dx.doi.org/10.1149/2162-8777/aca933]

[13]   Karol, V.; Prakash, C.; Sharma, A. Impact of magnesium content on various properties of Ba0.95-xSr0.05MgxTiO3 ceramic system synthesized by solid state reaction route. *Mater. Chem. Phys.,* **2021,** *271*, 124905.
[http://dx.doi.org/10.1016/j.matchemphys.2021.124905]

[14] Lalet, G.; Kurita, H.; Heintz, J. M.; Lacombe, G.; Kawasaki, A.; Silvain, J. F. Thermal expansion coefficient and thermal fatigue of discontinuous carbon fiber-reinforced copper and aluminum matrix composites without interfacial chemical bond. *J. Mater. Sci.,* **2013**, *49*(1), 397-402.

[15] Cho, S.H. Heat dissipation effect of Al plate embedded substrate in network system. *Microelectron. Reliab.,* **2008**, *48*(10), 1696-1702.
[http://dx.doi.org/10.1016/j.microrel.2008.04.018]

[16] Xia, L.; Jia, B.; Zeng, J.; Xu, J. Wear and mechanical properties of carbon fiber reinforced copper alloy composites. *Mater. Charact.,* **2009**, *60*(5), 363-369.
[http://dx.doi.org/10.1016/j.matchar.2008.10.008]

[17] Edie, D.D. The effect of processing on the structure and properties of carbon fibers. *Carbon,* **1998**, *36*(4), 345-362.
[http://dx.doi.org/10.1016/S0008-6223(97)00185-1]

[18] Ramesh, C.S.; Prasad, T.B. Friction and wear behavior of graphite-carbon short fiber reinforced Al-17%Si alloy hybrid composites. *J. Tribol.,* **2009**, *131*(1), 014501.
[http://dx.doi.org/10.1115/1.2991124]

[19] Liu, L.; Li, W.; Tang, Y.; Shen, B.; Hu, W. Friction and wear properties of short carbon fiber reinforced aluminum matrix composites. *Wear,* **2009**, *266*(7-8), 733-738.
[http://dx.doi.org/10.1016/j.wear.2008.08.009]

[20] Iijima, S. Helical microtubules of graphitic carbon. *Nature,* **1991**, *354*(6348), 56-58.
[http://dx.doi.org/10.1038/354056a0]

[21] Pérez-Bustamante, R.; Pérez-Bustamante, F.; Estrada-Guel, I.; Licea-Jiménez, L.; Miki-Yoshida, M.; Martínez-Sánchez, R. Effect of milling time and CNT concentration on hardness of CNT/Al2024 composites produced by mechanical alloying. *Mater. Charact.,* **2013**, *75*, 13-19.
[http://dx.doi.org/10.1016/j.matchar.2012.09.005]

[22] Aqel, A.; El-Nour, K.M.M.A.; Ammar, R.A.A.; Al-Warthan, A. Carbon nanotubes, science and technology part (I) structure, synthesis and characterisation. *Arab. J. Chem.,* **2012**, *5*(1), 1-23.
[http://dx.doi.org/10.1016/j.arabjc.2010.08.022]

[23] Endo, M.; Hayashi, T.; Kim, Y.A. Large-scale production of carbon nanotubes and their applications. *Pure Appl. Chem.,* **2006**, *78*(9), 1703-1713.
[http://dx.doi.org/10.1351/pac200678091703]

[24] Kok, M. Production and mechanical properties of Al2O3 particle-reinforced 2024 aluminium alloy composites. *J. Mater. Process. Technol.,* **2005**, *161*(3), 381-387.
[http://dx.doi.org/10.1016/j.jmatprotec.2004.07.068]

[25] Goh, C.S.; Wei, J.; Lee, L.C.; Gupta, M. Development of novel carbon nanotube reinforced magnesium nanocomposites using the powder metallurgy technique. *Nanotechnology,* **2006**, *17*(1), 7-12.
[http://dx.doi.org/10.1088/0957-4484/17/1/002]

[26] Chauhan, A.; Kumar, M.; Kumar, S. Fabrication of polymer hybrid composites for automobile leaf spring application. *Mater. Today Proc.,* **2022**, *48*, 1371-1377.
[http://dx.doi.org/10.1016/j.matpr.2021.09.114]

[27] Poirier, D.; Gauvin, R. X-ray microanalysis of porous materials using Monte Carlo simulations. *Scanning,* **2011**, *33*(3), 126-134.
[http://dx.doi.org/10.1002/sca.20259] [PMID: 21773976]

[28] Dusza, J.; Blugan, G.; Morgiel, J.; Kuebler, J.; Inam, F.; Peijs, T.; Reece, M.J.; Puchy, V. Hot pressed and spark plasma sintered zirconia/carbon nanofiber composites. *J. Eur. Ceram. Soc.,* **2009**, *29*(15), 3177-3184.
[http://dx.doi.org/10.1016/j.jeurceramsoc.2009.05.030]

[29] Kwon, H.; Estili, M.; Takagi, K.; Miyazaki, T.; Kawasaki, A. Combination of hot extrusion and spark plasma sintering for producing carbon nanotube reinforced aluminum matrix composites. *Carbon,* **2009**, *47*(3), 570-577.
[http://dx.doi.org/10.1016/j.carbon.2008.10.041]

[30] Chen, B.; Shen, J.; Ye, X.; Jia, L.; Li, S.; Umeda, J.; Takahashi, M.; Kondoh, K. Length effect of carbon nanotubes on the strengthening mechanisms in metal matrix composites. *Acta Mater.,* **2017**, *140*, 317-325.
[http://dx.doi.org/10.1016/j.actamat.2017.08.048]

[31] Barrena, M.I.; Gómez de Salazar, J.M.; Soria, A.; Cañas, R. Improved of the wear resistance of carbon nanofiber/epoxy nanocomposite by a surface functionalization of the reinforcement. *Appl. Surf. Sci.,* **2014**, *289*, 124-128.
[http://dx.doi.org/10.1016/j.apsusc.2013.10.118]

[32] Lahiri, D.; Bakshi, S.R.; Keshri, A.K.; Liu, Y.; Agarwal, A. Dual strengthening mechanisms induced by carbon nanotubes in roll bonded aluminum composites. *Mater. Sci. Eng. A,* **2009**, *523*(1-2), 263-270.
[http://dx.doi.org/10.1016/j.msea.2009.06.006]

[33] Kalaiselvan, K.; Murugan, N.; Parameswaran, S. Production and characterization of AA6061–B4C stir cast composite. *Mater. Des.,* **2011**, *32*(7), 4004-4009.
[http://dx.doi.org/10.1016/j.matdes.2011.03.018]

[34] Xue, Z.W.; Wang, L.D.; Zhao, P.T.; Xu, S.C.; Qi, J.L.; Fei, W.D. Microstructures and tensile behavior of carbon nanotubes reinforced Cu matrix composites with molecular-level dispersion. *Mater. Des.,* **2012**, *34*, 298-301.
[http://dx.doi.org/10.1016/j.matdes.2011.08.021]

[35] Shukla, A.K.; Nayan, N.; Murty, S.V.S.N.; Sharma, S.C.; Chandran, P.; Bakshi, S.R.; George, K.M. Processing of copper-carbon nanotube composites by vacuum hot pressing technique. *Mater. Sci. Eng. A,* **2013**, *560*, 365-371.
[http://dx.doi.org/10.1016/j.msea.2012.09.080]

[36] Chen, B.; Yang, J.; Zhang, Q.; Huang, H.; Li, H.; Tang, H.; Li, C. Tribological properties of copper-based composites with copper coated NbSe2 and CNT. *Mater. Des.,* **2015**, *75*, 24-31.
[http://dx.doi.org/10.1016/j.matdes.2015.03.012]

[37] Kwon, H.; Saarna, M.; Yoon, S.; Weidenkaff, A.; Leparoux, M. Effect of milling time on dual-nanoparticulate-reinforced aluminum alloy matrix composite materials. *Mater. Sci. Eng. A,* **2014**, *590*, 338-345.
[http://dx.doi.org/10.1016/j.msea.2013.10.046]

[38] Jiang, L.; Li, Z.; Fan, G.; Cao, L.; Zhang, D. The use of flake powder metallurgy to produce carbon nanotube (CNT)/aluminum composites with a homogenous CNT distribution. *Carbon,* **2012**, *50*(5), 1993-1998.
[http://dx.doi.org/10.1016/j.carbon.2011.12.057]

[39] Phuong, D.D.; Trinh, P.V.; An, N.V.; Luan, N.V.; Minh, P.N.; Khisamov, R.K.; Nazarov, K.S.; Zubairov, L.R.; Mulyukov, R.R.; Nazarov, A.A. Effects of carbon nanotube content and annealing temperature on the hardness of CNT reinforced aluminum nanocomposites processed by the high pressure torsion technique. *J. Alloys Compd.,* **2014**, *613*, 68-73.
[http://dx.doi.org/10.1016/j.jallcom.2014.05.219]

[40] Chen, B.; Kondoh, K.; Imai, H.; Umeda, J.; Takahashi, M. Simultaneously enhancing strength and ductility of carbon nanotube/aluminum composites by improving bonding conditions. *Scr. Mater.,* **2016**, *113*, 158-162.
[http://dx.doi.org/10.1016/j.scriptamat.2015.11.011]

[41] Borkar, T.; Hwang, J.; Hwang, J.Y.; Scharf, T.W.; Tiley, J.; Hong, S.H.; Banerjee, R. Strength versus ductility in carbon nanotube reinforced nickel matrix nanocomposites. *J. Mater. Res.,* **2014**, *29*(6),

761-769.
[http://dx.doi.org/10.1557/jmr.2014.53]

[42]   Guiderdoni, C.; Pavlenko, E.; Turq, V.; Weibel, A.; Puech, P.; Estournès, C.; Peigney, A.; Bacsa, W.; Laurent, C. The preparation of carbon nanotube (CNT)/copper composites and the effect of the number of CNT walls on their hardness, friction and wear properties. *Carbon,* **2013**, *58,* 185-197.
[http://dx.doi.org/10.1016/j.carbon.2013.02.049]

[43]   Sule, R.; Olubambi, P.A.; Sigalas, I.; Asante, J.K.O.; Garrett, J.C.; Roos, W.D. Spark plasma sintering of sub-micron copper reinforced with ruthenium–carbon nanotube composites for thermal management applications. *Synth. Met.,* **2015**, *202,* 123-132.
[http://dx.doi.org/10.1016/j.synthmet.2015.02.001]

[44]   Reinert, L.; Zeiger, M.; Suárez, S.; Presser, V.; Mücklich, F. Dispersion analysis of carbon nanotubes, carbon onions, and nanodiamonds for their application as reinforcement phase in nickel metal matrix composites. *RSC Advances,* **2015**, *5*(115), 95149-95159.
[http://dx.doi.org/10.1039/C5RA14310A]

[45]   Suárez, S.; Ramos-Moore, E.; Lechthaler, B.; Mücklich, F. Grain growth analysis of multiwalled carbon nanotube-reinforced bulk Ni composites. *Carbon,* **2014**, *70,* 173-178.
[http://dx.doi.org/10.1016/j.carbon.2013.12.089]

[46]   Suarez, S.; Lasserre, F.; Prat, O.; Mücklich, F. Processing and interfacial reaction evaluation in MWCNT/Ni composites. *physica status solidi,* **2014**, *211*(7), 1555-1561.
[http://dx.doi.org/10.1002/pssa.201431018]

[47]   Kim, K.T.; Eckert, J.; Menzel, S.B.; Gemming, T.; Hong, S.H. Grain refinement assisted strengthening of carbon nanotube reinforced copper matrix nanocomposites. *Appl. Phys. Lett.,* **2008**, *92*(12), 121901.
[http://dx.doi.org/10.1063/1.2899939]

[48]   Cha, S.I.; Kim, K.T.; Arshad, S.N.; Mo, C.B.; Hong, S.H. Extraordinary strengthening effect of carbon nanotubes in metal-matrix nanocomposites processed by molecular-level mixing. *Adv. Mater.,* **2005**, *17*(11), 1377-1381.
[http://dx.doi.org/10.1002/adma.200401933] [PMID: 34412442]

[49]   Daoush, W.M. Processing and characterization of CNT/Cu nanocomposites by powder technology. *Powder Metall. Met. Ceramics,* **2008**, *47*(9-10), 531-537.
[http://dx.doi.org/10.1007/s11106-008-9055-x]

[50]   Firkowska, I.; Boden, A.; Vogt, A.M.; Reich, S. Effect of carbon nanotube surface modification on thermal properties of copper–CNT composites. *J. Mater. Chem.,* **2011**, *21*(43), 17541-17546.
[http://dx.doi.org/10.1039/c1jm12671g]

[51]   Hwang, J.Y.; Lim, B.K.; Tiley, J.; Banerjee, R.; Hong, S.H. Interface analysis of ultra-high strength carbon nanotube/nickel composites processed by molecular level mixing. *Carbon,* **2013**, *57,* 282-287.
[http://dx.doi.org/10.1016/j.carbon.2013.01.075]

[52]   Kuzumaki, T.; Miyazawa, K.; Ichinose, H.; Ito, K. Processing of carbon nanotube reinforced aluminum composite. *J. Mater. Res.,* **1998**, *13*(9), 2445-2449.
[http://dx.doi.org/10.1557/JMR.1998.0340]

[53]   Hwang, J.Y.; Neira, A.; Scharf, T.W.; Tiley, J.; Banerjee, R. Laser-deposited carbon nanotube reinforced nickel matrix composites. *Scr. Mater.,* **2008**, *59*(5), 487-490.
[http://dx.doi.org/10.1016/j.scriptamat.2008.04.032]

[54]   Isaza, C.; Sierra, G.; Meza, J. M. A novel technique for production of metal matrix composites reinforced with carbon nanotubes. *J. Manuf. Sci. Eng.,* **2016**, *138*(2), 024501.
[http://dx.doi.org/10.1115/1.4030377]

[55]   Liu, Q.; Ke, L.; Liu, F.; Huang, C.; Xing, L. Microstructure and mechanical property of multi-walled carbon nanotubes reinforced aluminum matrix composites fabricated by friction stir processing. *Mater. Des.,* **2013**, *45,* 343-348.

[http://dx.doi.org/10.1016/j.matdes.2012.08.036]

[56] Chai, G.; Sun, Y.; Sun, J.J.; Chen, Q. Mechanical properties of carbon nanotube–copper nanocomposites. *J. Micromech. Microeng.,* **2008**, *18*(3), 035013.
[http://dx.doi.org/10.1088/0960-1317/18/3/035013]

[57] Chu, K.; Guo, H.; Jia, C.; Yin, F.; Zhang, X.; Liang, X.; Chen, H. Thermal properties of carbon nanotube-copper composites for thermal management applications. *Nanoscale Res. Lett.,* **2010**, *5*(5), 868-874.
[http://dx.doi.org/10.1007/s11671-010-9577-2] [PMID: 20672107]

[58] Kwon, H.; Takamichi, M.; Kawasaki, A.; Leparoux, M. Investigation of the interfacial phases formed between carbon nanotubes and aluminum in a bulk material. *Mater. Chem. Phys.,* **2013**, *138*(2-3), 787-793.
[http://dx.doi.org/10.1016/j.matchemphys.2012.12.062]

[59] Tokunaga, T.; Kaneko, K.; Horita, Z. Production of aluminum-matrix carbon nanotube composite using high pressure torsion. *Mater. Sci. Eng. A,* **2008**, *490*(1-2), 300-304.
[http://dx.doi.org/10.1016/j.msea.2008.02.022]

[60] Peng, X.; Luan, Z.; Ding, J.; Di, Z.; Li, Y.; Tian, B. Ceria nanoparticles supported on carbon nanotubes for the removal of arsenate from water. *Mater. Lett.,* **2005**, *59*(4), 399-403.
[http://dx.doi.org/10.1016/j.matlet.2004.05.090]

[61] Choi, H.J.; Shin, J.H.; Bae, D.H. The effect of milling conditions on microstructures and mechanical properties of Al/MWCNT composites. *Compos., Part A Appl. Sci. Manuf.,* **2012**, *43*(7), 1061-1072.
[http://dx.doi.org/10.1016/j.compositesa.2012.02.008]

[62] Cho, S.; Kikuchi, K.; Miyazaki, T.; Takagi, K.; Kawasaki, A.; Tsukada, T. Multiwalled carbon nanotubes as a contributing reinforcement phase for the improvement of thermal conductivity in copper matrix composites. *Scr. Mater.,* **2010**, *63*(4), 375-378.
[http://dx.doi.org/10.1016/j.scriptamat.2010.04.024]

[63] Guiderdoni, C.; Estournès, C.; Peigney, A.; Weibel, A.; Turq, V.; Laurent, C. The preparation of double-walled carbon nanotube/Cu composites by spark plasma sintering, and their hardness and friction properties. *Carbon,* **2011**, *49*(13), 4535-4543.
[http://dx.doi.org/10.1016/j.carbon.2011.06.063]

[64] Sule, R.; Olubambi, P.A.; Sigalas, I.; Asante, J.K.O.; Garrett, J.C. Effect of SPS consolidation parameters on submicron Cu and Cu–CNT composites for thermal management. *Powder Technol.,* **2014**, *258*, 198-205.
[http://dx.doi.org/10.1016/j.powtec.2014.03.034]

[65] Yamanaka, S. Fabrication and thermal properties of carbon nanotube/nickel composite by spark plasma sintering method *J. Jpn. I. Met. Mater.,* **2006**, *48*(9), 2506-2512.

[66] Nguyen, J.; Holland, T.B.; Wen, H.; Fraga, M.; Mukherjee, A.; Lavernia, E. Mechanical behavior of ultrafine-grained Ni–carbon nanotube composite. *J. Mater. Sci.,* **2014**, *49*(5), 2070-2077.
[http://dx.doi.org/10.1007/s10853-013-7897-1]

[67] Nam, D.H.; Kim, Y.K.; Cha, S.I.; Hong, S.H. Effect of CNTs on precipitation hardening behavior of CNT/Al–Cu composites. *Carbon,* **2012**, *50*(13), 4809-4814.
[http://dx.doi.org/10.1016/j.carbon.2012.06.005]

[68] Kim, K.T.; Cha, S.I.; Hong, S.H. Hardness and wear resistance of carbon nanotube reinforced Cu matrix nanocomposites. *Mater. Sci. Eng. A,* **2007**, *449-451*, 46-50.
[http://dx.doi.org/10.1016/j.msea.2006.02.310]

[69] Kim, K.T.; Eckert, J.; Liu, G.; Park, J.M.; Lim, B.K.; Hong, S.H. Influence of embedded-carbon nanotubes on the thermal properties of copper matrix nanocomposites processed by molecular-level mixing. *Scr. Mater.,* **2011**, *64*(2), 181-184.
[http://dx.doi.org/10.1016/j.scriptamat.2010.09.039]

[70] Suárez, S.; Soldera, F.; Oliver, C.G.; Acevedo, D.; Mücklich, F. Thermomechanical behavior of bulk Ni/MWNT composites produced *via* powder metallurgy. *Adv. Eng. Mater.*, **2012**, *14*(7), 499-502. [http://dx.doi.org/10.1002/adem.201200100]

[71] Rossi, P.; Suarez, S.; Soldera, F.; Mücklich, F. Quantitative assessment of the reinforcement distribution homogeneity in CNT/Metal composites. *Adv. Eng. Mater.*, **2015**, *17*(7), 1017-1021. [http://dx.doi.org/10.1002/adem.201400352]

[72] Carvalho, O.; Miranda, G.; Soares, D.; Silva, F.S. Carbon nanotube dispersion in aluminum matrix composites—Quantification and influence on strength. *Mech. Adv. Mater. Structures*, **2016**, *23*(1), 66-73. [http://dx.doi.org/10.1080/15376494.2014.929766]

[73] Uddin, S.M.; Mahmud, T.; Wolf, C.; Glanz, C.; Kolaric, I.; Volkmer, C.; Höller, H.; Wienecke, U.; Roth, S.; Fecht, H-J. Effect of size and shape of metal particles to improve hardness and electrical properties of carbon nanotube reinforced copper and copper alloy composites. *Compos. Sci. Technol.*, **2010**, *70*(16), 2253-2257. [http://dx.doi.org/10.1016/j.compscitech.2010.07.012]

[74] Chu, K.; Jia, C. C.; Li, W. S.; Wang, P. Mechanical and electrical properties of carbon-nanotub--reinforced Cu–Ti alloy matrix composites. *physica status solidi*, **2013**, *210*(3), 594-599. [http://dx.doi.org/10.1002/pssa.201228549]

[75] So, K.P.; Biswas, C.; Lim, S.C.; An, K.H.; Lee, Y.H. Electroplating formation of Al–C covalent bonds on multiwalled carbon nanotubes. *Synth. Met.*, **2011**, *161*(3-4), 208-212. [http://dx.doi.org/10.1016/j.synthmet.2010.10.023]

[76] Chu, K.; Jia, C.; Jiang, L.; Li, W. Improvement of interface and mechanical properties in carbon nanotube reinforced Cu–Cr matrix composites. *Mater. Des.*, **2013**, *45*, 407-411. [http://dx.doi.org/10.1016/j.matdes.2012.09.027]

[77] Lal, M. An alternative improved method for the homogeneous dispersion of CNTs in Cu matrix for the fabrication of Cu/CNTs composites. *Applied Nanoscience*, **2012**, *3*(1)

[78] Tu, J.P.; Yang, Y.Z.; Wang, L.Y.; Ma, X.C.; Zhang, X.B. Tribological properties of carbon-nanotub--reinforced copper composites. *Tribol. Lett.*, **2001**, *10*(4), 225-228. [http://dx.doi.org/10.1023/A:1016662114589]

[79] Varo, T.; Canakci, A. Effect of the CNT content on microstructure, physical and mechanical properties of Cu-based electrical contact materials produced by flake powder metallurgy. *Arab. J. Sci. Eng.*, **2015**, *40*(9), 2711-2720. [http://dx.doi.org/10.1007/s13369-015-1734-6]

[80] Suárez, S.; Ramos-Moore, E.; Mücklich, F. A high temperature X-ray diffraction study of the influence of MWCNTs on the thermal expansion of MWCNT/Ni composites. *Carbon*, **2013**, *51*(1), 404-409. [http://dx.doi.org/10.1016/j.carbon.2012.09.002]

[81] Xu, S.; Xiao, B.; Liu, Z.; Wang, W.; Ma, Z. "Microstructures and mechanical properties of CNT/Al composites fabricated by high energy ball-milling method," *Jinshu Xuebao. Chin Shu Hsueh Pao*, **2012**, *48*(7), 882-888. [http://dx.doi.org/10.3724/SP.J.1037.2012.00140]

[82] Rajkumar, K.; Aravindan, S. Tribological studies on microwave sintered copper–carbon nanotube composites. *Wear*, **2011**, *270*(9-10), 613-621. [http://dx.doi.org/10.1016/j.wear.2011.01.017]

[83] Kang, S.L. *Sintering: Densification*; Grain Growth and Microstructure, **2005**.

[84] Suárez, S.; Rosenkranz, A.; Gachot, C.; Mücklich, F. Enhanced tribological properties of MWCNT/Ni bulk composites – Influence of processing on friction and wear behaviour. *Carbon*, **2014**, *66*, 164-171. [http://dx.doi.org/10.1016/j.carbon.2013.08.054]

[85]   Gul, F.; Acilar, M. Effect of the reinforcement volume fraction on the dry sliding wear behaviour of Al–10Si/SiCp composites produced by vacuum infiltration technique. *Compos. Sci. Technol.,* **2004,** *64*(13-14), 1959-1970.
[http://dx.doi.org/10.1016/j.compscitech.2004.02.013]

[86]   Lakshmi, S.; Lu, L.; Gupta, M. *In situ* preparation of TiB2 reinforced Al based composites. *J. Mater. Process. Technol.,* **1998,** *73*(1-3), 160-166.
[http://dx.doi.org/10.1016/S0924-0136(97)00225-2]

[87]   Demir, A.; Altinkok, N. Effect of gas pressure infiltration on microstructure and bending strength of porous Al2O3/SiC-reinforced aluminium matrix composites. *Compos. Sci. Technol.,* **2004,** *64*(13-14), 2067-2074.
[http://dx.doi.org/10.1016/j.compscitech.2004.02.015]

[88]   Rehman, A.; Das, S.; Dixit, G. Analysis of stir die cast Al–SiC composite brake drums based on coefficient of friction. *Tribol. Int.,* **2012,** *51*, 36-41.
[http://dx.doi.org/10.1016/j.triboint.2012.02.007]

[89]   Surappa, M.K. Aluminium matrix composites: Challenges and opportunities. *Sadhana - Academy Proceedings in Engineering Sciences,* **2003,** *281–2*, pp. 319-334.

[90]   Garg, P.; Jamwal, A.; Kumar, D.; Sadasivuni, K.K.; Hussain, C.M.; Gupta, P. Advance research progresses in aluminium matrix composites: Manufacturing & applications. *J. Mater. Res. Technol.,* **2019,** *8*(5), 4924-4939.
[http://dx.doi.org/10.1016/j.jmrt.2019.06.028]

[91]   Hajjari, E.; Divandari, M.; Arabi, H. Effect of applied pressure and nickel coating on microstructural development in continuous carbon fiber-reinforced aluminum composites fabricated by squeeze casting. *Mater. Manuf. Process.,* **2011,** *26*(4), 599-603.
[http://dx.doi.org/10.1080/10426910903447311]

[92]   Aravind Senan, V.R.; Anandakrishnan, G.; Rahul, S.R.; Reghunath, N.; Shankar, K.V. An investigation on the impact of SiC/B4C on the mechanical properties of Al-6.6Si-0.4Mg alloy. *Mater. Today Proc.,* **2020,** *26*, 649-653.
[http://dx.doi.org/10.1016/j.matpr.2019.12.359]

[93]   Laha, T.; Chen, Y.; Lahiri, D.; Agarwal, A. Tensile properties of carbon nanotube reinforced aluminum nanocomposite fabricated by plasma spray forming. *Compos., Part A Appl. Sci. Manuf.,* **2009,** *40*(5), 589-594.
[http://dx.doi.org/10.1016/j.compositesa.2009.02.007]

[94]   Rohith, K.P.; Sajay Rajan, E.; Harilal, H.; Jose, K.; Shankar, K.V. Study and comparison of A356-WC composite and A356 alloy for an off-road vehicle chassis. *Mater. Today Proc.,* **2018,** *5*(11), 25649-25656.
[http://dx.doi.org/10.1016/j.matpr.2018.11.006]

[95]   Jiang, L.; Li, Z.; Fan, G.; Cao, L.; Zhang, D. Strong and ductile carbon nanotube/aluminum bulk nanolaminated composites with two-dimensional alignment of carbon nanotubes. *Scr. Mater.,* **2012,** *66*(6), 331-334.
[http://dx.doi.org/10.1016/j.scriptamat.2011.11.023]

[96]   Truong, H.T.X.; Lagoudas, D.C.; Ochoa, O.O.; Lafdi, K. Fracture toughness of fiber metal laminates: Carbon nanotube modified Ti–polymer–matrix composite interface. *J. Compos. Mater.,* **2014,** *48*(22), 2697-2710.
[http://dx.doi.org/10.1177/0021998313501923]

[97]   Lim, B.; Kim, C.; Kim, B.; Shim, U.; Oh, S.; Sung, B.; Choi, J.; Baik, S. The effects of interfacial bonding on mechanical properties of single-walled carbon nanotube reinforced copper matrix nanocomposites. *Nanotechnology,* **2006,** *17*(23), 5759-5764.
[http://dx.doi.org/10.1088/0957-4484/17/23/008]

[98]   Anh, N.N. Solar cell based on hybrid structural SiNW/ Poly(3,4 ethylenedioxythiophene): Poly(styrenesulfonate)/ graphene. **2020**. Available from: www.global-challenges.com

[99]   Bakshi, S.R.; Singh, V.; Seal, S.; Agarwal, A. Aluminum composite reinforced with multiwalled carbon nanotubes from plasma spraying of spray dried powders. *Surf. Coat. Tech.,* **2009**, *203*(10-11), 1544-1554.
[http://dx.doi.org/10.1016/j.surfcoat.2008.12.004]

[100] Mansoor, M.; Khan, S.; Ali, A.; Ghauri, K. M. Fabrication of aluminum-carbon nanotube nano-composite using aluminum-coated carbon nanotube precursor. *SAGE,* **2019**, *53*(28–30), 4055-4064.
[http://dx.doi.org/10.1177/0021998319853341]

[101] Kang, K.; Bae, G.; Kim, B.; Lee, C. Thermally activated reactions of multi-walled carbon nanotubes reinforced aluminum matrix composite during the thermal spray consolidation. *Mater. Chem. Phys.,* **2012**, *133*(1), 495-499.
[http://dx.doi.org/10.1016/j.matchemphys.2012.01.071]

[102] Liu, Z.Y.; Xiao, B.L.; Wang, W.G.; Ma, Z.Y. Developing high-performance aluminum matrix composites with directionally aligned carbon nanotubes by combining friction stir processing and subsequent rolling. *Carbon,* **2013**, *62*, 35-42.
[http://dx.doi.org/10.1016/j.carbon.2013.05.049]

[103] Liu, Z.Y.; Liu, Z.Y.; Liu, Z.Y.; Liu, Z.Y. Mechanical properties and interfacial phenomena in aluminum reinforced with carbon nanotubes manufactured by the sandwich technique. *J. Compos. Mater.,* **2016**, *51*(11), 1619-1629.
[http://dx.doi.org/10.1177/0021998316658784]

[104] Liu, Z.Y.; Xiao, B.L.; Wang, W.G.; Ma, Z.Y. Singly dispersed carbon nanotube/aluminum composites fabricated by powder metallurgy combined with friction stir processing. *Carbon,* **2012**, *50*(5), 1843-1852.
[http://dx.doi.org/10.1016/j.carbon.2011.12.034]

[105] Yang, X.; Liu, E.; Shi, C.; He, C.; Li, J.; Zhao, N.; Kondoh, K. Fabrication of carbon nanotube reinforced Al composites with well-balanced strength and ductility. *J. Alloys Compd.,* **2013**, *563*, 216-220.
[http://dx.doi.org/10.1016/j.jallcom.2013.02.066]

[106] Zhou, W.; Bang, S.; Kurita, H.; Miyazaki, T.; Fan, Y.; Kawasaki, A. Interface and interfacial reactions in multi-walled carbon nanotube-reinforced aluminum matrix composites. *Carbon,* **2016**, *96*, 919-928.
[http://dx.doi.org/10.1016/j.carbon.2015.10.016]

[107] Bi, S.; Xiao, B.L.; Ji, Z.H.; Liu, B.S.; Liu, Z.Y.; Ma, Z.Y. Dispersion and damage of carbon nanotubes in carbon nanotube/7055Al composites during high-energy ball milling process. *Acta Metall. Sin.,* **2021**, *34*(2), 196-204.
[http://dx.doi.org/10.1007/s40195-020-01138-5]

[108] Fu, X.; Jiang, J.; Jiang, X. Research progress in interfacial characteristics and strengthening mechanisms of rare earth metal oxide-reinforced copper matrix composites. *Materials,* **2022**, *15*(15), 5350.
[http://dx.doi.org/10.3390/ma15155350]

[109] Chen, B.; Shen, J.; Ye, X.; Imai, H.; Umeda, J.; Takahashi, M.; Kondoh, K. Solid-state interfacial reaction and load transfer efficiency in carbon nanotubes (CNTs)-reinforced aluminum matrix composites. *Carbon,* **2017**, *114*, 198-208.
[http://dx.doi.org/10.1016/j.carbon.2016.12.013]

[110] Raviathul Basariya, M.; Srivastava, V.C.; Mukhopadhyay, N.K. Microstructural characteristics and mechanical properties of carbon nanotube reinforced aluminum alloy composites produced by ball milling. *Mater. Des.,* **2014**, *64*, 542-549.
[http://dx.doi.org/10.1016/j.matdes.2014.08.019]

[111] Yoo, S.J.; Han, S.H.; Kim, W.J. Strength and strain hardening of aluminum matrix composites with randomly dispersed nanometer-length fragmented carbon nanotubes. *Scr. Mater.,* **2013**, *68*(9), 711-714.
[http://dx.doi.org/10.1016/j.scriptamat.2013.01.013]

[112] Wang, H. Synergistic strengthening effect of nanocrystalline copper reinforced with carbon nanotubes. *Scientific Reports,* **2016**, *6*(1), 1-8.
[http://dx.doi.org/10.1038/srep26258]

[113] Majid, M.; Majzoobi, G.H.; Noozad, G.A.; Reihani, A.; Mortazavi, S.Z.; Gorji, M.S. Fabrication and mechanical properties of MWCNTs-reinforced aluminum composites by hot extrusion. *Rare Met.,* **2012**, *31*(4), 372-378.
[http://dx.doi.org/10.1007/s12598-012-0523-6]

[114] Esawi, A.M.K.; Morsi, K.; Sayed, A.; Taher, M.; Lanka, S. The influence of carbon nanotube (CNT) morphology and diameter on the processing and properties of CNT-reinforced aluminium composites. *Compos., Part A Appl. Sci. Manuf.,* **2011**, *42*(3), 234-243.
[http://dx.doi.org/10.1016/j.compositesa.2010.11.008]

[115] Peng, T.; Chang, I. Mechanical alloying of multi-walled carbon nanotubes reinforced aluminum composite powder. *Powder Technol.,* **2014**, *266*, 7-15.
[http://dx.doi.org/10.1016/j.powtec.2014.05.068]

[116] Esawi, A.M.K.; Morsi, K.; Sayed, A.; Taher, M.; Lanka, S. Effect of carbon nanotube (CNT) content on the mechanical properties of CNT-reinforced aluminium composites. *Compos. Sci. Technol.,* **2010**, *70*(16), 2237-2241.
[http://dx.doi.org/10.1016/j.compscitech.2010.05.004]

[117] Jiang, L.; Fan, G.; Li, Z.; Kai, X.; Zhang, D.; Chen, Z.; Humphries, S.; Heness, G.; Yeung, W.Y. An approach to the uniform dispersion of a high volume fraction of carbon nanotubes in aluminum powder. *Carbon,* **2011**, *49*(6), 1965-1971.
[http://dx.doi.org/10.1016/j.carbon.2011.01.021]

[118] Huang, Y.; Ouyang, Q.; Zhang, D.; Zhu, J.; Li, R.; Yu, H. Carbon materials reinforced aluminum composites: A review. *Acta Metall. Sin.,* **2014**, *27*, 775-786.
[http://dx.doi.org/10.1007/s40195-014-0160-1]

[119] Serp, P.; Castillejos, E. Catalysis in carbon nanotubes. *ChemCatChem,* **2010**, *2*(1), 41-47.
[http://dx.doi.org/10.1002/cctc.200900283]

[120] Arai, S.; Suzuki, Y.; Nakagawa, J.; Yamamoto, T.; Endo, M. Fabrication of metal coated carbon nanotubes by electroless deposition for improved wettability with molten aluminum. *Surf. Coat. Tech.,* **2012**, *212*, 207-213.
[http://dx.doi.org/10.1016/j.surfcoat.2012.09.051]

[121] Hu, W.; Yanling, Z. Pretreatment and copper plating of carbon nanotubes by electroless deposition. *Biaomian Jishu,* **2019**, *48*(11), 211-218.

[122] Oh, S-I.; Lim, J-Y.; Kim, Y-C.; Yoon, J.; Kim, G-H.; Lee, J.; Sung, Y-M.; Han, J-H. Fabrication of carbon nanofiber reinforced aluminum alloy nanocomposites by a liquid process. *J. Alloys Compd.,* **2012**, *542*, 111-117.
[http://dx.doi.org/10.1016/j.jallcom.2012.07.029]

[123] Guo, B.; Zhang, X.; Cen, X.; Wang, X.; Song, M.; Ni, S.; Yi, J.; Shen, T.; Du, Y. Ameliorated mechanical and thermal properties of SiC reinforced Al matrix composites through hybridizing carbon nanotubes. *Mater. Charact.,* **2018**, *136*, 272-280.
[http://dx.doi.org/10.1016/j.matchar.2017.12.032]

[124] Aristizabal, K.; Katzensteiner, A.; Bachmaier, A.; Mücklich, F.; Suarez, S. Study of the structural defects on carbon nanotubes in metal matrix composites processed by severe plastic deformation. *Carbon,* **2017**, *125*, 156-161.

[http://dx.doi.org/10.1016/j.carbon.2017.09.075]

[125] Tong, L.; Sun, X.; Tan, P. *Effect of long multi-walled carbon nanotubes on delamination toughness of laminated composites,* **2008**, *42*(1), 5-23.
[http://dx.doi.org/10.1177/0021998307086186]

[126] Wang, L.; Choi, H.; Myoung, J.M.; Lee, W. Mechanical alloying of multi-walled carbon nanotubes and aluminium powders for the preparation of carbon/metal composites. *Carbon,* **2009**, *47*(15), 3427-3433.
[http://dx.doi.org/10.1016/j.carbon.2009.08.007]

[127] J., zhi Liao; M.J., Tan; I., Sridhar Spark plasma sintered multi-wall carbon nanotube reinforced aluminum matrix composites, *Materials & Design,,* **2010**, *31* (1).

[128] Choi, H.J.; Shin, J.H.; Bae, D.H. Grain size effect on the strengthening behavior of aluminum-based composites containing multi-walled carbon nanotubes. *Compos. Sci. Technol.,* **2011**, *71*(15), 1699-1705.
[http://dx.doi.org/10.1016/j.compscitech.2011.07.013]

[129] Tarlton, T.; Sullivan, E.; Brown, J.; Derosa, P.A. The role of agglomeration in the conductivity of carbon nanotube composites near percolation. *J. Appl. Phys.,* **2017**, *121*(8), 085103.
[http://dx.doi.org/10.1063/1.4977100]

[130] Neubauer, E.; Kitzmantel, M.; Hulman, M.; Angerer, P. Potential and challenges of metal-matri-composites reinforced with carbon nanofibers and carbon nanotubes. *Compos. Sci. Technol.,* **2010**, *70*(16), 2228-2236.
[http://dx.doi.org/10.1016/j.compscitech.2010.09.003]

[131] Zheng, L.; Sun, J.; Chen, Q. Carbon nanotubes reinforced copper composite with uniform CNT distribution and high yield of fabrication. *Micro & Nano Lett.,* **2017**, *12*(10), 722-725.
[http://dx.doi.org/10.1049/mnl.2017.0317]

[132] Datsyuk, V.; Kalyva, M.; Papagelis, K.; Parthenios, J.; Tasis, D.; Siokou, A.; Kallitsis, I.; Galiotis, C. Chemical oxidation of multiwalled carbon nanotubes. *Carbon,* **2008**, *46*(6), 833-840.
[http://dx.doi.org/10.1016/j.carbon.2008.02.012]

[133] Ci, L.; Ryu, Z.; Jin-Phillipp, N.Y.; Rühle, M. Investigation of the interfacial reaction between multi-walled carbon nanotubes and aluminum. *Acta Mater.,* **2006**, *54*(20), 5367-5375.
[http://dx.doi.org/10.1016/j.actamat.2006.06.031]

[134] Shin, S.E.; Choi, H.J.; Hwang, J.Y.; Bae, D.H. Strengthening behavior of carbon/metal nanocomposites. *Sci. Rep.,* **2015**, *5*(1), 16114.
[http://dx.doi.org/10.1038/srep16114] [PMID: 26542897]

[135] Bor, A.; Ichinkhorloo, B.; Uyanga, B.; Lee, J.; Choi, H. Cu/CNT nanocomposite fabrication with different raw material properties using a planetary ball milling process. *Powder Technol.,* **2018**, *323*, 563-573.
[http://dx.doi.org/10.1016/j.powtec.2016.06.042]

[136] Dahiya, M.; Khanna, V.; Anil Bansal, S. Effect of graphene size variation on mechanical properties of aluminium graphene nanocomposites: A modeling analysis. *Mater. Today Proc.,* **2022**.

[137] Gupta, P.; Ahamad, N.; Kumar, D.; Gupta, N.; Chaudhary, V.; Gupta, S.; Khanna, V.; Chaudhary, V. Synergetic effect of CeO_2 doping on structural and tribological behavior of $Fe-Al_2O_3$ metal matrix nanocomposites. *ECS J. Solid State Sci. Technol.,* **2022**, *11*(11), 117001.
[http://dx.doi.org/10.1149/2162-8777/ac9c92]

[138] Dahiya, M.; Khanna, V.; Anil Bansal, S. Aluminium-graphene metal matrix nanocomposites: Modelling, analysis, and simulation approach to estimate mechanical properties. *Mater. Today Proc.,* **2022**.

[139] Singh, K.; Bansal, S.A.; Khanna, V.; Singh, S. Effects of performance measures of non-conventional joining processes on mechanical properties of metal matrix composites. *Metal Matrix Composites,*

2022, (Aug), 135-165.
[http://dx.doi.org/10.1201/9781003194897-7]

[140] Khanna, V.; Kumar, V.; Bansal, S.A.; Prakash, C.; Ubaidullah, M.; Shaikh, S.F.; Pramanik, A.; Basak, A.; Shankar, S. Fabrication of efficient aluminium/graphene nanosheets (Al-GNP) composite by powder metallurgy for strength applications. *J. Mater. Res. Technol.,* **2023**, *22*, 3402-3412.
[http://dx.doi.org/10.1016/j.jmrt.2022.12.161]

CHAPTER 6

Biotribology: Recent advancements, Applications, Challenges and Future Directions

Harpreet Singh[1,*] and **Kirandeep Kaur**[2]

[1] *Ludhiana group of Colleges, Chaukimann, Punjab, India*

[2] *PSE Sales & SVCS/Distribution Associate, New Jersey, USA*

Abstract: Tribology deals with basic principles and understanding of three concepts: friction, wear, and lubrication. Now, bio tribology is one of the most exhilarating fields of tribological study. In this book chapter, the authors made efforts to review and provide brief thoughts about the various sections of the biotribology such as orthopedics, artificial implants, biomimetics, bio-lubricants, biomaterials, ocular tribology, skin tribology, haptics, dental tribology, sports tribology. Apart from these, biotribology deals with a few more exciting areas *i.e.*, in personal care like skin creams, cosmetics, *etc.*, and oral processing studies such as mouthfeel and taste perception.

This comprehensive review comes to a close with four studies, *i.e.*, bio-friction of the biological systems, tribology of medical and surgical devices, biocompatibility issues related to biomaterials, and critical aspects of bio-tribocorrosion. A critical review of bio-friction studies for the various biological systems is presented, and significant underlying tribological-lubrication mechanisms are also discussed.

The present emphasis and forthcoming advancements of the various medical and surgical instruments in context with the fundamental tribology principles and pertaining mechanisms for an efficient, versatile, and multi-functional bio-system will be discussed in this book chapter. Furthermore, major challenges faced by R&D officials and medical teams are discussed.

Biocompatibility and bio-tribo-corrosion of biomaterials are serious concerns in bio tribology. In-depth discussions of current trends, implementations, and their guidelines for the future are also included. In a nutshell, bio tribology studies can contribute noteworthy scientific, social, engineering, and healthcare benefits; the openings and possibilities are significant.

Keywords: Artificial implants, Biotribology, Bio-tribocorrosion, Bio-friction, Biomimetic, Bio-lubricants, Biomaterials, Challenges, Composites, Dentistry, Ocular tribology, Skin tribology, Sports tribology, Tribology.

---

* **Corresponding author Harpreet Singh:** Ludhiana Group of Colleges, Chaukimann, Punjab, India; E-mail: harpreetsingh6n2016@gmail.com

Virat Khanna, Prianka Sharma & Santosh Kumar (Eds.)

INTRODUCTION

Bio-tribology is the study of tribological aspects (friction, wear & lubrication) of biological systems or natural phenomena. Apart from this, it belongs to multidisciplinary fields Table **1** such as biomedical engineering, medical technology, nanotechnology, material engineering, *etc*. These things collectively diversify the impact on our daily life and research aspects. In general, it is not limited to implant replacement (Fig. **1**), but significantly affects sports equipment design and personal care products [1, 2].

Table 1. Brief about key subjects or areas that are included in the biotribology [1 - 40].

Major Application Sectors in the Biotribology	Area of Study
Joints (Synovial and cartilaginous) and ligaments	Basic types; applications; impairment and failure; replacement methods; risks and complications.
Tissue substitutes	Biological tissues; Biodegradable scaffolds; hydrogels; tissue-engineered skin substitutes; dental tissues; synthetic bone tissues, meniscus, and tendons.
Prostheses implant or Artificial implant	*Prostheses implant:* Craniofacial prostheses, extra/intra-oral prostheses, dental prostheses, neck Prostheses, breast prostheses, nipple prostheses, penile prostheses; Artificial *implant:* Neurosensory implantation, Cardiac replacements, Orthopedic inserts, Electric inserts, Contraception implants; Cosmetic implants, *etc*. Other applications: Articular joints, catheters, heart pumps, stents.
Biomimicry	Self-healing and bio-inspired materials; Development of surface based on surface tension Biomimetics; Bio-inspired technologies.
Locomotion	Microbial & bacterial motility; mechanics; slip mechanisms; locomotor effects; movement of microorganisms; study of motor skills.
Drug delivery	Controlled-release formulations; nanomedicine; biologic drug carriers; Oral dispersive medicines; injections; Modulated drug release; Targeted drug delivery; Bioavailability.
Ocular tribology	Anatomy; Targeting of the ocular surface and diseases; contact lenses; tear film; Disorders; wetting; Hydrogels; Micro-scale friction measurements (surface friction) and mechanisms; Diffusion phenomenon; various syndromes; Reverse engineering.
Skin tribology	Skin friction modeling; Damage and blistering mechanisms, bedsores, sweat lubrication; product-skin interactions; surface texturing and polymer coatings.
Haptics	Touch perception; Haptic Texture; surface texturing techniques; ergonomics; Haptic sensation; Surface haptic technology and Somatosensory system.
Tribology of personal care	Kinetic friction; Adhesion and wear phenomenon; Rheology; Multiple regression modeling; Skin and hair solutions; cosmetic treatments; Exfoliators.
Oral processing	Foodstuffs and beverages; mouth feel and texture perception; biomechanical functions; study of multiple physicochemical properties; sensory texture perception.

(Table 1) cont.....

Major Application Sectors in the Biotribology	Area of Study
Dental tribology	Wear of dental replacements; anchoring of biomechanical inserts; tribe-corrosion; restorative materials and implant design; biocompatibility; oral lubrication and the underlying mechanism; oral hygiene practice on dental tissues; Properties of dental tissues.
Tribology in Industry	Testing methodologies; Understanding of customer experience and ultimate guide to significant strategies; technological solutions; Energy saving technologies; Advanced manufacturing systems; Highly sophisticated diagnosis arrangements; Micro technologies for MEMS and Nanotechnologies for NEMS; Tribology networks; Mechanization; mass production; Automated production; Cyber physical systems; Friction studies on gears; bearings and lubricating oils; Additive technology; Computational and experimental tribology; EHD lubrication regime; contact mechanics and surface topographic investigations; condition monitoring, rheology; superlubricity; Tribotronics; Human machine interaction and concept of wearables.
Sports tribology	Reliable surfacing solutions; grip improving agents; low-friction characteristics; surface topography/texture tailoring; protruded scales and pattern on contacting surfaces; absorbing materials; study of adhesion properties; better tribological understanding of material-surface interaction; skin friction; ergonomics; material behavior.
Biolubricants	Renewable; Plant-derived or environmentally acceptable lubricants; Bio-corrosion and bio-fouling; Study of biodegradability; non-toxic and oxidative stability; tribo-reactive materials for automotive applications.
Tribology involved in natural phenomena	Geotribology; Fractures and discontinuities in rocks; Geo-mechanical modeling; microphysical models to map the underlying frictional phenomenon; Earthquakes; avalanches; landslip or other subsurface phenomena.

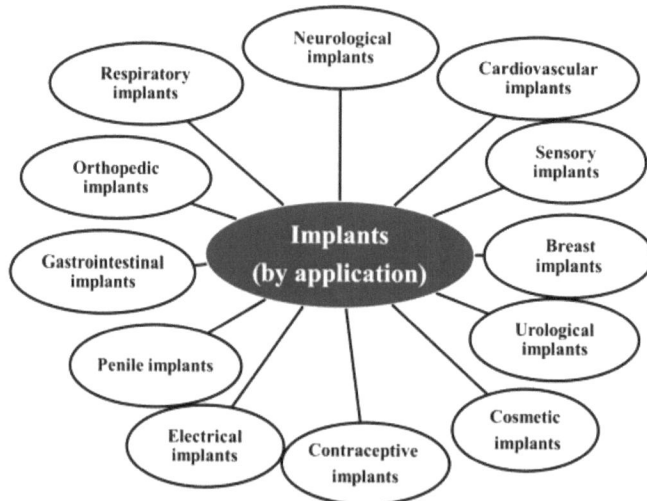

Fig. (1). Various type of implants (by application) [27, 80, 81, 87, 88].

Brief Report About Sub-sections of Biotribology

Orthopedics

Orthopedics is the division of surgery (more precisely includes both surgical and nonsurgical methods) that involves the musculoskeletal system and diagnosis and treatment of various diseases such as spine and degenerative, trauma, infections, tumors, and congenital disorders. Nowadays, the cure for sports injuries is a beautiful example/application of orthopedics. Modern orthopedics includes total hip replacement (Fig. **3**). Also, nanotechnological advancements in Orthopedics are key outcomes of the research field (Fig. **2**). Tribologists obtained that joint surfaces could be substituted by cemented implants or uncemented arthroplasty techniques to the bone. Knee replacements ought to be finished using a mobile bearing system. With these techniques, surgery would be less invasive and fabricate better-implanted components with more durability. In future directions, Orthopedic surgeons will aim to decrease the prescription of opioids and pain relief treatments for patients. Surface engineering techniques are designed for the manufacturing of biodegradable magnesium alloys [3, 4]. It is sensible to mention the basics of an implant and its functions. It is fabricated to substitute a misplaced biological structure, nominal sustenance to the injured tissues of the structure, or improvement in functionality. These artificial (or man-made) devices are generally made up of biomaterials of better functional, mechanical, biological, and chemical (reaction rate with cells to support the structure) and medicinal properties. Antibacterial coatings, release coatings (Ag, NO or polymeric/anti-biotic), photoactive antimicrobial coatings, traditional organic and inorganic coatings, functionalized coatings, nano-structural coatings, and drug-eluting nano-coatings are the featured biomedical coatings for the orthopedic applications. Various authors paid their valuable feedback and report on nanocomposites (classifications structure, composition, and biomechanical features). And most importantly, development means nanocomposite based bone-grafting [3 - 5].

Biomimetics

The mimicry of the models of inspiration, mechanisms, systems, and elements of nature for the solution of intricate human problems/systems or complications is called Biomimetics (or even, sometimes termed bio-mimicry). This novel technology provides solutions to biological systems at macro, micro, and sub-micro molecular level and nano scales. With this approach, researchers are able to solve engineering problems such as self-repairing abilities, self-assembly, environment (solar energy integration), and hydrophobicity. It is sensible to mention that the biological models could not be emulated precisely still; considerable outputs can be revived from such novel fabricated devices. The researchers highlighted the four challenges in biomimetic studies, namely, the design of micromechanical components (3D-MEMS), flow pumps (for lesser volumes of liquid), adhesive and lubricants, functional materials with crack redirecting features. Biomimetic design methodology (Lotus effect-inspired flexible, Bio-inspired tire, Riblet configurations, Gecko-inspired surfaces, Sandfish and Snake scaled skin effect) and bio-solution were proposed by bio-tribologists for the identification of problems in biomimetics. In recent advances, adhesion mechanistic studies and friction for the fluid-solid-inter-surface, wear performance and antiwear bio-inspired technologies are the primary research directions. Biophysics, biochemistry, biological surface texturing, bio-physical potency and mechanistic studies are the front-line interdisciplinary fields to reinforce the research and development activities in biomimetics tribology [6 - 10].

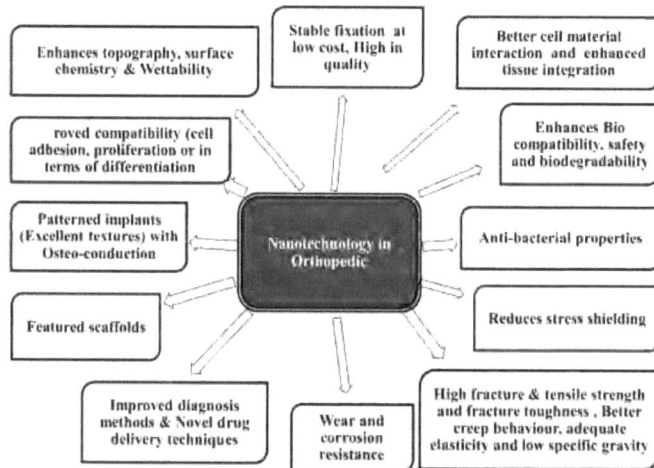

Fig. (2). Role of nanotechnology in Orthopedic applications [1 - 5, 27, 80, 81].

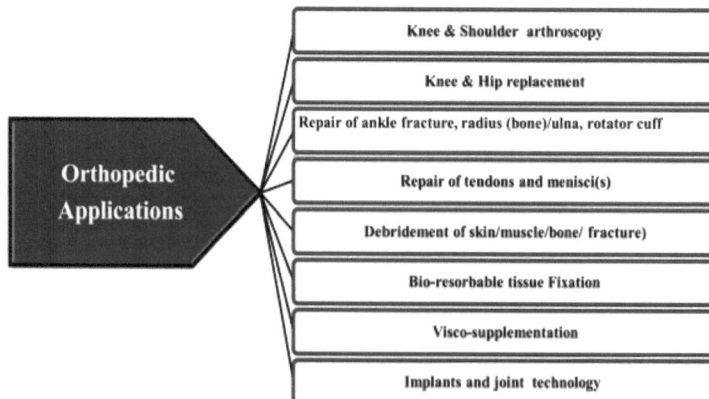

Fig. (3). Major Orthopedic applications [1 - 5, 27, 80, 81].

Biolubricants

Biolubricants are multi-functional fluids prepared from vegetable oils and downstream esters. Tribologists paid keen attention to these oils (coconut oil-based products) or blends due to their utilization in energy generation and security and for better environmental aspects. These technologies are widely applicable in the United States, Europe, and many other countries like Malaysia and Singapore. But in this direction, there are several disadvantages also, (kindly refer to (Fig. **4**). Table **2** list down the merits and de-merits and major application areas for the Biolubricants.

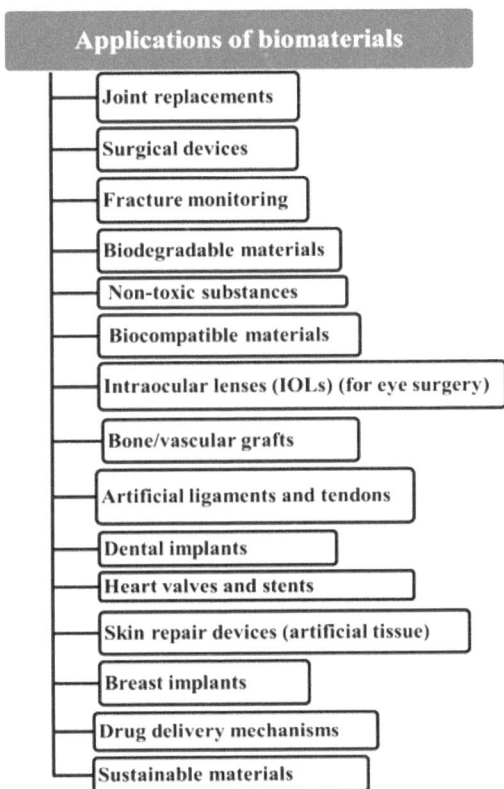

Applications of biomaterials

- Joint replacements
- Surgical devices
- Fracture monitoring
- Biodegradable materials
- Non-toxic substances
- Biocompatible materials
- Intraocular lenses (IOLs) (for eye surgery)
- Bone/vascular grafts
- Artificial ligaments and tendons
- Dental implants
- Heart valves and stents
- Skin repair devices (artificial tissue)
- Breast implants
- Drug delivery mechanisms
- Sustainable materials

Fig. (4). Major application of biomaterials [4, 15, 19, 84].

The tribologists are working in the direction of lubrication investigation mechanism studies. Various basic mechanisms are involved in bio lubrication, such as mending effect, tribofilm creation, surface property improvement and polishing effect. For nano-bio-lubrication, hardness of additive, size, proper distribution of additive in the fluid matrix, morphology (layered, spherical, polyhedral, Dahlia, elliptical or flaky), and basic mechanisms such as ball bearing, sliding or exfoliation are significant parameters for better lubrication characteristics. Hydrolytic stability, thermal stability, eco-toxicity, rheology, or wetting behavior are important properties for investigation and characterization For commercial viability, techno-economic analyses and lifecycle assessment play a vital role in the research and development of the novel bioproducts. Biodegradability is a serious issue in the development of biolubricants. In future directions, chemically modified biolubricants are gaining attention owing to novel lubrication mechanisms and performance characteristics [11 - 14].

Table 2. Merits, de-merits and major application areas for biolubricants [11 - 14].

Merits	Demerits	Major Application Areas
• High lubricity • High flash point • High viscosity index • Lower emissions • Better Skin compatibility • Better safety • High biodegradability • Customizable • Low contaminants • Noncorrosive • Excellent lubrication characteristics • Low COF • High anti-wear performance • Minimum specific fuel consumption • Sustainable • Eco-friendly • Low eco-toxicity • non-bioaccumulation	• High cost • Edible oils (debatable content) • High melting points • Low oxidation stability • Need advancement in the developing technologies • High acidity	• Metal working fluids • Hydraulic oils • Engine oils • Chemicals • Diesel fuels • Transmission fluids • Power transformer applications • Motor oils Others: • Cosmetics • Ant-corrosive coatings • Detergents • Petrochemical industries.

Biomaterials

Biomaterials guide or monitor interactions with numerous biological system components during any therapeutic or diagnostic process. More precisely, multi-disciplinary fields such as physical, biological, chemical sciences, tissue engineering and materials sciences are carefully studied to design biomaterials that are further utilized in medical (technology and regenerative science) and clinical sciences and diagnostic systems (Fig. **4**). It covers polymer production and property evaluation Table **3**, drug and gene vector design methodologies, host response characteristics, immune response and interaction biology and toxicology, and nanoscale self-aggregation. It has a great importance in bio-nanotechnology and tissue engineering.

Table 3. Typical properties need to be considered for the selection of biomaterials [15].

Bulk Properties	Surface Properties	Mechanical Properties
• Chemical Composition • Microstructure • Density • Electrical • Thermal Conductivity	• Wettability • Surface Energy • Surface Chemistry • Surface Texturing (Smooth/Rough), • Surface Tension • Surface Charge	• Elasticity • Ultimate tensile strength • hardness • Young's modulus • Elastic response to stresses • Toughness • Flexural rigidity

Bulk properties, surface properties, and mechanical properties must be monitored properly for the design and functionality of biomaterials Table **3**.

Ocular Tribology [16 - 25]

In context with the ocular surface, tribology principles apply to the motion or movement of the lid wipers against the globe and significant considerations on the mechanical interactions. These cases are solved by applying the concepts of the Stribeck curve. It is sensible to mention that a highly complex lubricating tear film is responsible for separating the sliding biological contacts. There exists a close relationship between the friction of eyelid juncture, wear of human tissues, and support of the hydrodynamic regime. Excessive wear induces patient symptomology. Now these days, the generation of very smooth surfaces (artificially implanted or fabricated), like contact lenses, is a quite daring task. Micro-tribological quantification provides a suitable platform for testing lubricious films. In the idealistic approach, the frictional coefficient approaches the minimal value, whenever there exists a transition regime. Also, the optimally designed tear film acts as a shock absorber for delicate tissues. However, in real-life scenarios living systems are stressed and affected by various parameters or factors, environmental aspects, humidity and moisture, temperature conditions, anatomy, lubrication regimes, pH and acidity, and osmotic concentration. Lid wiper epitheliopathy (LWE) is a serious concern, for those tribologists who are paying attention. Some authors are working in the direction of functional compatibility, safety, efficiency, frictional behavior and measurement (*in vitro* friction), lubrication mechanism, biomaterial (especially for contact lenses), and feasibility of the artificial tear formulations, rewetting drops and compatibility with contact lenses. Apart from this, major challenges and key opportunities in the field of soft tribology have been studied [16 - 25].

Haptics

The field of 'Haptics' considers any system of interaction involving touch. The key subsidiaries are haptic communication (communication *via* touching), perception (recognizing articles by touching), and technology (interfaces with the user through the sense of touch). In future directions, tactile perception and haptics during frictional contact, are essential to be analyzed to review skin contact comfort with daily needs and, thereby, the development of many advancements in robotics. In simpler words, haptics instruments might be equipped with tactile sensors that quantity/monitor various forces applied by the user on the interface. Gaming controllers, flightsticks, and driving wheels are the best examples of these devices. In Haptic technology, an understanding of touch by applying forces, vibrations, or movement to the user configures the creation of virtual articles in computational simulation and provides a means to monitor virtual objects, thereby improving remote sensing technologies. In recent trends through haptic therapy, programmable devices are designed for physical therapy exercises to heal and strengthen muscles and senses. This technology is not limited to the aforementioned fields but widens the approach to design and drafts models in automobile, aviation, space, smartphone, dentistry, electronic displays, and not but least in virtual reality [26 - 28].

Dental Tribology

Dentistry and tribology are facing critical challenges in designing and customizing oral care products. To accomplish this vision, extensive clinical and laboratory testing is needed. Wear dental implants are more common but researchers are now working with integrated technologies (especially nanotechnology) for lucid scientific understanding and profound insight into the wear and lubrication mechanisms (under the oral environment). Tribological aspects (improvement in tri-science, *i.e.*, friction, wear, and lubrication), and fragile tissue properties coupled with dental care, are the necessary practices. In a simpler way, the relative movement between teeth, implants, and restorative substances in the tribe-contact forms and the existence of food particles in the slurry form, saliva, and other external lubricating oils such as water, juice, and others are regarded as a dental tribological system. For the design of an ideal implant system, mechanical and biological integrity, corrosion, biocompatibility, friction, wear, and aesthetics are the pilot concerns. Composites and ceramics are predominant candidates to overcome bio-tribological issues. Composites are also categorized as the class of futuristic candidates. Tribo-corrosion behavior studies are an integral part of dental tribological advancement. Quantification erosion/corrosion is the pivotal factor for the quality of teeth health. It is sensible to mention that various mechanical properties such as hardness, Young's modulus of elasticity, and fracture toughness must be correlated with tooth wear in dental care product engineering and manufacturing [29].

Sports Tribology

In sports engineering, sports tribology is an integral part; it is applicable in designing and enhancing the performance of an individual sportsperson or even a team (Fig. **5**). In simpler words, sports engineers make collective efforts to give solutions to problems with adequate knowledge about tribology.

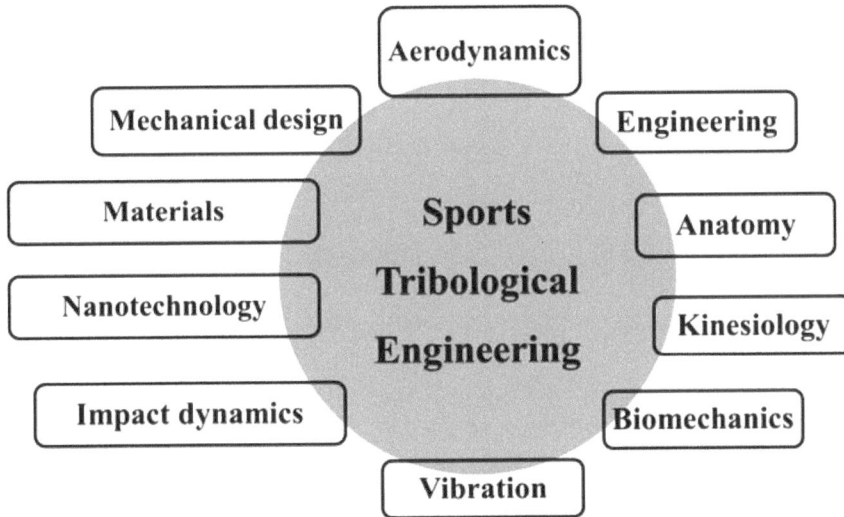

Fig. (5). Various fields involved in sports tribological engineering [1, 2, 31].

In this context, skin tribology is also equally important; designers aim to increase skin friction to enhance performance and sometimes prevent athletes from injuries. It is not that much easy, skin biomechanics must be strong enough to model skin friction with a close approximation [30, 31]

Skin Tribology

Basically, two major mechanisms are involved in skin friction *i.e.*, adhesion and deformation. It is most commonly applied to cricket, weightlifting, climbing, and gymnastics. Sometimes, boundary lubricant is also utilized for the concerned area. Roughness tailoring is also an important aspect, but most of the time, the problem arises when the environment involves a wet sports kit and wet skin. Also, by opting for innovative pitch constructs that involve advanced artificial playing surfaces upgrading from rubber granules, the skin friction coefficient might be reduced and intended to assist player sliding. Still, there exists a scope for the promotion of research to apprehend well the correlation between anatomy, physiology, and skin tribology. In skin tribology, skin friction measurement studies, skin friction injuries, skin failure modes, and skin-textile material interface modeling, are the most relevant studies that are highlighted by tribology officials. The smart biopolymers and 3D printing technology have been incorporated for skin tribological sectors. In skin tribology, the human skin acts as the interacting surface that is in relative motion. Researchers are trying to optimize the friction coefficient (skin) established upon a trial-and-error scheme and one's sense of touch. Precisely, skin tribology partially predicts skin friction (at the micro-scale). As tribology is involved, contact mechanics of product-skin interactions (at the asperity level) need to be understood. Empirical equations are also reported to model the friction outcomes. In this direction, surface texturing and polymer brush coatings are vital resources to monitor friction in the sliding members. Anti-slip coatings in the bathrooms, shaving, and clothing (textile) products are current facets of the research. To provide the best solution, surface, micro-mechanical, and material properties play key roles. The adhesion friction model and skin deformation mechanisms (including mechanical shearing) offer a background for the research, and still, there is scope for the analysis of the complex tribological behavior of human skin. Moreover, many authors revealed the significance and vivid understanding of the skin-pad interface as a system, frictional coefficient measurements, and slip mechanisms based on skin care products. In skin tribology, human skin-fabric interactions, the friction coefficient of the contact, textile material properties, and structure are important [31 - 33].

The Tribology of Personal Care

The tribology of personal care involves the design of hair-care products, the upgrading of nail polish chemicals, the formulation of skin creams, *etc*. Adhesion, lubrication, and wear phenomena are critical issues in this context. Tribological testing of these products is conducted on a mainly designed friction-wear monitor, by means of multiple-regression analysis (with sensory-panel data interpretation). Starved lubrication and fully lubricated modes are the most common operating conditions on friction (static and kinetic friction) measuring devices [34, 35].

Cardiovascular Tribology

In cardiovascular tribology, failure studies are significant. However, the authors attempt to achieve this aim by experimental, computational (fluid dynamics or evaluation of hemodynamic forces) or both of these analyses. Some authors are working on cardiovascular stent designs that include mechanical efficiency investigation, performance evaluation or FEM modeling. Basic challenges in cardiovascular tribology are integrated stent design and unavailability of durable devices. The study of vascular bio-tribological interaction (on macro and micro scale) is necessary to develop less invasive safer novel materials [36 - 38]. Authors tried to describe Bio-tribological mechanism on the basis of bio morphology, injury modes, frictional behaviour, creep behavior and stress conditions for composite cartilage, in their studies.

It is sensible to mention that authors have not reviewed reports on animal and plant tribology. This might be the next article by the respective authors.

BIO-FRICTION STUDIES [46 - 54]

In bio-friction, friction theories and mechanisms are applied to biological or living systems (Fig. 6). As we know that tribology is a tri-science that involves wear and lubrication too. Hence, tribological aspects and lubrication behavior are also studied thoroughly. In recent trends, the prediction of diseases and artificial replacements for natural living systems are considered.

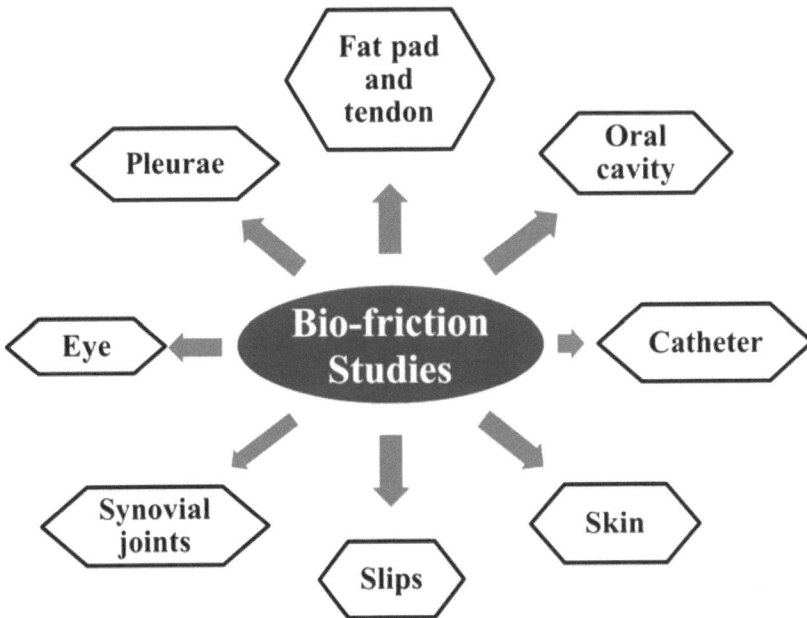

Fig. (6). Major bio-friction areas (emphasis on human health system) [46 - 51].

Synovial Joints

The natural synovial joint (basically, articular cartilage) is a load-bearing component and synovial fluid is the lubricant media inside the human system. In bio friction studies, research reports on the application of cartilage components under specific loading conditions and movements. Healthy synovial joints reveal a coefficient of friction of merely about 0.002. Hence, for the measurement of friction, highly sophisticated sensory friction devices are needed. Lubrication mechanism has various types (boundary/mixed/biphasic/ fluid-film lubrication). Three regimes (boundary, mixed, and full fluid) are well known, and here, the authors made an attempt to brief about biphasic lubrication. Biphasic lubrication deals with articular cartilage that belongs to both phases, solid as well as fluid. In the initial stages, the whole load is carried by the fluid component, and as time passes, on load is further transferred to the solid phase. Both the cases, solid and liquid, lower the friction. It is sensible to mention that boundary lubrication is more beneficial for long-duration of loading conditions. Some other mechanisms are also considered by the hydration lubrication mechanism, simply termed as the surface amorphous layer or the gel layer sometimes called the brush layer [23, 39 - 46].

Fat Pad and Tendon

The advantage (mechanically) of the musculoskeletal system is provided by fat pads. Fat pads are nothing but, masses of the adipose tissue (encapsulated form). Fat pads play a key function in minimizing friction in the biological system. Under severe conditions, high values in friction lead to abnormity and, consequently, pain. As per the research articles, the typical friction value between fat and bone from the bovine tissue is 0.01. Hydrodynamic lubrication is predominant in these cases. The authors further reveal that fat pads associated with tendons, might transfer forces (muscular) around the joint, enabling the movement of the joint. Higher friction values in a tendon cause cumulative trauma disorders (CTD). The tensile tendon may be able to perform elasto-hydrodynamically with a COF around 0.1, under physiological conditions. On the other hand, the compressive tendon's COF was computed as 0.008, in the mixed/boundary lubrication regime [23, 47 - 49].

Pleurae

An optimized level of friction is a necessity for the pleurae's normal functioning and curing of various diseases. Structurally, the pleurae comprise a membrane (double) with a monolayer of mesothelial cells, covering the lung (visceral pleura) and lining the chest wall (parietal pleura). Pleural (serous) fluid flows in between the membrane. Hence adequate lubrication and minimum value of shear stress are two significant conditions for the ease of breathing. Under severe conditions, higher friction may lead to damage to the tissue surfaces. More precisely, under normal working conditions, the friction characteristics are in line with boundary lubrication and must have a significant contact between the surfaces. It is sensible to mention that some of the experimental findings reveal the existence of a full-fluid film regime. In the friction-lubrication investigation study, under steady-state conditions, the frictional mapping was characterized to be dependable with evolution of regimes of the lubrication from mixed to fully developed hydrodynamic lubrication. Also, elastohydrodynamic lubrication at microscopic scales was also suggested for efficient lubricity and adequate frictional response among the pleural surfaces. The contact severities on the pleural surfaces and limited distortion encourage a hydrodynamic load support mechanism, and thereby separation occurs between the sliding surfaces. Boundary lubricant includes surface active phospholipids, whose functioning and characteristics are comparable to the synovial fluid [50 - 55].

Eye

The eye function depends on efficient lubrication and better anti-friction and anti-wear characteristics. Mucus gel and films (tear film) are also essential in creating eye lubricity. Under adverse conditions, lubrication loss and enhancement in coefficient of friction could lead to dry eye syndrome, higher conditions of shear stress values, and inflammation and harm to the anterior tissues. Artificial tear approaches are for dry eye syndrome and better health of the eye. Generally, the coefficient of friction lies between 0.025 to 0.075. Boundary lubrication and hydrodynamic lubrication mechanisms are the suggested lubrication regimes for the eye. Also, researchers are extensively working on theoretical lubrication modeling on contact lenses. It is sensible to mention that the elasto-hydrodynamic lubrication regime is accountable for contact lens friction and mechanics [23, 56 - 58].

Oral Cavity

The oral cavity is a relatively multifarious part as this includes both hard and soft tissues together with the temporomandibular joint (TMJ) (*i.e.*, synovial joint). From a tribological point of view, saliva is an important lubrication characteristic for normal functioning (during speech and food processing) of the oral cavity. Three sciences such as friction, rheology, and perception of foodstuffs, are equally important for the smooth functioning of the oral cavity [46, 59, 60].

Catheter

Catheters are utilized in medical sciences for diagnoses and interventions or sometimes in surgical instruments. In this process of insertion and manipulation, friction may arise which directly leads to shear stress that might cause harm to the tissue (natural). Coating techniques are applied to minimize friction. Anti-friction biomaterials are also developed. Moreover, the biomaterials (hydrophobic in nature) provide a lower frictional coefficient on the tissue, and specifically, hydrogel-coated catheters showed the lowermost frictional coefficients (static and kinetic) [23, 46, 61 - 65].

Skin

Friction mapping on the skin is important as it will help in understanding various processes and mechanisms such as the interaction of skin with other surfaces, its influences under different environments, effects of age and health conditions, and skin treatments using chemicals. The formation of blisters and ulcers is in close approximation to skin biomechanics and friction. Skin friction studies are an inherent part of the military and sports medicine. In addition to this, skin friction effects on touching, sensing (mechano-transduction techniques), sensation (mechanical sensation) and perception have been studied by various researchers. It is sensible to mention that studies must be conducted on skin deformation. Studies on human hair and shaving technologies are closely related to friction science [66 - 74].

Slips

Modern footwear advancement technologies precisely count the frictional coefficient between feet and footwear along with floor impact and the tendency of pedestrians to slip-and-fall. Hence, the biomechanics of slips are significant in footwear technologies. Significant factors involve footwear design, material considerations, and floral contact surfaces along with individual gait characteristics and various uncertainties must be included, such as natural variabilities among individual humans (age, mass, body configuration, walking speed, *etc.*) and various extrinsic parameters such as surface and shoe wear characteristics. In the following sections, the authors attempt to brief about the tribology of medical and surgical devices that include, artificial joints and teeth, dental replacements, cardiovascular instruments, contact lenses, artificial limbs, ocular systems, and skin systems. In the last section, we finalize the literature reports with recent and future advancements. Also, major tribological concerns are also discussed in the keynotes [23, 46, 75 - 77].

It is sensible to mention that authors are very much influenced by the key works on bio-friction studies by Jin and Dowson, (2016). Authors have attempted to brief the review on various biotribological systems.

Medical Devices [94, 95]

Artificial joints

Briefly, joints are of two types: relatively large motion mobility *i.e.* the hip, the knee, the shoulder, *etc.*, and smaller joints such as the ankle, the elbow, the wrist, and the finger. Also, implants of the spinal disc (total disc) and the temporomandibular joint (TMJ) prosthesis are equally important. It is sensible to mention that the joints facilitate complex 3-D motion and act as loading members. Major tribological concerns include serious concerns about the articulating surfaces, the connection between modular components, and the fixation to bone.

Articular Surfaces

Basically, biomaterials are utilized for the articular surfaces of the joints. Polymers, composites, and alloys are the dominating candidates in this direction. Anti-wear treatments, lower elastic modulus, and enhanced imaging quality are the basic consideration for ideal surfaces.

Modular Junctions

In the hip joint, modular head-neck combinations, and modular-neck stems are excellent examples of modular junctions. In the hip joint, modular head-neck combinations and modular-neck stems are excellent examples of modular junctions. These are significant due to the fact that they allow restoration of anatomy, optimization, and functionality (biomechanical) of the joints. Corrosion (fretting corrosion) is a serious problem in the advancement of clinical devices. It is possibly connected with the synergy of the mechanical and electrochemical effects [28]. Laboratory testing and computational modeling techniques are applied to overcome the tribe corrosion as well as wear problems.

Fixation

Cemented fixation or press-fitting (cement less fixation) are the basic methods of the implant fixations in artificial joint design. Fretting, corrosion, and wear are the dominating mechanisms in this direction. Surface finish and texture patterns are significant factors in this direction. It is sensible to mention that the micro-motion (or sometimes called strain) at the interface of implant-bone influences the stability (primary and leads to secondary) and adequate frictional response is needed to constrain the micro-motion [78 - 81].

Dental Artificial Tooth

Dental restorative substances are utilized to re-establish the role, coherence, and morphology of missing tooth assembly for dental restoration. Dental implants are used as the substitute for missing tooth structures to support these restorations. To overcome excessive friction and wear, currently, metals, alloys, ceramics, and composites or sometimes amalgam as materials are chiefly utilized for dental restorations and implants [27].

Dental Composite

High rate of aesthetics, bonding capability, and resin-based dental composites (encompass filler micro/nanoparticles in a polymer matrix, have been extensively utilized for restorative dentistry, recently. A major issue in dental composites is their weak anti-wear property. To tailor the anti-wear-corrosion, abrasive action, and material properties, various variables and components such as the size, shape, and hardness of the particle, additive content, inter-particle spacing, distribution of the additive, the interfacial bond strength, the nature and the surface hardness of the matrix phase, are significant in the design and manufacturing of the composite [82 - 84].

Dental Ceramics

High durability, improved aesthetic appeal and typical properties such as chemical and optical, made dental ceramics the first choice for dental restorations. However, researchers have to deal with brittleness and abrasive wear mechanism that leads to surface wear. High-toughness zirconia ceramic is dominating the market due to its toughness. It is sensible to mention that surface treatments are necessary for better degradation resistance, microstructure, and superior surface properties. Some researchers are working in the direction of advancement of mechanical properties such as fracture strength/resistance, thermal aging, and toughness.

Dental Amalgam

The research paid keen attention to selecting amalgam alloy (a mixture of mercury and powdered alloy) for dental restoration owing to its high anti-wear properties, durability, and ease of application in clinical and medical sciences. Some researchers believed that the high hardness value of the alloy may lead to enhanced abrasion resistance. Some serious issues such as its aesthetics, high toxicity, weak anti-corrosion properties, low value of fracture toughness and tensile strength, brittleness, *etc.* make its use critical and less feasible. At the end of the section, the authors reveal some of the recent works in the direction of tribology dental implants and orthodontic appliance (bracket-wire combinations) and their biocompatibility and bio-tribocorrosion. Still, a good interpretation of the basic underlying wear modes involved in the aforementioned medical and surgical devices is the major hurdle hampering the advancement of materials and tribological behavior.

Surgical Devices [94]

Briefly, surgical devices are classified on the basis of their functional usages such as cutting devices, grasping or holding devices, hemostatic forceps, retractors, clamps, distractors, accessories, and implants.

Invasive Grasper-tissue Interface

Invasive Surgery (Minimally) is an effective procedure in medical operations. It leads to loss of tissue disruption, and better and quick healing capability. However, these processes have some demerits too, such as ergonomically poor designs, difficulty in integration for haptic technologies, less precise force measurements (or simply say weak tool-tissue interaction and low operational efficiency). It is sensible to mention that researchers are working in the direction of measurement of forces and analyzing frictional response at the invasive grasper-tissue interface. These studies are important from the tribological point of view as well as in terms of the advancements in surgical instruments. The biochemistry, biotribology, and translational tissue engineering are the basic fields for the understanding of the tribology of cartilage. The shear-thinning behaviour/mechanism was also mentioned by the authors. However, this is not solely responsible for better tribological properties.

Interface Studies of Endoscopy and Oesophagus or Colon Interface

The prominent treatments (gastrointestinal endoscope, oesophagus or colon, and tissue repairing) in the digestive system and their advancement require an adequate amount of friction at the interface. In this direction, some researchers are studying the coefficient of friction of intestinal surface by viscoelastic model. The small intestine's friction trauma mechanism is another factor, that must be studied before surgery. It is sensible to mention that fundamental knowledge of gastrointestinal endoscopy, safety operations, and damage control can play a pivotal role in the clinical sciences.

Artificial Limb Stumps/Sockets

The friction between the linings and materials used in prosthetic sockets and the skin of the stump is much revolting in nature. High friction levels may cause various ailments like pressure ulcers, blisters, cysts, oedema, skin irritation, and dermatitis. Excessive levels may cause material degradation, and aging, and ultimately cause failure. Hence, the friction-wear-lubrication behaviour is very significant in the interface (limb skin-prosthetic socket), design and fitting. More precisely, by finite-element (FE) modelling approaches, one can able to measure the strains and stresses (for the internal soft tissue) throughout load-bearing conditions. Hence, an adequate frictional coefficient will provide stability and drop of slip with the general ability to sustain the loads. There exists a correlation between skin histological structure, surface topography, and coefficient of friction

Ocular Contact Lenses

Both the relative motion and loading are associated with the eye. Hence, tribology performs an eminent job in the successful functioning of the ocular contact lens. It is a well-known fact that lubricity behaviour between the contact lens, and the eye is unsympathetically significant. Tear film presence ensures fluid film lubrication that is further related to reduced friction and wear levels and smoother motion profiles. However, some authors reveal that boundary lubrication (brush-type), synergistic lubrication mechanism and a corresponding surface-brush boundary lubrication regime are present under certain conditions. Nowadays, novel soft contact lens materials with wetting agents are incorporated to minimize friction between the eye tribe contact pair. However, in this case, high levels of shear stresses may have a shallow impact on lubrication that virtually leads to wear. The elastic modulus, surface topography, material composition, and structure are significant considerations for better tribological functioning.

Cardiovascular Devices

Mechanical cardio-vascular assist devices exhibit relative movement between the components. Tribological fundamentals and extensive study of failure analysis of cardiovascular implants are crucial concerning the prototype of medical devices. The optimized lubricant (blood) film is available to evade blood damage and offers enough washouts. Friction, erosion, and wear of cardiovascular (heart valve or stents) devices are much more frequent. Novel coating approaches are established to reduce the development of thrombosis while at the same time enhancing the antiwear capability of the leaflets.

It is sensible to mention that the author's literature reports in the bio-tribology of medical and surgical instruments are the brief notes of the review articles by Jin *et al.* 2016 and Zhang *et al.* 2021.

BIOCOMPATIBILITY ISSUES RELATED TO BIOMATERIALS [4, 24, 63, 85 - 88]

Nowadays, various metal-based biomaterials such as stainless steel, titanium, magnesium alloys, and cobalt-chromium are immensely utilized due to good mechanical strength alloys, corrosion, wear, and biocompatibility for the implant materials. The basic problem lies in the selection of biomaterials is of their biocompatibility issues, higher degradation rate, poor inflammatory performance, more viable infections, low biochemical reaction response, weak anti-corrosion ability, poor mechanical properties (elastic modulus), poor wear performance, cell interactions and adhesion (biological point view) and shielding stress. Also, some researchers are working in the direction of bio-functionalization and surface treatments/modification of biomedical implants. Tribo-corrosion of the implants might affect the surface characteristics and biocompatibility. Various coatings such as carbon-based coatings (have good hydrophobicity character): a-C, diamond-like carbon, NCD, carbides and carbonitride, calcium phosphates, hydroxyapatite, and bioactive glass are applied in biomedical implants for betterment. Also, bio-resorbable-ceramic coatings have their advantages such as anti-wear/tear, high rate of biocompatibility, and anti-corrosive. Among, bio-degradable polymers, chitosan (CS) has excellent biocompatibility, biodegradability, wound healing, and antibacterial activity. On the other hand, basically, hydrogels (soft) are significant owing to their soft consistency, good stiffness, better viscoelastic behaviour, better immune response, high water content, porosity, better adhesion, and biocompatibility. Some biocompatible composites coating on various substrates, such as hydroxyapatite/PEEK composite coating on PEEK substrate, ZrO_2/PEEK coating on Ti6Al4V substrates, TiO_2/$MoSe_2$/Chitosan coating on Ti implants, is significant to address for the advancement of biomedical implants. Whereas, TiN coatings are known for better blood tolerability and biocompatibility. On the other hand, if researchers want to provide anti-oxidation ability, better hardness, anti-corrosive, and better structural stability along with biocompatibility, surface-modified coatings primed from ternary nitrides must be applied. Nano-TiN on Ti-6Al-4V and TiON coatings on Ti substrates leads to high biocompatibility with anti-wear ability, better haemoglobin compatibility, and superior biological activity. Pyrolytic carbon (PyC) coatings have good thrombo-resistant qualities with better biocompatibility. DLC-coated articles provide good surface coverage of cells and display high cell viability along with better biocompatibility. However, it is sensible to mention that chemical inertness, surface smoothness, and hydrophobicity are equally important for providing a high rate of compatibility with blood and minimizing platelet activation in contact with the blood, which could activate thrombosis. In the tribological evaluation of bio-implants, the study of materials, coatings and wear modes (Abrasive/adhesive/erosive/ corrosive/ fatigue), surface topography (surface roughness, texture, texturing geometry) and simulator design is significant and vital [24, 85 - 88]. It is sensible to mention that, according to the scope of the book, authors have listed a few composite-based applications for the various tribology segments Table **4**.

Table 4. Tribological investigation (based on composite usage/application) of various bio-tribological segments [1 - 40, 94 - 100].

Biotribology Segments	Composite Applications
Orthopedics & joint tribology	Poly(methyl methacrylate), poly(ether-ether-ketone), and ultra-high molecular weight polyethylene (UHMWPE) based polymer composites, UHMWPE with hydroxyapatite (HA) composites, alumina/zirconia toughened alumina composite ceramics (ZTA), Synthetic Polymeric composite materials, 3D Printed Polylactic-Acid (PLA) Green-Composite.
Tissue engineering	Bio-polymeric/natural materials or green composite materials.
Prostheses implant or Artificial implant	Glass-reinforced composites, continuous-fiber-reinforced polymers, PMMA composite, SiO_2 based nano composites, Carbon-fiber-reinforced, (PEEK), and UHMWPE based composites.
Bio-mimicry	CNT's based nano-composites, polymeric composites.
Locomotion	Fiber-elastomer composites, Actinomorphic soft robot with soft composite structures, Liquid metal-elastomer composites.
Drug delivery	Nanocellulose-based composite materials, Clay and Polymer-based composites, Interactive fibre-elastomer composites, Natural polysaccharide-based composites, engineered spider silk protein-based composites.
Ocular tribology	Reinforced composite materials for artificial corneas (transplantation).
Skin tribology	Utilization of **Polyamide (TPA) and polyurethane (TPU; PA6/diborontrioxide composites.**
Haptics technology	Ionic Polymer Metal Composites, Carbon fiber-reinforced polymer (CFRP), Silicon dioxide and PVC for actuator technology.
Dental tribology	Resin-based dental composites, PMMA-based denture composite, glass-polymer dental composite, Fiber reinforced dental composite, UDMA-based composites, Ti alloy – HAp composite, Graphene for Zirconia and Titanium Composites, Herbal composites, Polymethyl Methacrylate and Hydroxyapatite Composite.
Tribology in Industry	Polymer-based composites, Silicone rubber-based ceramizable composites.
Sport tribology	Polymeric composite materials, carbon fiber reinforced polymer composites, Ti_2AlC composite materials.
Biolubricants	Ceramic composite coating materials for biolubricants investigation.
Others	Graphene oxide-based composites, zeolite composite materials for the removal of pharmaceuticals and personal care products.

BIO-TRIBO-CORROSION

Bio-tribo-corrosion (multi-disciplinary science belongs to the synergetic pairing off mechanical and environmental effects, electrochemistry, material behaviour studies, surface-engineering, irreversible transformation science, mechanics, tribology, microbiology, oxidation, and corrosion) is an eminent area of research and has a practical influence on day-to-day life and also associated with the economic burdens. The intricacy of tribo-corrosion processes is understandable by specifying many variables (mechanical and corrosion) but, owing to the deficiency of integrated/efficient monitoring/measuring devices, the advancement is a little-bit hampered. This context will lead to the development of self-lubricating, self-repairing, and/or self-healing layered materials. Bio-trib--corrosion is an important aspect in terms of the biomedical point of view. Various experimental standardization techniques, micro-abrasion-corrosion maps, and tribo-electrochemical techniques are available to study the tribo-corrosion behavior of biomaterials. Bio-oils, biodegradable substances (metals, polymers, and composites), surface treatment technologies, anti-wear resistance biomedical implants, biological mechanisms/processes, and bio-tribo-nano corrosion are the current segments of the bio-tribo-corrosion. A better understanding of underlying mechanism/phenomenon, synergism, and antagonism, for complex interaction, inter/intra laboratory analysis and modeling (numerical/simulation/mathematical /software) are in the future directions of the bio-tribo-corrosion. Fundamental challenges in the study of bio-tribo-corrosion are inefficient non-synchronized simulating devices for integrated studies (or biological problems), few choices for tribular materials, and poor understanding of dynamics and synergism. However, recently, researchers have been studying seals, nuclear, lubricants, coatings and biomedical applications. In a future perspective, modeling, standardization, better synergy and antagonism and the need of multi-disciplinary platforms are salient hopes in bio-tribological works. Still, there is a lack of skilled/expert researchers who apply their knowledge to integrated biological-tribological works [89].

CHALLENGES

Wear testing using simulators is a time-consuming and costly task. Sometimes variation lies between the real outcome and the experimental results. Feasibility is also an issue in the practicality of artificial joints. Computational modelling itself is a stringent approach for biomaterials, and their combination, along with tribological constraints, plays a vital role in the development. Although, the studies are not fully integrated and coupled. The study of biomechanics and wear phenomenon are key studies, and major challenges lie in these approaches. The fracture exploration along with their alliance and transfer of forces through the fracture zone, are the important and basic considerations. Lubrication modelling and tribological simulation are equally significant in the relative sliding of the components. In soft tribology, multiple interdisciplinary fields such as poroelasticity, tribo-rheology, biotribology and microscopy have been studied to formulate biological problems with great care [90].

FUTURE DIRECTIONS

In the future direction, the authors want to describe that the medical devices must be designed per the unification with patients and doctors. As the patient anatomy is essential, hence, the progress of patient-specific implants might be introduced as per the roles, mainly as per the biomechanics of the loading system and functionality. Advancements in medical instruments might be profitable for a better understanding of natural physiology. For enhanced durability, robustness, easy repair, preservation and regeneration, wear-resistive systems might be the promising needs . Safety is a serious concern in this direction. Simulation, testing, and design cum manufacturing coupled with the biological environment would lead to the development of bio-tribologically effective medical and surgical instruments. Fluorescence microscopy (optical mode) for visualization (soft contact) of lubrication film in the cartilage contact area. Also, tribological features embedded with the haptic perception (grip and slipperiness) might be helpful in modelling of synovial joint. Also, some valuable efforts were made for the development of the correlation between the dynamic frictional coefficient and the perceived gliding quality of a surface. Now authors want to mention that practicability reports on complex structures, designing, and fabrication integrated with better typical properties (surface, mechanical chemical *etc.*) and integrity are still severe that need immense care and sophisticated instruments for testing and instrumentation. Fractography studies and mathematical, and numerical/ simulation modelling of biomedical coating must match targets [91 - 93].

CHAPTER SUMMARY

Biotribology belongs to tribological application and advancement in the technologies of biological sciences, mechanisms, and theories. The comprehensive review starts with keynotes on the various segments of bio tribology such as joint tribology, tissue substitute-based applications, implant tribology, biomimicry, drug delivery, ocular tribology, skin tribology, sports tribology, dental tribology, oral processing, haptics, bio lubrication *etc*. The next section deals with the bio-friction theories, mechanisms and lubrication regimes. In the bio-friction studies, authors have included recent reports and literature reviews on synovial joints, fat pad and tendon, pleurae, eye, oral cavity, catheter, skin and slip. The tribological application of medical and surgical instruments is described in a separate section. The authors finish with a brief discussion on biocompatibility and bio-tribe-corrosion of the biomaterials. In addition to this, various key challenges such as the development of tribological simulators, feasibility, compatibility and corrosiveness of the bio-tribological products, lack of understanding of biomechanics and wear phenomenon and numerical/ computational modelling, are mentioned in the other section. Finally, the authors quoted some future directions for the readers to gain maximum advantage of this review chapter.

REFERENCES

[1]   Available from: https://www.sciencedirect.com/journal/biotribology

[2]   Available from: https://www.elsevier.com

[3]   Simchi, A.; Tamjid, E.; Pishbin, F.; Boccaccini, A.R. Recent progress in inorganic and composite coatings with bactericidal capability for orthopaedic applications. *Nanomedicine,* **2011**, *7*(1), 22-39.
       [http://dx.doi.org/10.1016/j.nano.2010.10.005] [PMID: 21050895]

[4]   Chaichi, A.; Prasad, A.; Kootta Parambil, L.; Shaik, S.; Hemmasian Ettefagh, A.; Dasa, V.; Guo, S.; Osborn, M.L.; Devireddy, R.; Khonsari, M.M.; Gartia, M.R. Improvement of tribological and biocompatibility properties of orthopedic materials using piezoelectric direct discharge plasma surface modification. *ACS Biomater. Sci. Eng.,* **2019**, *5*(5), 2147-2159.
       [http://dx.doi.org/10.1021/acsbiomaterials.9b00009] [PMID: 33405717]

[5]   Shirdar, M.R.; Farajpour, N.; Shahbazian-Yassar, R.; Shokuhfar, T. Nanocomposite materials in orthopedic applications. *Front. Chem. Sci. Eng.,* **2019**, *13*(1), 1-13.
       [http://dx.doi.org/10.1007/s11705-018-1764-1]

[6]   Gebeshuber, I.C.; Stachelberger, H.; Austrian, T.; Kaplan-Straße, V. Tribology in biology: biomimetic studies across dimensions and across fields. *International Journal of Mechanical and Materials Engineering,* **2009**, *4*, 321-327.

[7]   Ivanović, L.; Vencl, A.; Stojanović, B.; Markovic, B. *Biomimetics Design for Tribological Applications*; Tribology in Industry, **2018**.
       [http://dx.doi.org/10.24874/ti.2018.40.03.11]

[8]   Dai, Z.; Tong, J.; Ren, L. Researches and developments of biomimetics in tribology. *Chin. Sci. Bull.,* **2006**, *51*(22), 2681-2689.
       [http://dx.doi.org/10.1007/s11434-006-2184-z]

[9]   Available from: https://en.wikipedia.org/wiki/Biomimetics

[10]   Available from: https://web.stanford.edu/group/mota/education/Physics%2087N%20Final%20Projects /Group%20Gamma/index.htm

[11]   Available   from: https://www.sciencedirect.com/topics/agricultural-and-biologic-l-sciences/biolubricants#:~:text=Biolubricants%20are%20functional%20fluids%20made,India%20for %20three%2Dwheeled%20vehicles

[12]   Saleem, M.S.; Khan, M.S.; Khan, S. *A Review on Tribological Performance of Nano Based Bio-Lubricants and its Applications.,* **2021**.

[13]   Narayana Sarma, R.; Vinu, R. Current status and future prospects of biolubricants: Properties and applications. *Lubricants,* **2022**, *10*(4), 70.
       [http://dx.doi.org/10.3390/lubricants10040070]

[14]   Available from: https://www.antala.uk/an-introduction-to-bio-lubricants/

[15]   Available from: https://en.wikipedia.org/wiki/Biomaterial

[16]   Moshirfar, M.; Pierson, K.; Hanamaikai, K.; Santiago-Cabán, L.; Muthappan, V.; Passi, S.F. Artificial tears potpourri: A literature review. *Clin. Ophthalmol.,* **2014**, *8*, 1419-1433.
       [PMID: 25114502]

[17]   Pucker, A.D. A review of the compatibility of topical artificial tears and rewetting drops with contact lenses. *Cont. Lens Anterior Eye,* **2020**, *43*(5), 426-432.
       [http://dx.doi.org/10.1016/j.clae.2020.04.013] [PMID: 32409235]

[18] Roba, M.; Duncan, E.G.; Hill, G.A.; Spencer, N.D.; Tosatti, S.G.P. Friction measurements on contact lenses in their operating environment. *Tribol. Lett.,* **2011**, *44*(3), 387-397.
[http://dx.doi.org/10.1007/s11249-011-9856-9]

[19] Samsom, M.L. *In Vitro Friction of Contact Lenses & Model Contact Lens Biomaterials: Effect of Proteoglycan 4,* **2017**.

[20] Tsai, H.J.; Horng, J.H.; Tan, C.M. Investigation of lubrication characteristics for a soft contact lens. *Key Eng. Mater.,* **2019**, *823*, 105-109.
[http://dx.doi.org/10.4028/www.scientific.net/KEM.823.105]

[21] Available from: https://www.ncbi.nlm.nih.gov

[22] Derler, S.; Gerhardt, L.C. Tribology of skin: Review and analysis of experimental results for the friction coefficient of human skin. *Tribol. Lett.,* **2012**, *45*(1), 1-27.
[http://dx.doi.org/10.1007/s11249-011-9854-y]

[23] Available from: https://www.cntribo.org

[24] Available from: https://link.springer.com

[25] Van Der Heide, E.; Zeng, X.; Masen, M.A. Skin tribology: Science friction? *Friction,* **2013**, *1*(2), 130-142.
[http://dx.doi.org/10.1007/s40544-013-0015-1]

[26] Available from: https://en.wikipedia.org/wiki/Haptics

[27] Zhou, Z.R.; Jin, Z.M. Biotribology: Recent progresses and future perspectives. *Biosurf. Biotribol.,* **2015**, *1*(1), 3-24.
[http://dx.doi.org/10.1016/j.bsbt.2015.03.001]

[28] Available from: https://en.wikipedia.org/wiki/Haptic_technology

[29] Zheng, Y.; Bashandeh, K.; Shakil, A.; Jha, S.; Polycarpou, A.A. Review of dental tribology: Current status and challenges. *Tribol. Int.,* **2021**.

[30] Available from: https://www.tribonet.org/news/sports-tribology/

[31] MacFarlane, M.J.; Theobald, P. Skin tribology in sport. *Biosurf. Biotribol.,* **2021**, *7*(3), 113-118.
[http://dx.doi.org/10.1049/bsb2.12015]

[32] D'Souza, B.; Kasar, A.K.; Jones, J.; Skeete, A.; Rader, L.; Kumar, P.; Menezes, P.L. A brief review on factors affecting the tribological interaction between human skin and different textile materials. *Materials,* **2022**, *15*(6), 2184.
[http://dx.doi.org/10.3390/ma15062184] [PMID: 35329636]

[33] Motamen Salehi, F.; Neville, A.; Bryant, M. Bio-tribology of incontinence management products: Additional complexities at the skin–pad interface. *Tribology - Materials, Surfaces & Interfaces,* **2018**, *12*(4), 193-199.
[http://dx.doi.org/10.1080/17515831.2018.1512784]

[34] Spencer, N.D.; Tysoe, W.T. *Weird and Wonderful Effects in Tribology.,* **2015**.
[http://dx.doi.org/10.1142/9789814656566_0009]

[35] Available from: https://www.stle.org

[36] Kareem, A.K., M.; Gabir, M.; Almoayed, O.M.; Ismail, A.; Taib, I.; Darlis, N.; Ali, I. A systematic review on cardiovascular stent and stenting failure: Coherent taxonomy, performance measures, motivations, open challenges and recommendations. *International Journal of Integrated Engineering.,* **2022**.

[37] Wagner, R.M.F.; Maiti, R.; Carré, M.J.; Perrault, C.M.; Evans, P.C.; Lewis, R. Bio-tribology of vascular devices: A review of tissue/device friction research. *Biotribology,* **2021**, *25*, 100169.
[http://dx.doi.org/10.1016/j.biotri.2021.100169]

[38]   Mozafari, H. *Bioresorbable Composite Stents for Enhanced Response of Vascular Smooth Muscle Cells.,* **2019**.

[39]   Hills, B.A.; Butler, B.D. Surfactants identified in synovial fluid and their ability to act as boundary lubricants. *Ann. Rheum. Dis.,* **1984**, *43*(4), 641-648.
[http://dx.doi.org/10.1136/ard.43.4.641] [PMID: 6476922]

[40]   Forster, H.; Fisher, J. The influence of loading time and lubricant on the friction of articular cartilage. *Proc. Inst. Mech. Eng. H,* **1996**, *210*(2), 109-119.
[http://dx.doi.org/10.1243/PIME_PROC_1996_210_399_02] [PMID: 8688115]

[41]   Dowson, D. Modes of lubrication in human joints. *Proceedings of the Institution of Mechanical Engineers,* **1966**, pp. 45-54.

[42]   Ateshian, G.A. A theoretical formulation for boundary friction in articular cartilage. *J. Biomech. Eng.,* **1997**, *119*(1), 81-86.
[http://dx.doi.org/10.1115/1.2796069] [PMID: 9083853]

[43]   Graindorge, S.; Ferrandez, W.; Jin, Z.; Ingham, E.; Grant, C.; Twigg, P.; Fisher, J. Biphasic surface amorphous layer lubrication of articular cartilage. *Med. Eng. Phys.,* **2005**, *27*(10), 836-844.
[http://dx.doi.org/10.1016/j.medengphy.2005.05.001] [PMID: 16046176]

[44]   Crockett, R.; Roos, S.; Rossbach, P.; Dora, C.; Born, W.; Troxler, H. Imaging of the surface of human and bovine articular cartilage with ESEM and AFM. *Tribol. Lett.,* **2005**, *19*(4), 311-317.
[http://dx.doi.org/10.1007/s11249-005-7448-2]

[45]   Goldberg, R.; Klein, J. Liposomes as lubricants: Beyond drug delivery. *Chem. Phys. Lipids,* **2012**, *165*(4), 374-381.
[http://dx.doi.org/10.1016/j.chemphyslip.2011.11.007] [PMID: 22119851]

[46]   Jin, Z.; Dowson, D. Bio-friction. *Friction,* **2013**, *1*(2), 100-113.
[http://dx.doi.org/10.1007/s40544-013-0004-4]

[47]   Theobald, P. Lubricating properties of the fat pad. In: *Encyclopedia of Tribology*; Springer: Berlin, **2013**. in press.
[http://dx.doi.org/10.1007/978-0-387-92897-5_1274]

[48]   Theobald, P.; Byrne, C.; Oldfield, S.F.; Dowson, D.; Benjamin, M.; Dent, C.; Pugh, N.; Nokes, L.D.M. Lubrication regime of the contact between fat and bone in bovine tissue. *Proc. Inst. Mech. Eng. H,* **2007**, *221*(4), 351-356.
[http://dx.doi.org/10.1243/09544119JEIM176] [PMID: 17605392]

[49]   Evans, R.B. Managing the injured tendon: Current concepts. *J. Hand Ther.,* **2012**, *25*(2), 173-190.
[http://dx.doi.org/10.1016/j.jht.2011.10.004] [PMID: 22326362]

[50]   Finley, D.J.; Rusch, V.W. Anatomy of the Pleura. *Thorac. Surg. Clin.,* **2011**, *21*(2), 157-163.
[http://dx.doi.org/10.1016/j.thorsurg.2010.12.001] [PMID: 21477764]

[51]   Lai-Fook, S.J. Pleural mechanics and fluid exchange. *Physiol. Rev.,* **2004**, *84*(2), 385-410.
[http://dx.doi.org/10.1152/physrev.00026.2003] [PMID: 15044678]

[52]   Loring, S.H.; Brown, R.E.; Gouldstone, A.; Butler, J.P. Lubrication regimes in mesothelial sliding. *J. Biomech.,* **2005**, *38*(12), 2390-2396.
[http://dx.doi.org/10.1016/j.jbiomech.2004.10.012] [PMID: 16214486]

[53]   Loring, S.H.; Butler, J.P. Pleural lubrication and friction in the chest. In: *Encyclopedia of Tribology*; Springer: Berlin, **2013**. in press
[http://dx.doi.org/10.1007/978-0-387-92897-5_1305]

[54]   Bodega, F.; Sironi, C.; Porta, C.; Pecchiari, M.; Zocchi, L.; Agostoni, E. Mixed lubrication after rewetting of blotted pleural mesothelium. *Respir. Physiol. Neurobiol.,* **2013**, *185*(2), 369-373.
[http://dx.doi.org/10.1016/j.resp.2012.09.003] [PMID: 22982215]

[55]   Hills, B.A. Graphite-like lubrication of mesothelium by oligolamellar pleural surfactant. *J. Appl. Physiol.,* **1992**, *73*(3), 1034-1039.
[http://dx.doi.org/10.1152/jappl.1992.73.3.1034] [PMID: 1400014]

[56]   Rennie, A.C.; Dickrell, P.L.; Sawyer, W.G. Friction coefficient of soft contact lenses: Measurements and modeling. *Tribol. Lett.,* **2005**, *18*(4), 499-504.
[http://dx.doi.org/10.1007/s11249-005-3610-0]

[57]   Ngai, V.; Medley, J.B.; Jones, L.; Forrest, J.; Teiehroeb, J. Friction of contact lenses: Silicone *versus* conventional hydrogels. *Tribology and Interface Engineering Series,* **2005**, *48*, 371-379.
[http://dx.doi.org/10.1016/S0167-8922(05)80039-2]

[58]   Jones, M.B.; Fulford, G.R.; Please, C.P.; McElwain, D.L.S.; Collins, M.J. Elastohydrodynamics of the eyelid wiper. *Bull. Math. Biol.,* **2008**, *70*(2), 323-343.
[http://dx.doi.org/10.1007/s11538-007-9252-7] [PMID: 18066629]

[59]   Goh, S.M. Tribology of foods. In: *Encyclopedia of Tribology*; Springer: Berlin, **2013**. in press
[http://dx.doi.org/10.1007/978-0-387-92897-5_1301]

[60]   Zhou, Z.R.; Zheng, J. Oral tribology. *Proc. Inst. Mech. Eng., Part J J. Eng. Tribol.,* **2006**, *220*(8), 739-754.
[http://dx.doi.org/10.1243/13506501JET145]

[61]   Luboz, V.; Zhai, J.; Odetoyinbo, T.; Littler, P.; Gould, D.; How, T.; Bello, F. Guidewire and catheter behavioural simulation. *Stud. Health Technol. Inform.,* **2011**, *163*, 317-323.
[PMID: 21335811]

[62]   Lawrence, E.L.; Turner, I.G. Materials for urinary catheters: A review of their history and development in the UK. *Med. Eng. Phys.,* **2005**, *27*(6), 443-453.
[http://dx.doi.org/10.1016/j.medengphy.2004.12.013] [PMID: 15990061]

[63]   Nickel, J.C.; Olson, M.E.; Costerton, J.W. *In vivo* coefficient of kinetic friction: Study of urinary catheter biocompatibility. *Urology,* **1987**, *29*(5), 501-503.
[http://dx.doi.org/10.1016/0090-4295(87)90037-9] [PMID: 3576867]

[64]   Khoury, A.E.; Olson, M.E.; Villari, F.; Costerton, J.W. Determination of the coefficient of kinetic friction of urinary catheter materials. *J. Urol.,* **1991**, *145*(3), 610-612.
[http://dx.doi.org/10.1016/S0022-5347(17)38405-7] [PMID: 1997718]

[65]   Kazmierska, K.; Szwast, M.; Ciach, T. Determination of urethral catheter surface lubricity. *J. Mater. Sci. Mater. Med.,* **2008**, *19*(6), 2301-2306.
[http://dx.doi.org/10.1007/s10856-007-3339-4] [PMID: 18071872]

[66]   Hills, R.J.; Unsworth, A.; Ive, F.A. A comparative study of the frictional properties of emollient bath additives using porcine skin. *Br. J. Dermatol.,* **1994**, *130*(1), 37-41.
[http://dx.doi.org/10.1111/j.1365-2133.1994.tb06879.x] [PMID: 8305314]

[67]   Dowson, D. A tribological day. *Proc. Inst. Mech. Eng., Part J J. Eng. Tribol.,* **2009**, *223*(3), 261-273.
[http://dx.doi.org/10.1243/13506501JET557]

[68]   Cichowitz, A.; Pan, W.R.; Ashton, M. The heel. *Ann. Plast. Surg.,* **2009**, *62*(4), 423-429.
[http://dx.doi.org/10.1097/SAP.0b013e3181851b55] [PMID: 19325351]

[69]   Egawa, M.; Oguri, M.; Hirao, T.; Takahashi, M.; Miyakawa, M. The evaluation of skin friction using africtional feel analyzer. *Skin Res. Technol.,* **2002**, *8*(1), 41-51.
[http://dx.doi.org/10.1034/j.1600-0846.2002.080107.x] [PMID: 12005119]

[70]   Hung, C.; Dubrowski, A.; Gonzalez, D.; Carnahan, H. Surface exploration using instruments: The perception of friction. *Stud. Health Technol. Inform.,* **2007**, *125*, 191-193.
[PMID: 17377264]

[71]   Shao, F.; Chen, X-J.; Barnes, C.J.; Henson, B. A novel tactile sensation measurement system for qualifying touch perception. *Proc. Inst. Mech. Eng. H,* **2010**, *224*(1), 97-105.

[http://dx.doi.org/10.1243/09544119JEIM658] [PMID: 20225461]

[72]   Skedung, L.; Danerlöv, K.; Olofsson, U.; Michael Johannesson, C.; Aikala, M.; Kettle, J.; Arvidsson, M.; Berglund, B.; Rutland, M.W. Tactile perception: Finger friction, surface roughness and perceived coarseness. *Tribol. Int.,* **2011**, *44*(5), 505-512.
[http://dx.doi.org/10.1016/j.triboint.2010.04.010]

[73]   Zahouani, H.; Vargiolu, R.; Hoc, T. Bio tribology of tactile perception: Effect of mechano-transduction. In: *Encyclopedia of Tribology*; Springer: Berlin, **2013**. in press
[http://dx.doi.org/10.1007/978-0-387-92897-5_1318]

[74]   Bhushan, B.; Wei, G.; Haddad, P. Friction and wear studies of human hair and skin. *Wear,* **2005**, *259*(7-12), 1012-1021.
[http://dx.doi.org/10.1016/j.wear.2004.12.026]

[75]   Clarke, J.D.; Lewis, R.; Carré, M.J. Tribology in daily life: Footwear-surface interactions in pedestrian slips. In: *Encyclopedia of Tribology*; Springer: Berlin, **2013**. in press
[http://dx.doi.org/10.1007/978-0-387-92897-5_1299]

[76]   Redfern, M.S.; Cham, R.; Gielo-Perczak, K.; Grönqvist, R.; Hirvonen, M.; Lanshammar, H.; Marpet, M.; Pai, C.Y.C., IV; Powers, C. Biomechanics of slips. *Ergonomics,* **2001**, *44*(13), 1138-1166.
[http://dx.doi.org/10.1080/00140130110085547] [PMID: 11794762]

[77]   Chang, W.R.; Grönqvist, R.; Leclercq, S.; Myung, R.; Makkonen, L.; Strandberg, L.; Brungraber, R.J.; Mattke, U.; Thorpe, S.C. The role of friction in the measurement of slipperiness, Part 1: Friction mechanisms and definition of test conditions. *Ergonomics,* **2001**, *44*(13), 1217-1232.
[http://dx.doi.org/10.1080/00140130110085574] [PMID: 11794765]

[78]   Gilbert, J.L.; Buckley, C.A.; Jacobs, J.J. *In vivo* corrosion of modular hip prosthesis components in mixed and similar metal combinations. The effect of crevice, stress, motion, and alloy coupling. *J. Biomed. Mater. Res.,* **1993**, *27*(12), 1533-1544.
[http://dx.doi.org/10.1002/jbm.820271210] [PMID: 8113241]

[79]   Zhang, H.Y.; Blunt, L.A.; Jiang, X.Q.; Fleming, L. T.; Barrans, S.M. The influence of bone cement type on production of fretting wear on the femoral stem surface: a preliminary study. *Clin. Bio mech.,* **2012**, *27*, 666-672.

[80]   Pulido, L.; Rachala, S.R.; Cabanela, M.E. Cementless acetabular revision: past, present, and future. *Int. Orthop.,* **2011**, *35*(2), 289-298.
[http://dx.doi.org/10.1007/s00264-010-1198-y] [PMID: 21234562]

[81]   Harrison, N.; McHugh, P.E.; Curtin, W.; Mc Donnell, P. Micromotion and friction evaluation of a novel surface architecture for improved primary fixation of cementless orthopaedic implants. *J. Mech. Behav. Biomed. Mater.,* **2013**, *21*, 37-46.
[http://dx.doi.org/10.1016/j.jmbbm.2013.01.017] [PMID: 23455331]

[82]   Zhou, Z.R.; Zheng, J. Tribology of dental materials: A review. *J. Phys. D Appl. Phys.,* **2008**, *41*(11), 113001.
[http://dx.doi.org/10.1088/0022-3727/41/11/113001]

[83]   Arsecularatne, J.A.; Chung, N.R.; Hoffman, M. An *in vitro* study of the wear behaviour of dental composites. *Biosurf. Biotribol.,* **2016**, *2*(3), 102-113.
[http://dx.doi.org/10.1016/j.bsbt.2016.09.002]

[84]   Venhoven, B.A.M. Influence of filler parameters on the mechanical coherence of dental restorative resin composites. *Biomaterials,* **1996**, *7*, 735-740.

[85]   Link, J.M.; Salinas, E.Y.; Hu, J.C.; Athanasiou, K.A. The tribology of cartilage: Mechanisms, experimental techniques, and relevance to translational tissue engineering. *Clin. Biomech. (Bristol, Avon),* **2019**.
[PMID: 31676140]

[86]   Available from: https://mdpi-res.com

[87] Amirtharaj Mosas, K.K.; Chandrasekar, A.R.; Dasan, A.; Pakseresht, A.; Galusek, D. Recent advancements in materials and coatings for biomedical implants. *Gels,* **2022,** *8*(5), 323. [http://dx.doi.org/10.3390/gels8050323] [PMID: 35621621]

[88] Shen, G.; Fang, F.; Kang, C. Tribological performance of bioimplants: A comprehensive review. *Nanotechnology and Precision Engineering,* **2018,** *1*, 107-122.

[89] Mathew, M.T.; Srinivasa Pai, P.; Pourzal, R.; Fischer, A.; Wimmer, M.A. Significance of tribocorrosion in biomedical applications: Overview and current status. *Adv. Tribol.,* **2009,** *2009*, 1-12. [http://dx.doi.org/10.1155/2009/250986]

[90] Pitenis, A.A.; Urueña, J.M.; McGhee, E.O.; Hart, S.M.; Reale, E.R.; Kim, J.; Schulze, K.D.; Marshall, S.L.; Bennett, A.I.; Niemi, S.R.; Angelini, T.E.; Sawyer, W.G.; Dunn, A.C. Challenges and opportunities in soft tribology. *Tribology - Materials, Surfaces & Interfaces,* **2017,** *11*(4), 180-186. [http://dx.doi.org/10.1080/17515831.2017.1400779]

[91] Gulsen, A.; Merve, G.; Meltem, P. Biotribology of cartilage wear in knee and hip joints review of recent developments. *IOP Conf. Series Mater. Sci. Eng.,* **2018,** *295*, : 012040.. [http://dx.doi.org/10.1088/1757-899X/295/1/012040]

[92] Singh, K.; Khanna, A.; Rosenkranz, V. Panorama of physico-mechanical engineering of graphene-reinforced copper composites for sustainable applications. *Materials Today Sustainability,* **2023,** *24*, 100560. [http://dx.doi.org/10.1016/j.mtsust.2023.100560]

[93] Čípek, P.; Rebenda, D.; Nečas, D.; Vrbka, M.; Křupka, I.; Hartl, M. *Visualization of Lubrication Film in Model of Synovial Joint*; Tribology in Industry, **2019**.

[94] Jin, Z.M.; Zheng, J.; Li, W.; Zhou, Z.R. Tribology of medical devices. *Biosurf. Biotribol.,* **2016,** *2*(4), 173-192. [http://dx.doi.org/10.1016/j.bsbt.2016.12.001]

[95] Zhang, X.; Zhang, Y.; Jin, Z. A review of the bio-tribology of medical devices. *Friction,* **2022,** *10*(1), 4-30. [http://dx.doi.org/10.1007/s40544-021-0512-6]

[96] Dahiya, M.; Khanna, V.; Anil Bansal, S. Effect of graphene size variation on mechanical properties of aluminium graphene nanocomposites: A modeling analysis. *Mater. Today Proc.,* **2022,** (Jul) [http://dx.doi.org/10.1016/j.matpr.2022.07.259]

[97] Gupta, P.; Ahamad, N.; Kumar, D.; Gupta, N.; Chaudhary, V.; Gupta, S.; Khanna, V.; Chaudhary, V. Synergetic effect of ceo $_2$ doping on structural and tribological behavior of fe-al $_2$ o $_3$ metal matrix nanocomposites. *ECS J. Solid State Sci. Technol.,* **2022,** *11*(11), 117001. [http://dx.doi.org/10.1149/2162-8777/ac9c92]

[98] Dahiya, M.; Khanna, V.; Anil Bansal, S. Aluminium-graphene metal matrix nanocomposites: Modelling, analysis, and simulation approach to estimate mechanical properties. *Mater. Today Proc.,* **2022,** (Nov) [http://dx.doi.org/10.1016/j.matpr.2022.10.181]

[99] Singh, K.; Bansal, S.A.; Khanna, V.; Singh, S. Effects of performance measures of non-conventional joining processes on mechanical properties of metal matrix composites. *Metal Matrix Composites,* **2022,** (Aug), 135-165. [http://dx.doi.org/10.1201/9781003194897-7]

[100] Khanna, V.; Kumar, V.; Bansal, S.A.; Prakash, C.; Ubaidullah, M.; Shaikh, S.F.; Pramanik, A.; Basak, A.; Shankar, S. Fabrication of efficient aluminium/graphene nanosheets (Al-GNP) composite by powder metallurgy for strength applications. *J. Mater. Res. Technol.,* **2023,** *22*, 3402-3412. [http://dx.doi.org/10.1016/j.jmrt.2022.12.161]

CHAPTER 7

A Review on Reinforcement and Its Effect on Aluminium-Based Composites

Anupam Thakur[1,*], Virat Khanna[1] and Qasim Murtaza[2]

[1] *Department of Mechanical Engineering, MAIT, Maharaja Agrasen University, H.P., India*

[2] *Mechanical Department, Delhi Technological University, Delhi, India*

Abstract: In today's world, there are different materials that are already used in certain applications and have been performing well. As the need and the complications in certain areas have been progressively discovered, there is a wide requirement for materials' research that has a combined effect of more than one property which is a limitation of monolithic materials. To have such an effect, the fabrication of certain materials having a well-tailored blend of properties as per the reinforcement is used to form a composite. A review has been carried out for the various research works and the effort is being made to summarize the effect of mono and hybrid reinforcements on the materialistic properties as compared to the base material.

Keywords: Aluminium, Composites, Hardness, MMC, Tribological test.

INTRODUCTION

As material science has advanced in many ways, there is a need for advancement in material research as the application areas have become very competitive in nature. This thus requires a lot of research on metals and non-metals thus leading to different analytical studies to find the optimum use of the material in a certain field of application. During the research, it has been found that various Metal Matrix Composites (MMCs) combine the properties of two different materials which lead to the production of a new material which is basically an alloy. This alloy has a higher percentage of the base material and a much lower percentage of the alloying materials that are combined with it. The main idea behind this type of material production is to majorly have the properties of the base metal but due to the application areas in certain fields, these properties somewhat need an alteration. These small changes in properties can be made with the help of alloying elements being used to increase the tensile strength, and compressive strength and alter wear properties to a great extent. As we see, the introduction of

---

* **Corresponding author Anupam Thakur:** Department of Mechanical Engineering, MAIT, Maharaja Agrasen University, H.P., India; E-mail: anupamthakur@hotmail.com

Virat Khanna, Prianka Sharma & Santosh Kumar (Eds.)

an alloying element leads to a composite, similarly the formation of a metal composite sometimes also called MMC in the case of a metal alloy. Thus an MMC is created when another alloying element is added to a base metal to positively alter its properties. This alteration makes it useful in various fields like aerospace, marine and automotive areas (Fig. **1**). The term 'composite' is usually referred to describe a material in which distinct phases can be seen easily under a microscope, rather than at the atomic level. These distinct phases contribute to the formulation of a well-tailored material in which the properties of a composite differ from those of the individual element constituting it. In other words, the strength of one element rectifies the weakness of the alloying element. It is observed that the combination of these materials should be carefully calculated to exhibit distinctive mechanical and thermal properties. In recent years, the use of bio-reinforced composites has also gained some interest in research. The utilization of bio-waste material such as groundnut shell ash, fly ash, sugarcane bagasse ash, *etc.* has been explored which could be also be an additional reinforcement beyond the elemental reinforcement (Fig. **2**). This approach leads to the reuse of industrial waste and concurrently reduces the cost of the composite production.

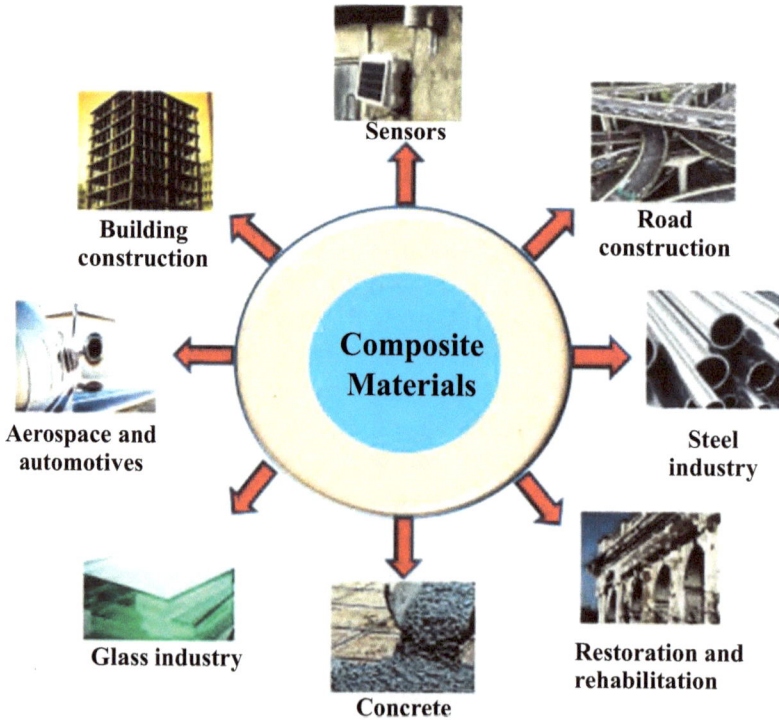

Fig (1). Use of composite materials in various fields.

It is observed that various types of reinforcements are used in the development of the MMCs. These can be present in the form of fibers referred to as fiber-reinforced composites, laminar sheets, known as laminar composites or in particle form called particulate composites. The MMCs thus developed in the various categories have shown various positive results. The density of the material reduces by 20-40%. Despite achieving lower density, higher strength is obtained which can be up to four times the material strength to density ratio. Higher directional mechanical property is observed, particularly in terms of the tensile properties of the material. Thus, the developed MMCs have greater strength than the pure steel or aluminium with greater fatigue endurance. The toughness of the material is also not compromised; in fact, it surpasses that of ceramics and glasses. Machining these MMCs is also easy with the various methods, making them applicable in various fields. The corrosive properties of these composites are found to be superior as compared to the regular metals.

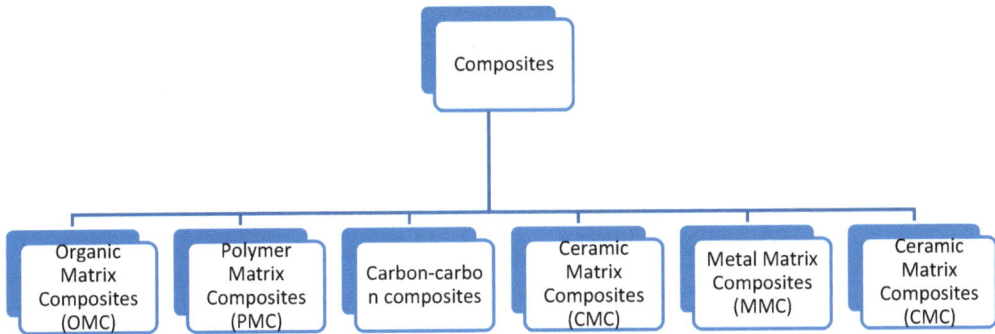

Fig (2). Classification of composites.

Metal matrix composites combine the properties of two different metals. Aluminium matrix composites (AMCs) are also types of composites that find their application with various available alloying elements available, usually ceramic materials (TiC, SiC, B_4C, Gr, Al_2O_3, WC, TiO_2 *etc.*). The process of making these AMCs can be executed with the help of powder metallurgy or liquid metallurgy. Additionally, due to the lightweight nature of aluminium, it is used in various moving parts such as automobile components like engine heads, pistons, piston rings, shafts, drives *etc.* Considering this, the addition of a self-lubricating material needs to be studied to overcome the wear properties of aluminium, and materials like graphite can be considered. Due to the above-mentioned points, research on composite appears very promising, leading to the discovery of better AMCs having much superior properties.

The ratio of reinforcement being introduced in the metal matrix is in terms of volume % or weight %, varying from 1-30% for tribological applications as compared to the metal. In the various composites, the addition of the reinforcement could be in the form of single/mono reinforcement or hybrid reinforcement. If single reinforcement is used, then properties such as tensile strength, hardness, wear resistance, toughness, *etc.* change, which depends on the type of reinforcement used for composite formation. A small amount of added reinforcement changes the properties of the composite compared to the parent metal, like hardness, toughness, brittleness, fracture toughness, fatigue strength, *etc.* Ceramic materials like TiC, SiC, and B_4C are found to increase the tensile strength of the parent aluminium and increase the hardness. If we use the reinforcement having high hardness, then the MMC formed will also be hard. While this may improve the wear resistance of the composite, it may lead to wear out the counter surface due to increased hardness and surface wear may be seen as a result. To overcome this issue, Hybrid MMCs are formed, which have two reinforcements present inside the metal matrix. Generally, in hybrid MMCs, there is more than one reinforcement, and the combination of these reinforcement addresses the limitation of the single reinforcement. The hybrid MMC thus combines the effect of both properties obtained from individual reinforcements (E.g. Si, Gr, MoS_2, etc). This results in a better composite material that can overcome the limitation of single reinforcement, especially in terms of wear resistance under varying load, speed, and temperature conditions. Thus, hybrid MMCs offer a solution to the limitation observed in single MMCs.

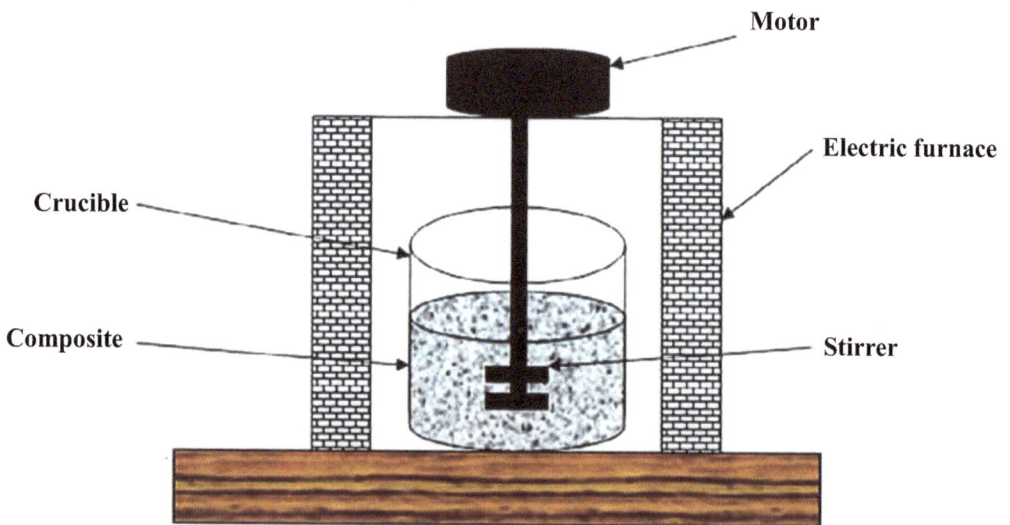

Fig (3). Stir casting setup [29].

A lot of research aiming at the synthesis and characterization of the various composites has been seen in the entire research community related to composites. Various grades of materials have been chosen, and metal matrix composites have been developed with varying percentages of composites using carbides and some naturally occurring reinforcements. A focus has been made on various aluminium composites being developed and researched with varying percentages of weight reinforcement. Depending on the type of reinforcements and the various destructive analyses performed, a large number of results have been found. A focus has been made to summarize some of these in the form of a review which are as under. Various natural reinforcements like groundnut shell powder have been studied [1] on the effect of using groundnut shell powder in different ratios with silicon in matrix metal. Mechanical properties of the same were studied stating that the lower percentage contribution of groundnut shell powder up to 4%wt. has resulted in lowering the hardness, UTS and specific strength of composite but a further increase of groundnut shell powder weight percentage increased fracture toughness. The same effect of groundnut shell ash was studied in aluminium composite, and an increase in mechanical strength and wear properties was reported [2]. The effect of agro waste particles is also seen, which also depicts good mechanical properties in terms of mechanical strength. The agro waste can be used in the form of small micro-sized particulates [3]. It was also confirmed through some research that an appreciable amount of hardness can be attained while maintaining the low density of the material compared to the parent metal of the alloy [4]. Not only naturally occurring materials can be used as reinforcements, but other metals and non-metals can also be used to achieve the same results by liquid metallurgy. Mechanical and tribological properties of the same are studied and verified by various research works [5]. Stir casting is the most common technique which can be used to form an MMC (Fig. **3**). The effect of Al_2O_3 [6] on the preparation of 6061Al-Al_2O_3 MMCs has been studied by stir casting. The evaluation of mechanical and wear properties has been conducted, reinforcement varied from 6 to 12%wt. in steps of 3%wt., the hardness of the prepared composites increased with increasing wt% of Al_2O_3 particulates with a higher tensile strength of composites, while ductility of composites was less when compared to cast 6061Al. The effect of TiO_2 (0-12%) in Al- 15% Sic% using powder metallurgy was also studied [7]. The tribological wear test concluded that the wear resistance of the hybrid composite was enhanced compared to the earlier MMC, with an increase in micro hardness as the percentage of TiO_2 in reinforcement increased. The results were verified using the XRD, SEM, and EDS test which quantifies the same. The oxide layer formation is the reason for the less contact area between the sliding surfaces thus improving the surface morphology. The effect of the various reinforcements on the Al 6061, 2124 has also been reported which depicts the result of good wear resistance attained with higher thermal resistance and hardness as compared to pure aluminium. The results of the study depicted the beneficial use of B_4C (4-12%) in Al2124, 6061 specifically for increased hardness and improved wear resistance properties [8]. The study also gives the scope for the further study of B_4C-based composites for wear study with another type of reinforcement. A study on the rice husk ash (RHA) extracted pure silica was confirmed by FESEM, and XRD tests [9]. The use of Si (2.5-3%wt.) with Gr (2.5-2%wt.) resp. in aluminium alloy has been seen, increasing the hardness of the MMC fabricated [10]. The thermal conductivity of the developed hybrid MMC has shown considerable reduction. It has been proposed that the combination of Si 3%wt. and Gr. 2%wt. could be used for pistons and other automobile parts for better mechanical properties.

Effect of Carbide reinforcements, Silicon and Graphite

Studies have been carried out on fabricating hybrid MMC of AZ91D-SiC-Gr. and conducted a wear study for it. SiC varied in two levels 3%wt. and 5%wt. and Gr in three levels 0%wt., 3%wt., and 5%%wt. Bottom pouring stir casting was used for fabricating the samples of the same. A mean wear rate of 0.0147 mm^3/min was obtained with the least wear for Si 3% and Gr 5% with AZ 91D. L_{18} orthogonal array was used for performing the tribological test, and the optimal condition for wear was found to be SiC 5%wt., Gr 5%wt., load-10N, dist 500m and speed 1m/s [11]. The increase in hardness was evident with SiC increment whereas Gr was seen to decrease the same as its contribution increased, but Gr proved to play a vital role of solid lubricant for wear. To study the comparative effect of these self-lubricating materials, a comparative study on the mechanical as well as tribological effect of Al MMC is performed. The composites comprising B_4C, Si, and Gr with aluminium matrix were formed [12]. An average mesh size of 220 was used for the reinforcements. Three samples were created, S_1- AlLM0-SiC6%+Gr4%, S_2-Al LM0-Sic10%, and S_3- Al LM0- B_4C-5%. Bending strength was found to be maximum for S_3 whereas minimum for S_2. The impact test stated that S_1 absorbed maximum energy of 56J with the least for S_3. Hardness was maximum for S_3 of 88HV, whereas the least for S1 of 27.26 HV. The effect of graphite was found to impact the wear rate inversely with the increase in sliding distance, sliding velocity, and load. The MRR of the composite increased from 60.20 to 87.53 mm^3 /min. with surface roughness varying from 3.2 to 4.1 μm with graphite addition. It is also seen that the corrosion resistance of the composite also improves with the addition of graphite.

Based on the various aspects of reinforcement that are possible and being used in the fabrication of an MMC, it has been seen that mainly fabrication is done with the stir casting process, and the reason being the simplicity of the machine, its availability in the manufacturing industry and economic process. A pie chart is shown in Fig. (**4**) [13]. A discussion on the various types of reinforcement and their effect on the matrix of parent alloy in the form of mechanical properties have also been reported.

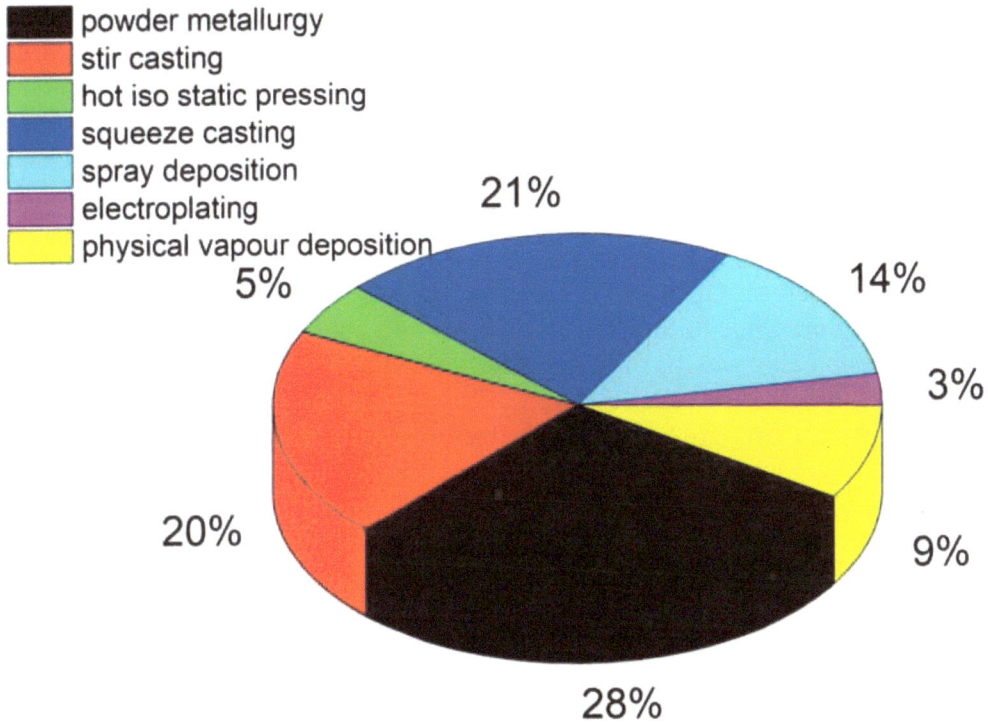

Fig (4).   Pie chart for various routes for MMC fabrication [13].

On increasing the combined percentage of SiC and Gr up to 2.5%, the hardness, and degradation increases on further weight percentage increments. Porosity was found to be the reason for this degradation. Emphasis on the Gr particles (Fig. **4**) for the wear study is seen in various research works [14]. They investigated the structure density and wear properties of the nanocomposite formed using graphite nanoplatelet with aluminium powder of 99.7% purity and 10-15 microns by powder sintering compaction process (Fig. **5**). The percentage of graphite was set as 0.1, 0.3, 0.5 and 1% wt with ethanol solvent for mixing in Al powder. The composite sample having 0.5% GNP had increased hardness and wear resistance. Out of the two methods adopted for reinforcement addition in powder aluminium, *i.e.* EDBS (sodium dodecyl benzene sulfate) and EC (ethyl cellulose), EC had better dispersion of reinforcement.

The effect of graphite on the sliding wear of the AA7075 alloy has been studied. Graphite varied between 5-20% (16-20 microns Gr particles) in the aluminium matrix. Pin-on-disc method was used for the wear test. The wear rate showed a downward trend up to 5%wt. and then started to increase as Gr % increased further. The coefficient of friction also showed the same trend. Mechanical tests revealed a gradual decrease in hardness from 117 BHN to 85 BHN, and wear loss was found to be 2.5 times less compared to base Al 7075 at 5% Gr inclusion. Tensile strength was also seen to increase in reinforcement up to 5% and then start to decrease again [15]. A similar study on different grades of Al 2024 with Gr is performed. Reinforcements used were graphite (0 to 6% wt. an increment of 2%)/ SiC (2%fixed)/ Si_3N_4 (2%fixed) and fabricated using stir casting machine. Physical properties and wear tests were performed on the fabricated samples. A significant effect of Gr % was seen on the wear test. Surface morphology confirmed the presence of the 2D carbon layer as a solid lubricant. Gr particle size was 99 microns, SiC was 10 microns and Si_3N_4 was 44 microns. Along with these, 2% Mg was added to increase the wettability among the phases. Micro-hardness, UTS, and impact strength were seen to decrease with graphite weight percentage increment. The wear rate was seen to improve a lot using Gr as reinforcement. Other fabrication processes of MMC like spray forming are also seen [16]. Using the spray forming, MMC of Al/Gr is synthesized followed by rolling at 250°C to refine the grain structure. Gr particles of 25-35 microns were used for the fabrication. 3%wt. Gr was used with an extra 0.3%wt. for the assurity of Gr in the aluminium matrix. EDS images confirmed the presence of Gr phases. The accumulation of Gr with the Si particle phase is seen. The rolling process reduced the porosity of the matrix. Tensile and ductile properties increased as the percentage of rolling decreased up to 80. The tensile test results conclude that the failure of composite shifts to a ductile nature as the rolling percent is increased [17]. The fabrication of Al (LM13, LM30)-Si alloy with Gr (3% wt. particle size 63-120 microns) is done and alloys were formed under two categories *i.e.* near the eutectic temperature of Al-Si alloy (LM 13), and another hyper eutectic alloy(LM30) using the stir casting setup. Magnesium as 1%wt. was added prior to Gr addition to increase the wettability of Gr particles followed by aged heat treatment. Wear test was performed to study the wear rate in dry as well as lubricated conditions. It was seen that the wear rate reduced for the fabricated composite and was even better for heat-treated composite. Heat-treated LM13–Gr composite had a coefficient of friction of 0.059, LM30 –Gr had a coefficient of friction of 0.071 whereas non-heat treated LM 13-Gr had 0.103 and LM30 had 0.140. UTS was found to decrease in both cases generally, but after the heat treatment, the UTS improved compared to base alloy [18]. Dry sliding wear of the Aluminium MMC with SiC/Gr was carried on A356 with SiC/Gr reinforcement. The fabrication of composite was performed using compact casting. The wear rate was first seen to increase when graphite was included up to 1%wt. and then reduced as Gr was increased up to 4%wt. SiC particles of 40 microns size and graphite particles of 200-800 microns were used. SEM images clearly show the different phases present in the matrix. Wear rate was found to be the same for Al-Si(10%)-Gr-1% and Al-Si10% alloy of 5.44 mm^3/min and 5.86 mm^3/min respectively, whereas for Al-Si(10%)-Gr(3%), the wear rate decreased to 1.37 mm^3/min [19]. The coefficient of friction, on the other hand, was seen to increase from 0.4 (for pure A356) to 0.69 (for Al-Si (10%)-Gr (3%)) due to the presence of SiC particle protrusion. Investigation of the wear and mechanical properties of Al7075-SiC/Marble dust/Gr hybrid MMC has also been done [20]. The range of reinforcement was kept as SiC (0–8%wt. interval of 2%) / marble dust (8–0%wt. interval of 2%) and graphite (3%wt. fixed). Gr particles of 97.5 microns, marble dust and SiC 20 microns size were used. SiC 6%wt. -Marble dust 2%wt. -Gr3%wt. showed the best wear properties. It has been found that with the introduction of graphite as a reinforcement, the hardness of the alloy decreases. Fabrication of a hybrid composite Al LM30-Gr was performed to study the same. The comparison of the prepared composites was done based on the fabrication technique *i.e.* gravity die-cast and vertical centrifuge casting. Gr (60-micron size) particles were used as 5% by weight. Both methods successfully incorporated Gr into the matrix; the SEM images verify the same. Tribological tests report a significant improvement in the surface morphology of weared surface. The hardness of gravity die casted MMC was reported to vary from 119-130 HB, whereas for centrifuge, it varied from 119 to 90 HB [21].

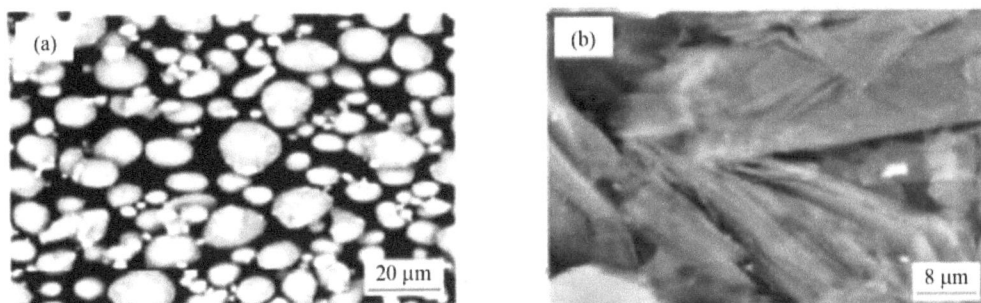

Fig (5). (a) Al powder, (b) Graphite nanoplatlets.

Effect of other Carbide Reinforcements on Wear/Mechanical Properties

Influence of factors such as material, graphite particle size, vol./wt. %, sliding speed and load have been carried out. It has been found that lower ranges of Gr %wt. had different results in the sliding test compared to the high amount of Gr %wt. Ductile properties reduce with an increase in Gr wt%. Coarse-size particles tend to squeeze and spread throughout the contact area and result in increased wear resistance of the fabricated composite. Overall, the graphite inclusion tends to increase the resistance to wear for a fabricated AMC. In a similar study, having Al (K series) as the base matrix, fabricated out the Al-Si (16%wt.)-Ni (5% wt.)-Gr (5% wt.) using stir casting setups and performed the tribological studies under different conditions. Steel specimens were used to compare the results with the fabricated composites [22, 23]. An abnormal increase in friction coefficient was seen for steel specimen as the load increased beyond 700 N, whereas the composite formed did not show any such change in the friction coefficient up to 990N under limited lubrication conditions. The effect of B_4C reinforcement (4%wt.) for the Al7075 series with porcelain weight percentage from 0 to 16% in steps of 4% is also studied [24]. Tribological tests were conducted to find out the wear rate along with mechanical strength tests for strength analysis. The results depict the decrease in density with the increase of porcelain in the Al/B_4C composite. FESEM images were also captured to study the surface morphology for the wear test conducted. Best wear results were summarized for 12% wt. porcelain inclusion. The tensile strength increase of 141 MPA to 199 MPA has been seen as the %wt. of reinforcement increased from 0 to 12%. Fabrication of Al7050- B_4C and Gr keeping the B_4C percentage as 7.5 and varying the Gr in the range of 2.5 to 10% in steps of 2.5% has been performed . K_2TiF_6 (potassium titanium fluoride) was added to the molten Al 7050 matrix to remove the porosity, having good bonding in the matrix with the reinforcements. The study aimed to investigate the effect of Gr particles on mechanical wear and corrosion analysis for various applications. It has been found that Gr reduces hardness if increased in large quantity, but its value decreases from 95 to 64HRB. The wear rate was increased due to the Gr addition without increasing the B_4C content from 60.20 mm^3/min. to 87.53 mm^3/min [9]. However, it was found that the Gr addition prevents less propagation to the formation of cracks as compared to the base alloy.

Research has been conducted on the inclusion of fly ash particulates, ranging from 0 to 12% by weight in steps of 2% in the Al 6063 composite [25]. The investigation is particularly relevant for the automotive and aerospace industries, where the utilization of such composites is eminent due to their lightweight characteristic and enhanced mechanical properties. The compact casting technique was used for the composite fabrication. Bulk density, apparent porosity measurements, Charpy test, Vickers micro hardness measurements, (FESEM), (EDS) were carried out for further evaluation. Porosity and micro hardness increase were directly related to the fly ash particles with an inverse relationship with bulk density and impact energy, demonstrating the effect of particle size increase on these properties.

A study was performed on Al6061 for B_4 C (5, 10, 20% vol.) reinforcement inclusion. Coarse AlB_2 and Al_3BC precipitates, novel structures of dispersed Al_4C_3/(Ti, Cr)B_2 layer with TiB_2/MgO precipitates were generated at the B_4C -Al interface with Ti, Mg, and Cr add-on in the Al6061 alloy. A thermodynamically stable and continuous (Ti, Cr)B_2 layer on the Al_4C_3 layer was formed [26]. The EPMA analysis of 5% B_4C (Fig. **6**) has been carried out which shows the interface formation at the Al- B_4C boundary. Due to the presence of Ti in the aluminium, the coarse Al-B_4C precipitates were not seen in the structure. Also, the Cr and Mg diffused concentrations were not seen at the interface. From the analysis, it has been found that the Al and B_4C first react to form the Al_4C_3 layer followed by the even diffusion of Ti, Cr particles in the B_4C to form the thermodynamically stable Ti, CrB_2 on the prior Al_4C_3 layer formed. Fig. (**7**) shows the SAD analysis results which state the complex phases of Tib_2 (Titanium Boride), CrB_2 (Chromium boride) and Al_4C_3. A large amount of TiB_2 is generated at the Ti-Al6063 interfacial layer alloying to B_4C. Fig. (**7c**) shows the same precipitates. The schematic diagram showing the various interface layers and the reaction is shown in Fig. (**7f**).

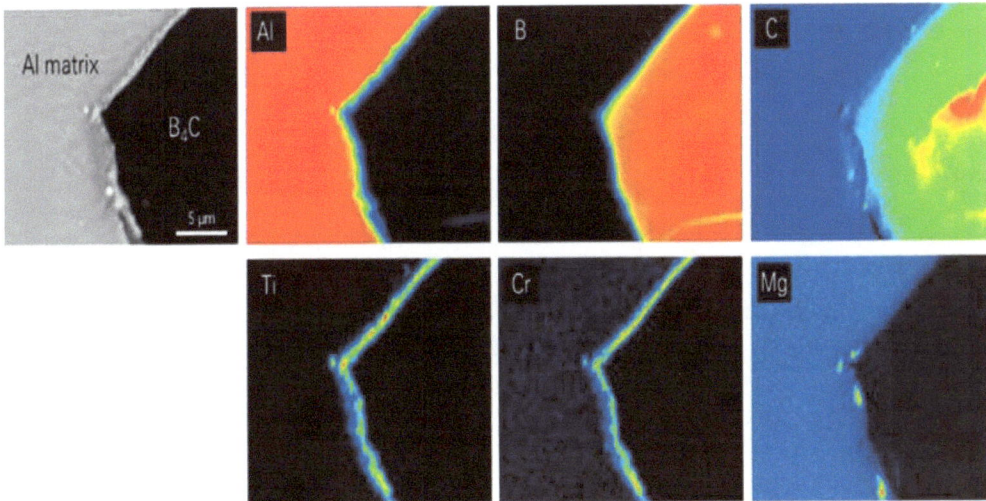

Fig. (6). EPMA analysis of AL-B_4Cinterface [26].

Experiments have been performed to study MMCs developed by mixing the Al-Mg-Si alloy with fly ash, SiO_2, Al_2O_3, Fe_2O_3, CaO, LOI(0,5,10,15,20%wt.) [27]. The various mechanical tests performed show the best result for 10% wt fly ash as compared to other concentrations used. To the same metal alloy, 5% fly ash + B_4C (2.5, 5, 7.5% wt.) was added by the stir casting process. The results show the use of 5% fly ash and 5% B_4C conc. as the best combination with 38.6% higher compression strength, 18.7% higher tensile strength, and 11.3% higher hardness compared with the unreinforced Al-Mg-Si heat-treated alloy.

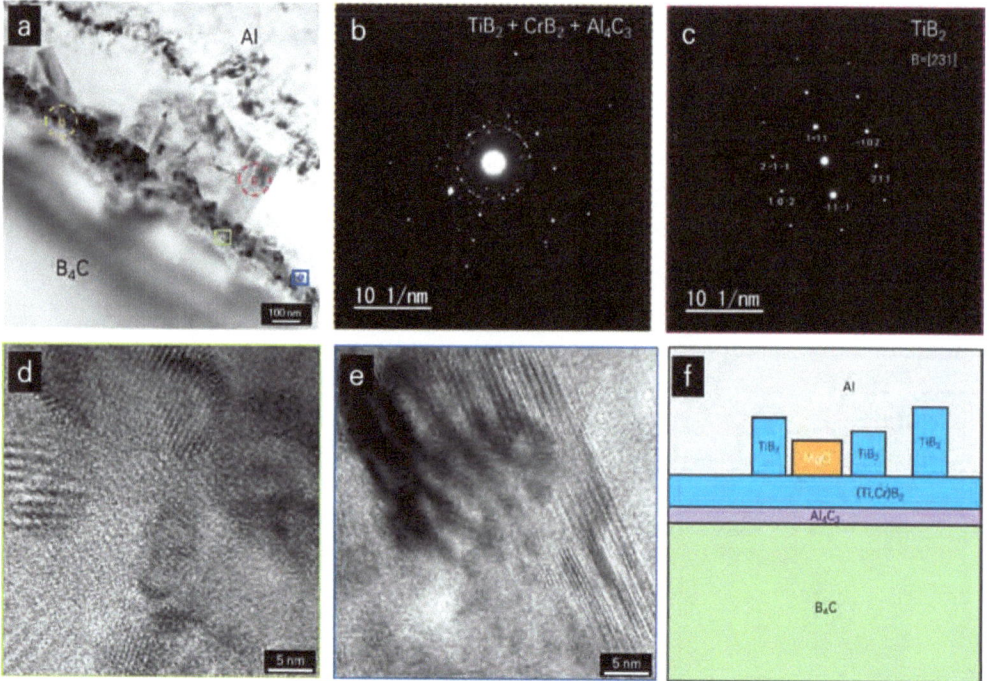

Fig. (7). Phase images at the AL-B$_4$C interface. **(a)** AL-B$_4$C TEM images **(b)** Selective area Diffraction (SAD) at the b and c target points in the coating layer. **(c,d,e)** Magnified images of Ti(Cr)B$_2$-Al and Al$_4$C$_3$/TiCrB$_2$. **(f)** Diagram showing the Al-B$_4$C interface and the various phases seen [26].

Synthesis of a hybrid composite of Al7075 with B$_4$C (3, 6, 9%) and 3% BN reinforcement using stir casting methods [28] was performed. Tensile strength increased from 221 to 280 MPA, compressive strength increased from 495 to 590 MPA, and hardness increased from 60 to 100 HRC. The corrosion rate decreased by 18.5% for 3 to 6%wt. B$_4$C reinforcement inclusion and 22.4% for 6-9%wt. inclusion of B$_4$C reinforcement change. A corrosion rate of 0.00111311 mm/year was obtained for Al7075 + 9%B$_4$C + 3%BN. Some reviews on aluminium-based composites developed through various casting techniques available with different reinforcements to improve the tribological behaviour of the MMC [29]. Some agro-based waste, borides, carbides and nitrides are conventional reinforcements for this purpose (Fig. **8**). The comparison of stir casting, investment casting, centrifugal casting, die casting, squeeze casting and spark plasma sintering has also been made. A direct relation in the W (Tungsten) and Si carbides has been related to mechanical properties increment.

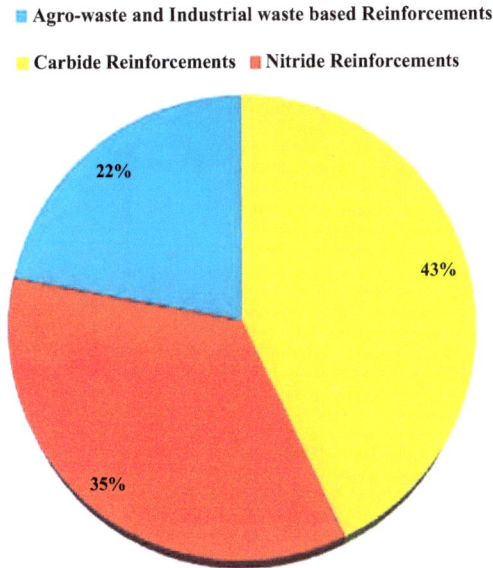

Fig. (8). Percentage of reinforcement utilization in MMC fabrication [29].

The Al6063 composite alloy matrix exhibited an augmentation in hardness upon the incorporation of the particle reinforcement. B_4C was used as 5%wt. for the composite fabrication, having a mesh size of 325 [30]. The stir casting technique was employed for the fabrication of this MMC, and microscopic images reveal a well-dispersed arrangement of B_4C particulates in the aluminium matrix. Hexachloroethane tablets were added into the molten metal to mitigate porosity in the composite during solidification. The hardness of the MMC increased from 74.40 to 95.08 BHN with 10%wt. B_4C concentration. Al-Li 8090 alloy composite fabrication is performed incorporating micro B_4C particles of 30 μm in 2, 5, 10%wt. configuration by the aid of stir casting [31]. The use of K_2TiF_6 has been evident in increasing the wettability of the particles aiming to create a well-dispersed MMC of B_4C in Al8090. SEM analysis was performed to study the morphological structure of the composites fabricated. Microhardness reached a maximum of 120 HV for Al-B_4C 2%wt. (14% increase), while UTS was found a max of 154.23 MPa (12% increase) for Al-B_4C5%wt. The wear rate was lowered to a minimum of 0.00056mm^3/minute (74% decrease) for Al-B_4C 2%wt. Notably, the clustering of B_4C particles has been seen, evident through the electron microscopic images taken, at higher concentrations level of the reinforcement of 5 and 10%wt (Fig. **9**). Thus, it can be stated that, B_4C inclusion up to 4%wt. can be beneficial, preventing the agglomeration and coagulation of particles.

(a)

(b)

(c)

(d)

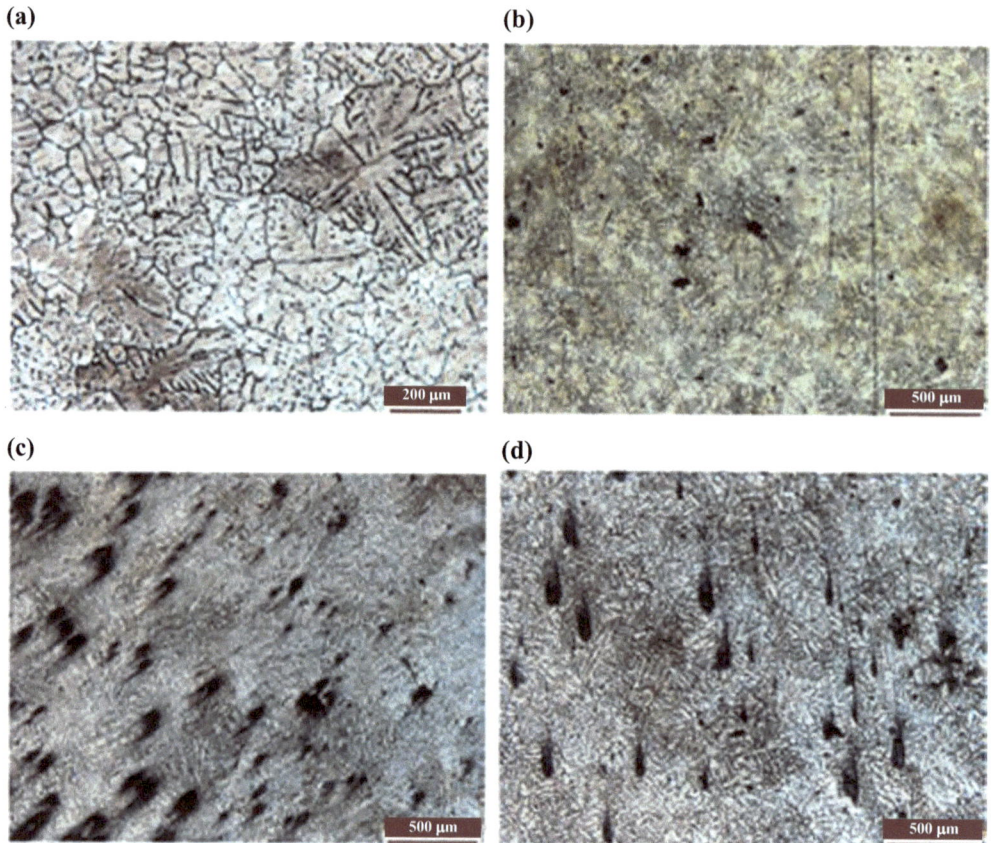

Fig. (9). Microstructure of Al 8090 with varying percent of B_4C in Alloy (a) 0%wt. B_4C (b) 2%wt. B_4C (c) 5%wt. B_4C (d) 10%wt.B_4C [31]

From Table **1**, we can summarize the various MMCs being developed from different categories of aluminium with their respective application areas. The composites fabricated when tested mechanically, show the property change in terms of hardness, yield strength and ultimate tensile strength as an effect of reinforcement inclusion in the aluminium matrix. A summary has been provided for the various monolithic and hybrid MMCs being fabricated along with their respective casting technique and mechanical properties quantification.

Table 1. Uses of various Al-based MMC with %ge of reinforcement used in different fields.

Aluminium Category	Reinforcement	Percentage	Casting Technique	Result	Application Area	Authors
Al A356	4%wt. nano Al_2O_3	4%wt.	SC	The compressive strength up to 631 MPa is obtained	Automobile sector, industries	Parizi *et al.* [32].

(Table 1) cont.....

Aluminium Category	Reinforcement	Percentage	Casting Technique	Result		Application Area	Authors
Al A356	SiC + Graphite	Fixed 10%wt., 1%wt. respectively	CC	Yield strength up to 169 MPa.	Modulus of elasticity reached up to 81.3 GPa.	Spacecraft	Aleksandar et al. [32]
Al	SiC	15%vol.	Sq.C	139 Brinell harness is obtained.	Ultimate tensile strength of 240 MPa is obtained.	Automotive sector in pistons	Onat [33]
Al A356	MicroTiB$_2$	0.5%vol.	SC	Yield strength increased upto271 MPa	Ultimate tensile strength of 309 MPa is obtained	Drum Brake lining	Akbari et al. [34]
Al 6061	Si$_3$N$_4$	10% vol.	SC	Vickers Hardness has increased to 87 VHN.	Ultimate tensile strength of 201 MPa is obtained.	Magneto parts, appliance fittings.	Ramesh et al. [35]
2009Al	CNT	1,3%wt. (2 levels)	FSP	Yield strength has improved up to 384 MPa	Ultimate tensile strength of 477 MPa is achieved.	Automobile and aerospace industry	Liu et al. [36]
Al 2024	Gr + SiC	5%wt., 20%wt. respectively	PM	Density of 2.94 g/cm^3 is obtained in MMC	Brinell Hardness is obtained of 62.98 BHN.	Aerospace industry	Ravindran et al. [37]
Al 7075	SiC+ Gr	7%wt., 3%wt. respectively	SC	Density of 2.8, 2.79 g/cm^3 resp. is obtained in MMC	Hardness was measured upto175 Hv, 219 Hv for both resp.	Transmission part in signal towers	Dhiman et al. [38]
Al 6061	AlN	20%wt.	SC	Ultimate tensile strength of 241 MPa	Mic.hardnessupto91 VHN is measured.	Electronic devices	Murugan et al. [39]
Al 6082	TiC	24%vol.	FSP	Mic.hardnessup to 150 VHN was measured	Ultimate tensile strength of 382 MPa.	Elevated temp. Areas and application	Murugan et al. [40]
Al 6061-T6	SiC + Gr	8,4%vol. respectively	FSP	Ultimate tensile strength of 218.9 MPa is obtained	Yield strength was increased up to 185 MPa.	Automobile sector	Aruri et al. [41].
Al 7009	SiC	25%wt.	DC	Vickers Hardness has increased to 138 VHN.		Ground-based military systems	Rao et al. [42].
Al A356	4%wt. nano Al$_2$O$_3$	4%wt.	SC	The compressive strength up to 631 MPa is obtained.		Automobile sector, industries	Parizi et al. [43].

Abbreviation: SC-Stir casting, Sq.C- Squeeze casting, CC- Compact casting, PM- Powder metallurgy, FSP-Friction stir processing, DC-Die casting.

SiC is one of the mostly used ceramic reinforcements, known for excellent wear resistance as well as corrosion with lower thermal conductivity at the same time.

The hardness of this ceramic based composite is of higher order and has good wear resistance. It can be seen from Table **2** that SiC has already been studied thoroughly for the same and shows an increase in hardness, in the range of 100-150 HV, the wear resistance is also improved of the MMC formed in the order of 0.1-0.2mm^3/min using this.

From Table **2** and Table **3**, it can be seen that there is a considerable change in the mechanical properties of the composite in terms of tensile strength compared to the base alloy. The hardness is seen to increase with the addition of B$_4$C and this leads to better wear resistance as well. Also, on the other hand, it is seen that the addition of solid lubricants such as Gr and Si has improved the wear resistance and friction coefficient of the MMC [76 - 80]. Thus it can be summarized that carbide reinforcement like SiC and B$_4$C enhances the hardness of the composite, whereas to mitigate the wear, the use of Gr can be made for a hybrid composite fabrication and impart a self-lubricating property to the contact surface.

Table 2. Other various compositions categorized according to the reinforcements and mechanical results.

Aluminium Grade	Reinforcement	Result /Mechanical Properties	Authors.
Al 2024	B$_4$C 4% wt.	Tensile strength and hardness improved.	Rebba *et al.* [44].
AA 2024	B$_4$C/Gr	Wear losses reduced for AA2024/B$_4$C MMC compared to pure AA2024.	Vijayaraghavan *et al.* [45].
Al2024	10%wt. SiC+5%wt. Gr	Hardness value increased as compared to the base metal.	Louis *et al* . [46]
Al2024	Sic + Gr	Tensile strength + yield strength increased.	Fenoel *et al.* [47].
Al6061	Nano Al$_2$O$_3$	Composite was fabricated using ultrasonic assist. Squeeze casting to avoid particulate accumulation in the matrix.	Singh *et al* . [48]
AA7075	Sic+ Al$_2$O$_3$ Nano powder.	Mono and hybrid composite were fabricated which show a hardness increase of 63 to 81%, respectively.	Kannan *et al* . [49]
LM25	Al$_2$O$_3$(2%wt.), B$_4$C(3%wt.).	Tensile strength was found to be 54.6 MPa with a hardness of 52.8 BHN.	Ramnath *et al* . [50]
Al6061	12 wt% Al$_2$O$_3$	Tensile strength was found to be195MPa with a hardness of 185 BHN.	Ravi *et al* . [51]
AA6061	10%wt. B$_4$C	Tensile strength was found to be145 MPa with a hardness of 68 HV.	Kumar *et al* . [52]

(Table 2) cont.....

Aluminium Grade	Reinforcement	Result /Mechanical Properties	Authors.
Al6082	10%wt. WC	Tensile strength was found to be190N/mm^2 with a hardness of 83 BHN.	Kaushik *et al* . [53]
AA6063	AA -12%wt. Al$_2$O$_3$ + 1% Gr	Tensile strength was found to increase up to 351.6 MPa.	Pardeep *et al* . [54]
AA5083	10%wt. SiC	Tensile strength was found to be 253 MPa with a hardness of 76 BHN.	Sulaiman *et al.* [55].
Al 6061	8%wt. SiC	Tensile strength was found to be 298 MPa with a hardness of %1HRB.	Maurya *et al* . [56]
A 356	5%wt. SiC	Tensile strength was found to be 309.8 MPa with a hardness of 104 BHN.	Dwivedi *et al* . [57]

Table 3. Comparative study of various Gr/B$_4$C reinforced composites.

Matrix	Reinforcement	Reinforcement %ge	Hardness	Result	Remarks	Authors
Al6061	Gr	Gr-4%wt.	51.83 BHN	112.33 MPA UTS	-	Swami *et al.* [58]
AA6061	B$_4$C	B$_4$C -12%wt.	80.8 HV	215 MPA Tensile strength	Added K$_2$TiF$_6$ salt for homogeneity and wettability of B$_4$C.	Kalaiselvan *et al.* [59].
Al6061	SiC/Gr	Sic-10/ Gr (-2,5,8%wt.)	59,55,49 HV	wear loss0.09,0.38,1.09,0.71g respectively	Gr% increase results in hardness decrease.	Guo *et al.* [60]
Al6082	Sic/Gr	Sic-5%wt. /Gr-5%wt.	49HRW	0.3964mm3/min	-	Kaushik *et al.* [61]
Pure Al	B$_4$C	B$_4$C -10%wt.	51HV	132 MPA Tensile strength	-	Mazaheri *et al.* [62].
Al6106	Gr/SiC	Gr- 9/Sic-(10,20, 30,40%wt.)	63,69,73,75 BHN	.085(Only Graphite), wear loss of 0.19,0.1,0.18,0.23mm^3 /min resp.	-	Mahdhavi *et al.* [63].
Pure AL	SiC/Gr	SiC- 5/G (5,10%wt.)	53BHN,51HN	0.0069,0.0061g reduction in weight after wear test	Wear loss reduced by 21.79% by Gr introduction in the matrix.	Mahdhavi *et al.* [64].
Pure Al	6.3Sn/Gr	Gr-(1.6,2.4,3.4,5.6,8.4%wt.)	116,109,104,99.6,96.2BHN and Tensile strength 147.3,152.3,158.4165.3,173.2 MPa	-	-	Ravindran *et al.* [65]

(Table 3) cont.....

Matrix	Reinforcement	Reinforcement %ge	Hardness	Result	Remarks	Authors
Al-Si10 Mg alloy	Gr/ Al$_2$O$_3$	Gr-3/ Al$_2$O$_3$-(3,6, 9%wt.)	126,129,133 HV and tensile strength172.1,190,201 MPa tensile strength, 0.003,0.0029,0.0027 mm3/min.	-	Tensile strength raised by 31% using Gr-3% and Al$_2$0$_3$ -9%.	Srivastava *et al.* [66]
Pure Al	B$_4$C	B$_4$C -8%wt.	54HV	155 MPA UTS	-	Radhika *et al.* [67]
Al-Si10 Mg alloy	Gr/Al$_2$O$_3$	Gr-3/ Al$_2$O$_3$-(3,6, 9%wt.)	126,129,133 HV and tensile strength172.1,190,201 MPa tensile strength, 0.003,0.0029,0.0027 mm3/min.	-	Tensile strength raised by 31% using Gr-3% and Al$_2$0$_3$ -9%.	Harichandran *et al.* [68].
Pure Al	B$_4$C	B$_4$C – 8%wt.	54HV	155 MPA UTS	-	Shabani *et al.* [69].
AA6061	B$_4$C	B$_4$C -15%wt.	80VHN	260 UTS	-	Mazahery *et al.* [70].
AA2024	B$_4$C	B$_4$C -30%wt.	120 BHN	115 UTS	-	Mazahery *et al.* [71]
AA7075	B$_4$C	B$_4$C -20%wt.	210 BHN	340 UTS	-	
Pure AL	B$_4$C	B$_4$C -10%wt.	51 HV	132 UTS	-	
Pure AL	B$_4$C	B$_4$C - 15%wt.	77BHN	210 UTS	Porosity of 1.8% found	Mazaher *et al.* [72].
AA6061	B$_4$C	B$_4$C -15%wt.	97BHN	270 UTS	-	Mazaher *et al.* [73].
Al2024	Gr	Gr-5%wt.	48 BHN	0.119g of wear loss found in sliding test as compared to 0.111 g of pure Al2024.	Gr. Decrease the hardness of the Al.	Ravindran *et al.* [74]
Al6061	Gr	Gr-4%wt.	46.8 HV	160 UTS, 0.00021mm^3/min	-	Sharma *et al.* [75]

CONCLUSION

Various composites being developed have been studied, and a concise review has been drafted, which states that MMCs reinforced with B$_4$C up to 5%wt. have increased hardness in comparison to the base aluminium. Conversely, the addition of solid lubricants, such as Graphite and Si leads to better wear resistance as compared to the base alloy. It is found that the stir casting process is the most economical and effective in developing MMC having homogenous dispersion of the reinforcement in the metal matrix. As compared to the monolithic composites, the hardness of the hybrid MMC has increased, in terms of VHN and BHN, UTS with enhanced wear resistance. The inclusion of magnesium, from 1 to 1.5%wt. is seen to be most beneficial for increasing the wettability of the reinforcement with aluminum alloys. K$_2$TiF$_6$ is also seen to have enhanced wettability of the reinforcement with the aluminum metal matrix verified through EDAX, XRD and SEM analysis. The thermal stability of the composite fabricated is also enhanced by increasing the reinforcement percentage and offers better mechanical performance of the material in terms of wear application where temperature is a crucial factor.

REFERENCES

[1] Alaneme, K.K.; Bodunrin, M.O.; Awe, A.A. Microstructure, mechanical and fracture properties of groundnut shell ash and silicon carbide dispersion strengthened aluminium matrix composites. *J. King Saud Univ. Sci. Eng. Sci.,* **2018,** *30,* 96-103.

[2] Dwivedi, P.; Maurya, M.; Maurya, K.; Srivastava, K.; Sharma, S.; Saxena, A. Utilization of groundnut shell as reinforcement in development of aluminum based composite to reduce environment pollution: A review. *Evergreen,* **2020,** *7*(1), 15-25.
[http://dx.doi.org/10.5109/2740937]

[3] Olufunmilayo, O.J.; Kunle, O.B. Agricultural waste as a reinforcement particulate for aluminum metal matrix composite (AMMCs): A review. *Fibers,* **2019,** *7,* 33.
[http://dx.doi.org/10.3390/fib7040033]

[4] Lakshumu, N.A.; Sudarshan, B.; Hari, K.K. Study on mechanical behavior of groundnut shell fiber reinforced polymer metal matrix composities. *Int. J. Eng. Res. Technol.,* **2013,** *2*(2)

[5] Mohanavela, V.; Suresh, K.S.; Sathisha, T.; Adithiyaac, T.; Mariyappand, K. Microstructure and mechanical properties of hard ceramic particulate reinforced AA7075 alloy composites. *Mater. Today Proc.,* **2018,** *5,* 26860-26865.
[http://dx.doi.org/10.1016/j.matpr.2018.08.168]

[6] Bharath, V. Preparation of 6061Al-Al2O3 MMC's by stir casting and evaluation of mechanical and wear properties. *3rd International Conference on Materials processing and Characterization (ICMPC 2014), Procedia Materials Science,* **2014,** *6,* pp. 1658-1667.

[7] Kumar, C.A.V.; Rajadurai, J.S. Influence of rutile (TiO2) content on wear and microhardness characteristics of aluminium-based hybrid composites synthesized by powder metallurgy. *Trans. Nonferrous Met. Soc. China,* **2016,** *26*(1), 63-73.
[http://dx.doi.org/10.1016/S1003-6326(16)64089-X]

[8] Akhileshwar, N.; Soren, S.; Navneet, K.; Kaushal, D.R. A comprehensive review on mechanical properties of Al-B4C stir casting fabricated composite. *Mater. Today,* **2019,** *21*(3), 1432-1435.
[http://dx.doi.org/10.1016/j.matpr.2019.09.172]

[9] Ranjith, R.; Giridharan, P. K.; Velmurugan, C.; Chinnusamy, C. Formation of lubricated tribo layer, grain boundary precipitates, and white spots on titanium-coated graphite–reinforced hybrid composites. *J. Aust. Ceram. Soc.,* **2018,** *55,* 645-655.

[10] Saini, S.; Gupta, A.; Mehta, A.J.; Pramanik, S. Rice huskextracted silica reinforced graphite/aluminium matrixhybrid composite. *J. Therm. Anal. Calorim.,* **2020.**
[http://dx.doi.org/10.1007/s10973-020-10404-8]

[11] Sandeep, K.K.; Rajeev, Verma; Sandeep, S.K.; Archana, T.; Thakur, R.S. Optimization and effect of reinforcements on the sliding wearbehavior of self-lubricating AZ91D-SiC-Gr hybrid composite. In: *Silicon*; Springer, **2020.**
[http://dx.doi.org/10.1007/s12633-020-00523-0]

[12] Rajesh, J.H.N.; Sankaranarayanan, R.; Tharmaraj, R. A comparative study of the mechanical and tribological behavioursof different aluminium matrix–ceramic composites. *J. Braz. Soc. Mech. Sci. Eng.,* **2019,** *41,* 330.
[http://dx.doi.org/10.1007/s40430-019-1831-7]

[13] Venkateshwar, P.R.; Suresh, G.K.; Mohana, D.K.; Raghavendra, H.R. Mechanical and wear performances of aluminiumbased metal matrixcomposites: A review. *J. Bio- Tribo-Corros.,* **2020,** *6,* 83.
[http://dx.doi.org/10.1007/s40735-020-00379-2]

[14]   Baig, Z.; Mamat, O. Mazli MustaphaAsad Mumtaz, Sadaqat Ali, and Mansoor Sarfraz,Surfactant-decorated graphite nanoplatelets (GNPs) reinforced aluminumnanocomposites: sintering effects on hardness and wear. *Int. J. Miner. Metall. Mater.,* **2018**, *25*(6), 704.
        [http://dx.doi.org/10.1007/s12613-018-1618-3]

[15]   Baradeswaran, A.; Elaya, P.A. Effect of graphite on tribological and mechanical properties of AA7075 composites. *Tribol. Trans.,* **2015**, *58*(1), 1-6.
        [http://dx.doi.org/10.1080/10402004.2014.947663]

[16]   Bhaskar, S.; Kumar, M. Effect of graphite particulates on sliding wear performance of hybrid AA2024 alloy composites. *J. Mater. Eng. Perform.,* **2021**, *30*(6), 3976-3989.
        [http://dx.doi.org/10.1007/s11665-021-05677-5]

[17]   Chourasiya, S.K.; Gautam, G.; Singh, D. Performance enhancing of spray formed al/graphite alloy compositeby rolling. *Met. Mater. Int.,* **2019**.
        [http://dx.doi.org/10.1007/s12540-019-00547-1]

[18]   Das, S.; Prasad, S.V. Microstructure and wear of cast (Al-Si ALLOY)-graphitecomposites. *Wear,* **1989**, *133*, 173-187.

[19]   Vencl, A.; Vučetić, F.; Bobić, B. Tribologicalcharacterisation in dry sliding conditions of compocastedhybrid A356/SiCp/Grp composites with graphite macroparticles. *Int. J. Adv. Manuf. Technol.,* **2018**.
        [http://dx.doi.org/10.1007/s00170-018-2866-0]

[20]   Kumar, A.; Mukesh, K.; Bhavna, P. Investigations on mechanical and sliding wear performance of AA7075 - SiC/Marble Dust/Graphite Hybrid Alloy Composites UsingHybrid ENTROPY -VIKOR Method. In: *Silicon*; Springer, **2021**.
        [http://dx.doi.org/10.1007/s12633-021-00996-7]

[21]   Senthil, M.S.; Rajan, T.P.D. Characterization of graphite-reinforced lm30-aluminiu mmatrix composite processed through gravity and verticalcentrifugal casting processes. *J. Inst. Eng. India Ser. D,* **2020**.
        [http://dx.doi.org/10.1007/s40033-020-00242-1]

[22]   Emad, O.A Influences of graphite reinforcement on the tribological propertiesof self-lubricating aluminum matrix composites for greentribology, sustainability, and energy efficiency—a review. *Int. J. Adv. Manuf. Technol.,* **2015**, *83*, 325-346.
        [http://dx.doi.org/10.1007/s00170-015-7528-x]

[23]   Omrani, E. Effect of graphite particles on improving tribological propertiesAl-16Si-5Ni-5Graphite self-lubricating composite under fullyflooded and starved lubrication conditions for transportationapplications. *Int. J. Adv. Manuf. Technol.,* **2015**, *87*, 929-939.
        [http://dx.doi.org/10.1007/s00170-016-8531-6]

[24]   Aherwar, A.; Patnaik, A.; Pruncu, C.I. Effect of B4C and waste porcelain ceramic particulate reinforcements on mechanical and tribological characteristics of high strength AA7075 based hybrid composite. *J. Mater. Res. Technol.,* **2020**, *9*(5), 9882-9894.

[25]   Mohammed Razzaq, A.; Majid, D.; Ishak, M.; Basheer, U. Effect of fly ash addition on the physical and mechanical properties of AA6063 alloy reinforcement. *Metals,* **2017**, *7*(11), 477.
        [http://dx.doi.org/10.3390/met7110477]

[26]   Lee, D.; Kim, J.; Lee, S.K.; Kim, Y.; Lee, S.B.; Cho, S. Experimental and thermodynamic study on interfacial reaction of B4C–Al6061 composites fabricated by stir casting process. *J. Alloys Compd.,* **2021**, *859*, 157813.
        [http://dx.doi.org/10.1016/j.jallcom.2020.157813]

[27]   Saravana Kumar, M.; Vasumathi, M.; Rashia Begum, S. Influence of B4C and industrial waste fly ash reinforcement particles on the micro structural characteristics and mechanical behavior of aluminium (AleMgeSi-T6) hybrid metal matrix composite. *J. Mater. Res. Technol.,* **2021**, *15*, 1201-1216.

[28]   Ramadoss, N.; Pazhanivel, K.; Anbuchezhiyan, G. Synthesis of B4C and BN reinforced Al7075 hybrid composites using stir casting method. *J. Mater. Res. Technol,* **2020**, *9*(3), 6297-6304.

[29]   Bharti, C.; Singh, A.; Rahul, R.; Sharma, D.; Dwivedi, S.P. A critical review of aluminium based composite developed by various casting technique with different reinforcement particles to enhance tribo-mechanical behaviour. *Mater. Today Proc.,* **2021**, *47*, 4092-4097.
[http://dx.doi.org/10.1016/j.matpr.2021.06.366]

[30]   Naik, T.P.; Gairola, S.; Patowari, P.K. Wire electrical discharge machining of AA6063/B4C composite fabricated by stir-casting process. *Mater. Today: Proc.,* **2021**, *46*, 10845-10853.
[http://dx.doi.org/10.1016/j.matpr.2021.01.806]

[31]   Patil, C.S.; Ansari, M.I.; Selvan, R.; Thakur, D.G. Influence of micro B4C ceramic particles addition on mechanical and wear behavior of aerospace grade Al-Li alloy composites. *Sadhana,* **2021**, *46*(1), 11.
[http://dx.doi.org/10.1007/s12046-020-01543-7]

[32]   Vencl, A.; Bobic, I. Mechanical and welding properties of A6082-SiC-ZrO2 hybrid composite fabricated by stir and squeeze casting. *J. Alloys Compd.,* **2010**, *506*(2), 631-639.

[33]   Onat, A. Mechanical and dry sliding wear properties of silicon carbide particulate reinforced aluminium–copper alloy matrix composites produced by direct squeeze casting method. *J. Alloys Compd.,* **2010**, *489*(1), 119-124.
[http://dx.doi.org/10.1016/j.jallcom.2009.09.027]

[34]   Karbalaei Akbari, M.; Baharvandi, H.R.; Shirvanimoghaddam, K. Tensile and fracture behavior of nano/micro TiB2 particle reinforced casting A356 aluminum alloy composites. *Mater. Des.,* **2015**, *66*, 150-161.
[http://dx.doi.org/10.1016/j.matdes.2014.10.048]

[35]   Ramesh, C.S.; Keshavamurthy, R.; Channabasappa, B.H.; Pramod, S. Tribological behavior of thin electroplated and chemically deposited Ni-P coatings on copper substrates. *Tribol. Int.,* **2010**, *43*(3), 623-634.
[http://dx.doi.org/10.1016/j.triboint.2009.09.011]

[36]   Liu, Z.Y.; Xiao, B.L.; Wang, W.G.; Ma, Z.Y. Modelling of carbon nanotube dispersion and strengthening mechanisms in Al matrix composites prepared by high energy ball milling-powder metallurgy method. *Carbon,* **2012**, *50*(5), 1843-1852.
[http://dx.doi.org/10.1016/j.carbon.2011.12.034]

[37]   Ravindran, P.; Manisekar, K.; Rathika, P.; Narayanasamy, P. A study of porosity effect on tribological behavior of cast Al A380M and sintered Al 6061 alloys. *Mater. Des.,* **2013**, *45*, 561-570.
[http://dx.doi.org/10.1016/j.matdes.2012.09.015]

[38]   Kumar, R.; Dhiman, S. A study of sliding wear behaviors of Al-7075 alloy and Al-7075 hybrid composite by response surface methodology analysis. *Mater. Des.,* **2013**, *50*, 351-359.
[http://dx.doi.org/10.1016/j.matdes.2013.02.038]

[39]   Ashok Kumar, B.; Murugan, N. Metallurgical and mechanical characterization of stir cast AA6061-T6–AlNp composite. *Mater. Des.,* **2012**, *40*, 52-58.
[http://dx.doi.org/10.1016/j.matdes.2012.03.038]

[40]   Thangarasu, A.; Murugan, N.; Dinaharan, I.; Vijay, S.J. Synthesis and characterization of titanium carbide particulate reinforced AA6082 aluminium alloy composites *via* friction stir processing. *Arch. Civ. Mech. Eng.,* **2015**, *15*(2), 324-334.
[http://dx.doi.org/10.1016/j.acme.2014.05.010]

[41]   Aruri, D.; Adepu, K.; Adepu, K.; Bazavada, K. Wear and mechanical properties of 6061-T6 aluminum alloy surface hybrid composites [(SiC+Gr) and (SiC+Al2O3)] fabricated by friction stir processing. *J. Mater. Res. Technol.,* **2013**, *2*(4), 362-369.
[http://dx.doi.org/10.1016/j.jmrt.2013.10.004]

[42]   Rao, R.N.; Das, S.; Mondal, D.P.; Dixit, G. RETRACTED: Effect of heat treatment on the sliding wear behaviour of aluminium alloy (Al–Zn–Mg) hard particle composite. *Tribol. Int.,* **2010**, *43*(1-2), 330-339.
       [http://dx.doi.org/10.1016/j.triboint.2009.06.013]

[43]   Ezatpour, H.R.; Torabi-Parizi, M.; Sajjadi, S.A. Microstructure and mechanical properties of extruded Al/Al2O3 composites fabricated by stir-casting process. *Trans. Nonferrous Met. Soc. China,* **2013**, *23*(5), 1262-1268.
       [http://dx.doi.org/10.1016/S1003-6326(13)62591-1]

[44]   Bhargavi, R.N. Studies on mechanical properties of 2024 Al – B4c composites. *Int. J. Adv. Mater. Manuf. Charact.,* **2014**, *4*(1), 42-46.

[45]   Vijayaraghavan, K.; Arul, K.A. Analysis on aluminium metal matrix composites with boron carbide and graphite. *Int. J. Innov. Res. Sci. Eng. Technol.,* **2016**, 5.

[46]   Marlon, J.L. Fabrication, testing and analysis ofaluminium 2024 metal matrix composite. *IJRAME,* **2014**, *2*, 29.

[47]   Mayuresh, S.R.S. Microstructures and mechanical properties of 2024Al/Gr/SiC hybrid composites fabricated by vacuum hot pressing. *Trans. Nonferrous Met. Soc.,* **2016**, *26*(5), 1259-1268.

[48]   Mayuresh Singh, R.S. Development and analysis of al-matrix nano composites fabricated by ultrasonic assisted squeezecasting process. *Mater. Today Proc.,* **2015**, *2*, 3697.
       [http://dx.doi.org/10.1016/j.matpr.2015.07.146]

[49]   Kannan, C.; Ramanujam, R. Comparative study on the mechanical and microstructural characterisation of AA 7075 nano and hybrid nanocomposites produced by stir and squeeze casting. *J. Adv. Res.,* **2017**, *8*(4), 309-319.
       [http://dx.doi.org/10.1016/j.jare.2017.02.005] [PMID: 28386480]

[50]   Vijaya Ramnath, B.; Elanchezhian, C.; Jaivignesh, M.; Rajesh, S.; Parswajinan, C.; Siddique, A.G.A. Evaluation of mechanical properties of aluminium alloy–alumina–boron carbide metal matrix composites. *Mater. Des.,* **2014**, *58*, 332-338.
       [http://dx.doi.org/10.1016/j.matdes.2014.01.068]

[51]   Ravi, B.; Naik, B.B.; Prakash, J.U. Characterization of aluminum matrix composite (AA 6061/B4C) Fabricated by stir casting technique. *Mater. Today Proc.,* **2015**, *2*(4-5), 2984-2990.
       [http://dx.doi.org/10.1016/j.matpr.2015.07.282]

[52]   Saravana Kumar, A.; Sasi Kumar, P.; Sivasankaran, S. Synthesis and mechanical behavior of AA6063-X Wt. % Al2O3-1% Gr (X1/43, 6, 9 and 12 Wt.%) hybrid composites. *Proc Eng,* **2014**, *97*, 951.
       [http://dx.doi.org/10.1016/j.proeng.2014.12.371]

[53]   Kaushik, N.C.; Rao, R.N. Effect of grit size on two body abrasive wear of Al 6082 hybrid composites produced by stir casting method. *Tribol. Int.,* **2016**, *102*, 52-60.
       [http://dx.doi.org/10.1016/j.triboint.2016.05.015]

[54]   Sharma, P.; Paliwal, K.; Garg, R.K.; Sharma, S.; Khanduja, D. A study on wear behaviour of Al/6101/graphite composites. *J. Asian Ceram. Soc.,* **2017**, *5*(1), 42-48.
       [http://dx.doi.org/10.1016/j.jascer.2016.12.007]

[55]   Sulaiman, S; Marjom, Z; Ismail, MIS Effect of modifier on mechanical properties of aluminium silicon carbide (Al- SiC) composites. *Proc Eng.,* **2017**, *2017*(184), 773.

[56]   Maurya, M.; Maurya, N.K.; Bajpai, V. Effect of SiC reinforced particle parameters in the development of aluminium based metal matrix composite. *EVERGREEN Joint. J. Novel Carb. Resour. Sci. Green Asia Strat.,* **2019**, *06*, 200.

[57]   Dwivedi, S.P.; Sharma, S.; Mishra, R.K. Microstructure and mechanical properties of A356/SiC composites. *Prog. Mater. Sci.,* **2014**, *6*, 1524.

[58] Swamy, A.R.K.; Ramesha, A.; Kumar, G.B.V.; Prakash, J.N. Effect ofparticulate reinforcements on the mechanical properties ofAl6061-WC and Al6061-gr MMCs. *J. Miner. Mater. Charact. Eng.,* **2011,** *10*(12), 1141-1152.
[http://dx.doi.org/10.4236/jmmce.2011.1012087]

[59] Kalaiselvan, K.; Murugan, N.; Parameswaran, S. Production and characterization of AA6061–B4C stir cast composite. *Mater. Des.,* **2011,** *32*(7), 4004-4009.
[http://dx.doi.org/10.1016/j.matdes.2011.03.018]

[60] Guo, M.L.T.; Tsao, C.Y.A. Tribological behavior of self-lubricating aluminium/SiC/graphite hybrid composites synthesized by the semi-solid powder-densification method. *Compos. Sci. Technol.,* **2000,** *60*(1), 65-74.
[http://dx.doi.org/10.1016/S0266-3538(99)00106-2]

[61] Kaushik, N.C.; Rao, R.N. The effect of wear parameters and heat treatment on two body abrasive wear of Al–SiC–Gr hybrid composites. *Tribol. Int.,* **2016,** *96*, 184-190.
[http://dx.doi.org/10.1016/j.triboint.2015.12.045]

[62] Mazaheri, Y.; Meratian, M.; Emadi, R.; Najarian, A.R. Comparison of microstructural and mechanical properties of Al–TiC, Al–B4C and Al–TiC–B4C composites prepared by casting techniques. *Mater. Sci. Eng. A,* **2013,** *560*, 278-287.
[http://dx.doi.org/10.1016/j.msea.2012.09.068]

[63] Mahdavi, S.; Akhlaghi, F. Effect of SiC content on the processing, compaction behavior, and properties of Al6061/SiC/Gr hybrid composites. *J. Mater. Sci.,* **2011,** *46*(5), 1502-1511.
[http://dx.doi.org/10.1007/s10853-010-4954-x]

[64] Ravindran, P.; Manisekar, K.; Narayanasamy, R.; Narayanasamy, P. Tribological behaviour of powder metallurgy-processed aluminium hybrid composites with the addition of graphite solid lubricant. *Ceram. Int.,* **2013,** *39*(2), 1169-1182.
[http://dx.doi.org/10.1016/j.ceramint.2012.07.041]

[65] Srivastava, S.; Mohan, S.; Srivastava, Y.; Shukla, A.J. Study ofthe wear and friction behavior of immiscible as cast-Al-Sn/graphite composite. *IJMER,* **2012,** *2*(2), 25-42.

[66] Radhika, N.; Subramanian, R.; Venkat Prasat, S.; Anandavel, B. Dry sliding wear behaviour of aluminium/alumina/graphite hybrid metal matrix composites. *Ind. Lubr. Tribol.,* **2012,** *64*(6), 359-366.
[http://dx.doi.org/10.1108/00368791211262499]

[67] Harichandran, R.; Selvakumar, N. Effect of nano/micro B4C particles on the mechanical properties of aluminium metal matrix composites fabricated by ultrasonic cavitation-assisted solidification process. *Arch. Civ. Mech. Eng.,* **2016,** *16*(1), 147-158.
[http://dx.doi.org/10.1016/j.acme.2015.07.001]

[68] Radhika, N.; Subramanian, R.; Venkat Prasat, S.; Anandavel, B. Dry sliding wear behaviour of aluminium/alumina/graphite hybrid metal matrix composites. *Ind. Lubr. Tribol.,* **2012,** *64*(6), 359-366.
[http://dx.doi.org/10.1108/00368791211262499]

[69] K., Singh; V., Khanna; A., Rosenkranz; V., Chaudhary, Sonu; G., Singh; S., Rustagi Panorama of physico-mechanical engineering of graphene-reinforced copper composites for sustainable applications, *Materials Today Sustainability.,* **2023,** *24*, 100560.
[http://dx.doi.org/10.1016/j.mtsust.2023.100560]

[70] Shabani, M.O.; Mazahery, A. Good bonding between coated B4C particles and aluminum matrix fabricated by semisolid techniques. *Russ. J. Non-Ferrous Met.,* **2013,** *54*(2), 154-160.
[http://dx.doi.org/10.3103/S1067821213020120]

[71] Mazahery, A.; Shabani, M.O. Sol–gel coated B4C particles reinforced 2024 Al matrix composites. *Proc InstMech Eng. Part L,* **2012,** *226*(2), 159-169.

[72] Mazahery, A.; Shabani, M.O.; Salahi, E.; Rahimipour, M.R.; Tofigh, A.A.; Razavi, M. Hardness and tensile strength study on Al356–B$_4$C composites. *Mater. Sci. Technol.,* **2012,** *28*(5), 634-638.

[http://dx.doi.org/10.1179/1743284710Y.0000000010]

[73]   Mazahery, A.; Shabani, M.O. Existence of good bonding between coated B4C reinforcement and al matrix *via* semisolid techniques: Enhancement of wear resistance and mechanical properties. *Tribol. Trans.,* **2013**, *56*(3), 342-348.
[http://dx.doi.org/10.1080/10402004.2012.752552]

[74]   Ravindran, P.; Manisekar, K.; Vinoth, S.K.; Rathika, P. Investigation of microstructure and mechanical properties of aluminum hybrid nano-composites with the additions of solid lubricant. *Mater. Des.,* **2013**, *51*, 448-456.
[http://dx.doi.org/10.1016/j.matdes.2013.04.015]

[75]   Sharma, P.; Paliwal, K.; Garg, R.K.; Sharma, S.; Khanduja, D. A study on wear behaviour of Al/6101/graphite composites. *J. Asian Ceram. Soc.,* **2017**, *5*(1), 42-48.
[http://dx.doi.org/10.1016/j.jascer.2016.12.007]

[76]   Dahiya, M.; Khanna, V.; Anil Bansal, S. Effect of graphene size variation on mechanical properties of aluminium graphene nanocomposites: A modeling analysis. *Mater. Today Proc.,* **2022**, (Jul)
[http://dx.doi.org/10.1016/j.matpr.2022.07.259]

[77]   Gupta, P.; Ahamad, N.; Kumar, D.; Gupta, N.; Chaudhary, V.; Gupta, S.; Khanna, V.; Chaudhary, V. Synergetic effect of CeO_2 doping on structural and tribological behavior of $Fe-Al_2O_3$ metal matrix nanocomposites. *ECS J. Solid State Sci. Technol.,* **2022**, *11*(11), 117001.
[http://dx.doi.org/10.1149/2162-8777/ac9c92]

[78]   Dahiya, M.; Khanna, V.; Anil Bansal, S. Aluminium-graphene metal matrix nanocomposites: Modelling, analysis, and simulation approach to estimate mechanical properties. *Mater. Today Proc.,* **2022**, (Nov)
[http://dx.doi.org/10.1016/j.matpr.2022.10.181]

[79]   Singh, K.; Bansal, S.A.; Khanna, V.; Singh, S. Effects of performance measures of non-conventional joining processes on mechanical properties of metal matrix composites. *Metal Matrix Composites,* **2022**, (Aug), 135-165.
[http://dx.doi.org/10.1201/9781003194897-7]

[80]   Khanna, V.; Kumar, V.; Bansal, S.A.; Prakash, C.; Ubaidullah, M.; Shaikh, S.F.; Pramanik, A.; Basak, A.; Shankar, S. Fabrication of efficient aluminium/graphene nanosheets (Al-GNP) composite by powder metallurgy for strength applications. *J. Mater. Res. Technol.,* **2023**, *22*, 3402-3412.
[http://dx.doi.org/10.1016/j.jmrt.2022.12.161]

CHAPTER 8

Hybrid Glass Fiber Reinforced Composites: Classification, Fabrication and Applications

Rahul Mehra[1], **Satish Kumar**[1,*] and **Santosh Kumar**[1,*]

[1] *Department of Mechanical Engineering, Chandigarh Group of Colleges, Landran, Mohali, Punjab, India*

Abstract: The need to develop and use materials that are both much lighter and stronger than current materials but are also more energy-efficient has been felt due to the ongoing depletion of resources and the rising demand for component efficiency. Composites are the best available suitable materials due to their excellent ultra-light weight and outstanding strength characteristics. They have great energy absorption capacity, high stiffness, high fracture toughness, and low thermal expansionin addition to being highly strong in effect and light in weight. Today, composites are being used in an increasing number of technical fields, from the automotive to aviation.

Keywords: Applications, Composites, Fabrication Techniques, Fiber reinforced polymer, Mechanical Properties.

INTRODUCTION

Materials with high strength, stiffness, hardness, and other properties are needed to meet the constantly growing need of new advanced technologies. Composites are becoming increasingly popular because of their reduced weight, increased durability and outstanding wear resistance. Composites are resilient to a variety of environmental stresses. On a macro scale, these materials often have multiple chemically or physiologically distinct component constituents. They are divided by a distinct interface based on their respective compositions. Composites are formed up with one or more constant phases and one or more discontinuous phases [1].

Classification of Composites

As indicated in Fig. (**1**), composites can be split into two groups: those composed of natural fibres, such as wood, palm, and bamboo, and those made of artificial

---
* **Corresponding authors Satish Kumar and Santosh Kumar:** Department of Mechanical Engineering, Chandigarh Group of Colleges, Landran, Mohali, Punjab, India; E-mails: satish.4310@cgc.edu.in and santoshdgc@gmail.com

fibres, such as glass and carbon. Products made from synthetic composites can be distinctive, flexible, and of high quality, which is not possible with materials from other traditional sources. High-durability artificial composites are utilised gradually for a variety of applications in demanding operating environments. Materials with special mechanical qualities, a lower specific weight, and greater resistance to external influences are needed to meet fundamental industrial needs. Moreover [2 - 9], enhanced mechanical characteristics in composites are necessary to reach and maintain certain safety standards and cost-effectiveness.

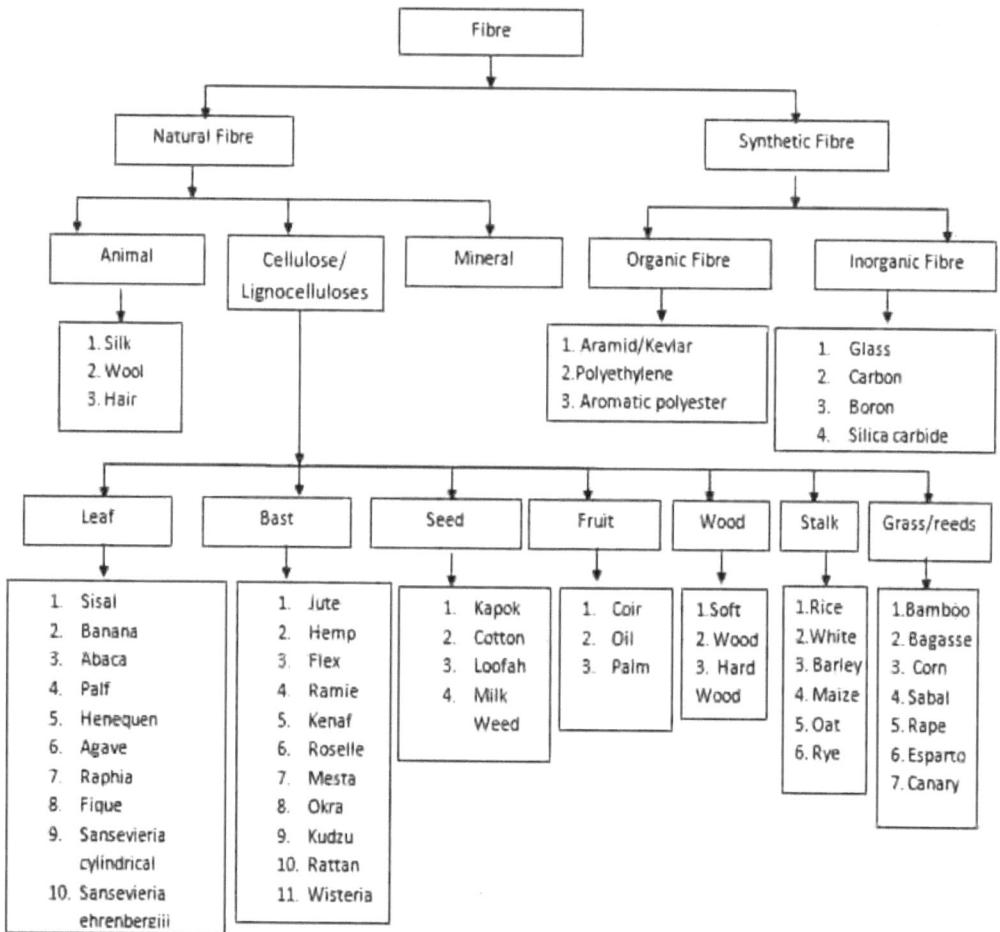

Fig. (1). Classification of various fibre materials [12].

Various synthetic composites used in polymers are either thermoplastics or thermosetting plastic polymers. In these composites, the cross-linked polymeric chain is used which ensures that the structure should be strong and rigid. Due to this rigidity, it makes the polymer almost impossible to melt or dissolve, whereas thermosetting plastics start melting when exposed to various solvents. They also own the characteristic of moulding under thermal conditions [10 - 12].

Moreover, an epoxy resin is used to connect these polymers together. Epoxy resin typically consists of linear epoxy resins and a curing agent such as poly [13 - 22] bisphenol-A diglycidylether) DGEBA and triethylenetetramine (TETA), as shown in Fig. (2), which reflects the repeated number of units, often between two and 25.

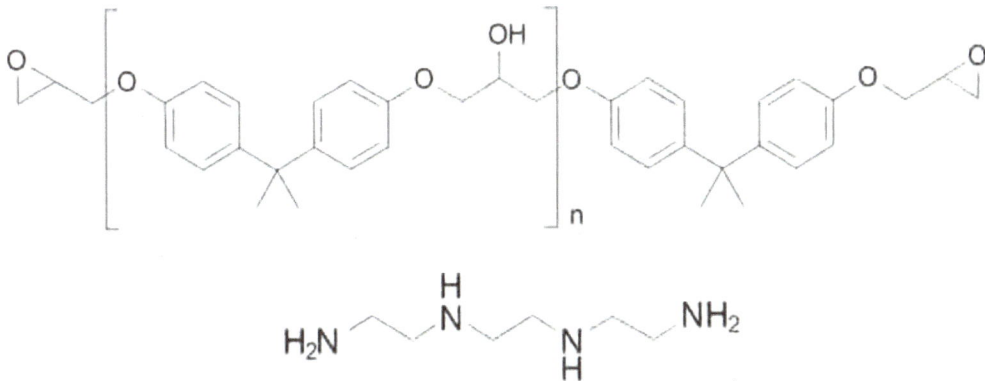

Fig. (2). Structure of DGEBA and TETA polymers with an as the repititive unit.

These groups interact with one another to cure the matrix and create a hard, stiff structural unit. A cross-linked structure made of DGEBA and TETA is shown in Fig. (3).

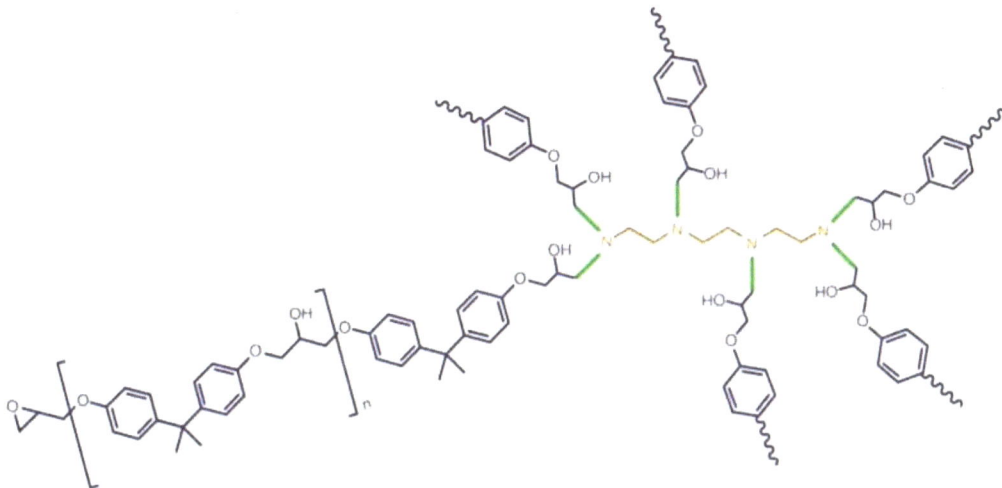

Fig. (3). An example of cross linked chemical structure of DGEBA and TETA.

The polymer ratio and epoxy group utilised during fabrication determine the physical attributes and cross-linking density. After interacting with all of the groups, the resin can be fully cured and have the maximum density; at this point, it can be seen as a single large molecule with highly advanced mechanical characteristics. The typical engineering composite fibre is separated into three categories: polymer matrix composites, MMC and CMC.

Polymer Matrix Composites (PMCs)

These matrix materials are being utilized throughout several sectors. For many industrial applications, the increased mechanical qualities are just insufficient, especially given their poor fatigue and shear resistance. These shortcomings are addressed by employing a variety of reinforcements, such as ceramics, glass, *etc.*, and combining them properly and appropriately with epoxy resin. High temperatures and pressures are not necessary for PMCs. Moreover, basic tools are needed to create PMCs. This leads to the fast development of PMCs and their easy adoption for industrial applications. PMCs have a high elastic modulus and are less fragile than other materials. Fig. (**4**) illustrates how PMCs are further separated into two main types *i.e.* FRP and PRP.

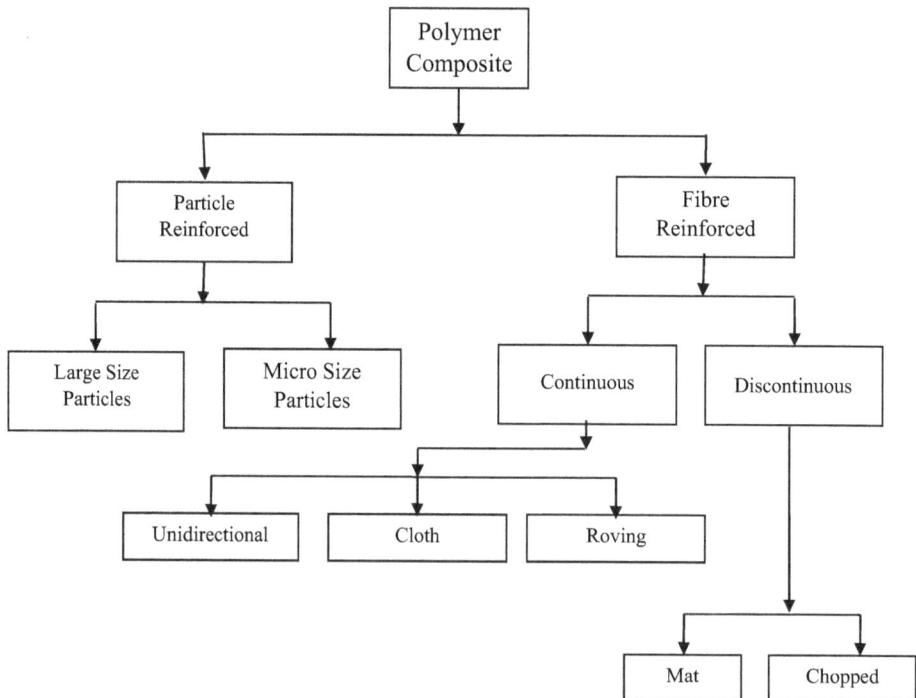

Fig. (4). Categorization of polymer compositives based on its reinforcement.

Particle Reinforced Polymer

The main source of strength in these composites comes from the majority of man-made fibres that are cemented together to offer the essential mechanical qualities. In addition, filler material is added to provide the product with unique qualities and to ease the production process, which lowers the cost. According to Fig. (5), typical long and short reinforcing materials include glass and carbon fibres as well as those constructed of silicon, titanium, beryllium, and a variety of other natural fibres.

Similar to this, common matrix materials include vinyl ester, polyester, polyurethane, epoxy, phenol resin, and epoxy. Epoxy and polyesters are among those that are employed in industrial sectors the most.

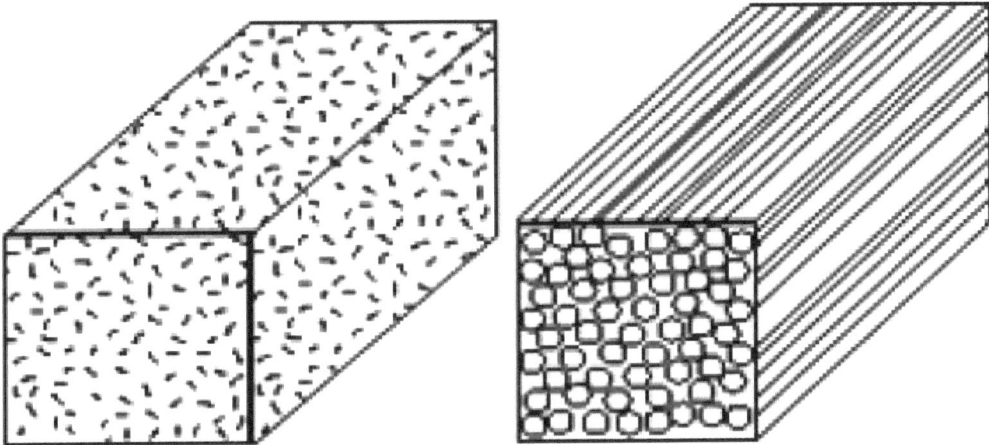

Fig. (5). Compositives with short long reinforcement fibers [22].

Fibre Reinforced Polymer

Fiber reinforced polymers are made of fibres whose length is higher than their diameter (l/d). The aspect ratio (l/d) is higher for continuous fibres than for discontinuous fibres.

Comparatively speaking, continuous fibres have a regular preferred orientation versus discontinuous fibres. The irregular fibres are dispersed at random. Fibers that are continuous and discontinuous are shown in Fig. (**6a** and **6b**). They include chopped and mat for discontinuous FRPs as well as unidirectional (UD), woven fabric and roving form with a 30-degree helical filament winding. It is possible to obtain the appropriate mechanical characteristics with a big volume of fibre by stacking several continuous fibres with varied orientations by as much as 60 to

70%. Fig. (7) shows the epoxy laminate, as well as the orientation and cross-section of the laminate.

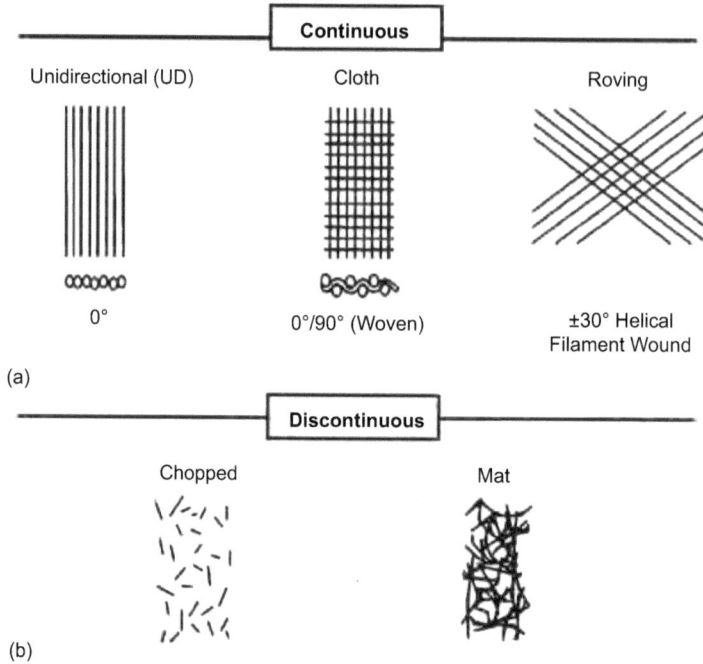

Fig. (6). (**a**) Continues fibre (**b**) disontinues fibre [23].

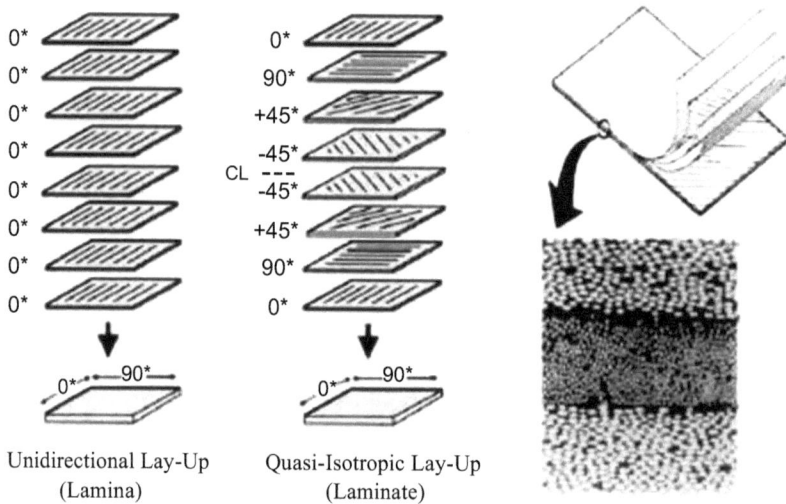

Fig. (7). Orientation and cross section of epoxy lamina [24].

When all of the plies are piled in the same orientation in a single ply, the layup is referred to as a lamina.

Moreover, it shields the FRP from environmental deterioration that causes exterior harm. The tight relationship exists between fibre and matrix enables to transfer the load from the matrix to fiber. Several loadings are depicted in Fig. (8) depending on the fibre orientation.

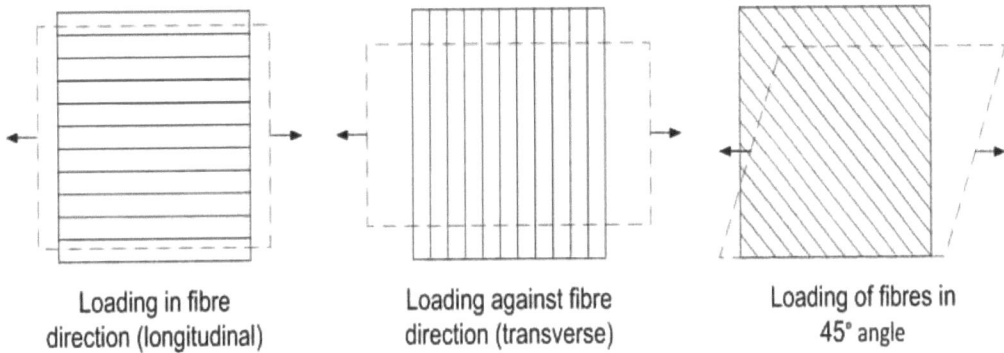

| Loading in fibre direction (longitudinal) | Loading against fibre direction (transverse) | Loading of fibres in 45° angle |

Fig. (8). Various loadings dependings on the orientation of fibre.

Due to their unique qualities of being extremely lightweight and having a high level of durability, hybrid glass fibre-reinforced plastics (GFRPs) are employed. These polymers are created using the previously mentioned epoxy resins and are made up of both long and short fibres that are incorporated in matrix material.

As required by ASTM standards, numerous writers have investigated the impact of different mechanical qualities such as tensile, compressive, and bending strengths [25 - 35]. Advanced reinforcements such as TiC, milled carbon, SiCp, and particles of graphite have been shown to increase the wear properties, microstructures, tensile, impact, and flexural strengths of composites, according to several researchers [36 - 40].

Manufacturing Methods for Hybrid Fibre Reinforced Plastics (FRPs)

This technique is utilized to make these composites, as shown in Fig. (9), including bladder moulding, compression, hand layup, stir casting, and others.

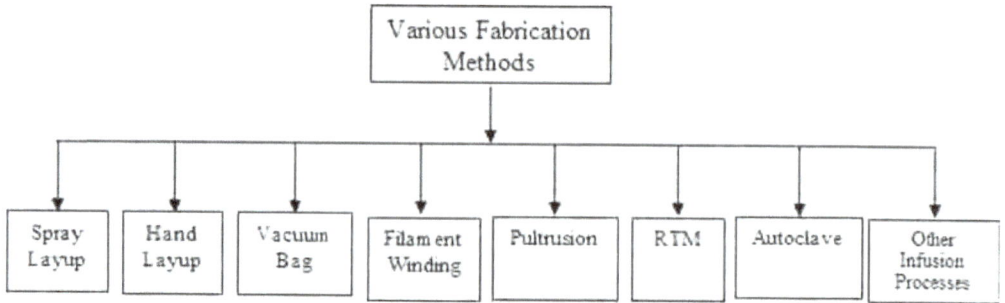

Fig. (9). Various methods of fabrication of gFRP.

Spray Lay-up Method

As shown in Fig. (**10**), the process entails spraying either separately or concurrently roving glass as the primary reinforcement and polyester as the resin on the secondary layer.

Panels that are minimally laden and enclosed constructions are made using this technique. The most frequent items made with the spray layup process are shower trays, bathtubs and fixtures, caravan bodies, *etc*.

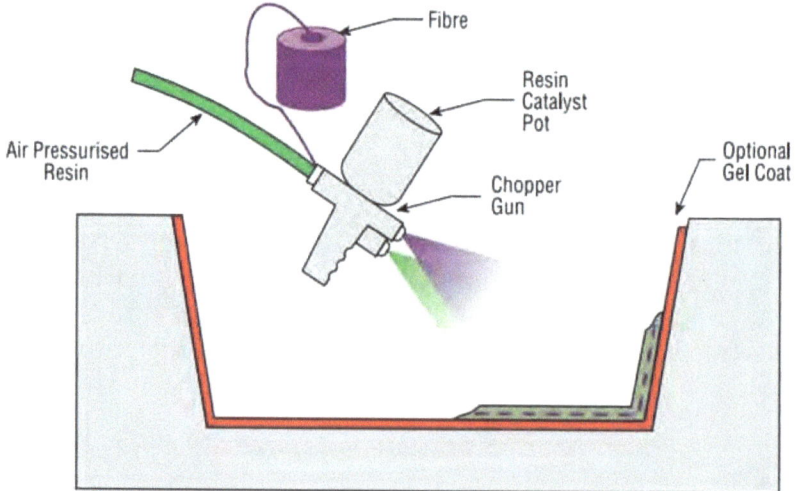

Fig. (10). Spray lay-up method [85].

Hand Lay-up

With this method, the resin is painted or sprayed over glass fibre while preserving normal atmospheric conditions. Glass fibre is manually placed in an open mould. In order to prevent the matrix from failing and the collapse of the entire composite, air and bubble formation that has become trapped among the fibre sheets must be prevented with the use of rollers. Usually, this technique is used to create the blades of wind turbines and boats. In Fig. (**11**), the hand layup method is displayed.

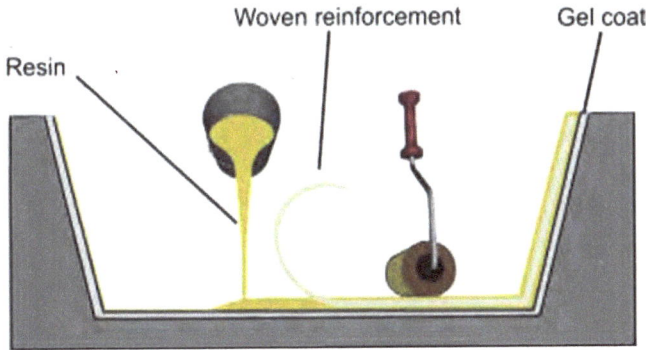

Fig. (11). Hand lay-up method [86].

Vacuum Bag Technique

The only difference between this approach and the hand layup technique is how the vacuum bag is used (Fig. **12**).

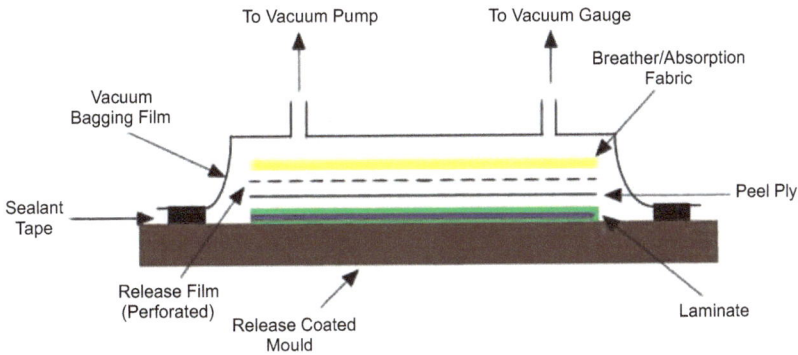

Fig. (12). Vacuum bag method [87].

Filament Winding

After going through a number of steps, the winding is completed in this phase. After passing *via* a moving carriage, fibre from the creel enters the resin bath. The method for winding filament is shown in Fig. (**13**).

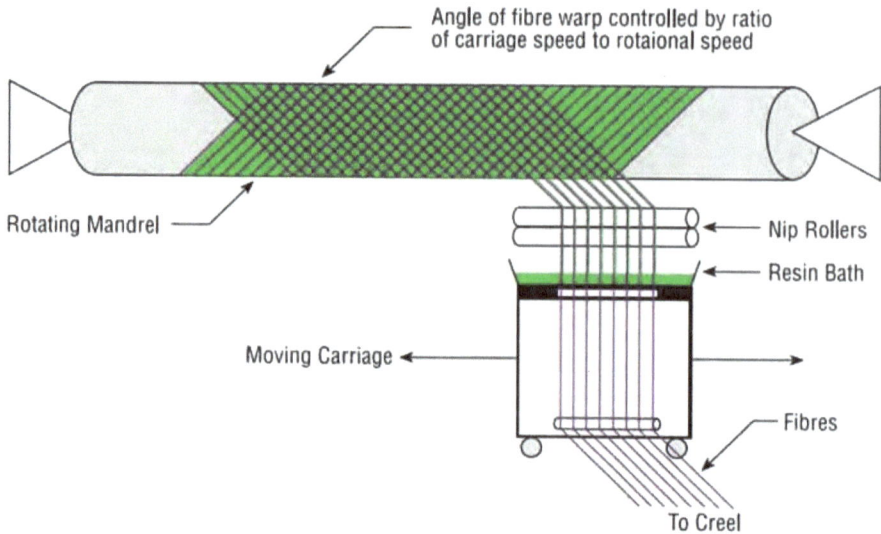

Angle of fibre warp controlled by ratio of carriage speed to rotaional speed

Rotating Mandrel

Nip Rollers

Resin Bath

Moving Carriage

Fibres

To Creel

Fig. (13). Filament winding method [84].

The rate of the feed mechanism, which regulates the ratio of the carriage speed to the rotation speed, is in charge of controlling the angular movement.

Pultrusion

In this procedure, fibre is drawn out of the fibre rack's creel and transferred with the aid of the working guides to the cloth rack. The fibre is forced through a die in a resin solution, as shown in Fig. (**14**).

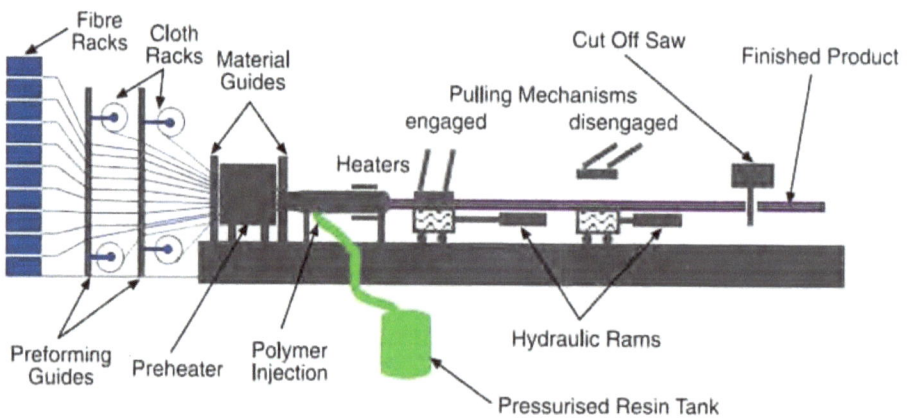

Fibre Racks
Cloth Racks
Material Guides
Cut Off Saw
Finished Product
Pulling Mechanisms
engaged
disengaged
Heaters
Preforming Guides
Preheater
Polymer Injection
Hydraulic Rams
Pressurised Resin Tank

Fig. (14). Pultrusion method [88].

Resin Transfer Molding (RTM)

As shown in Fig. (**15**), this technique clamps reinforcement in the form of dry cloth between two moulding tools. Once the fibre is thoroughly soaked, the resin is subsequently injected under pressure, temperature, vacuum, or none of these conditions.

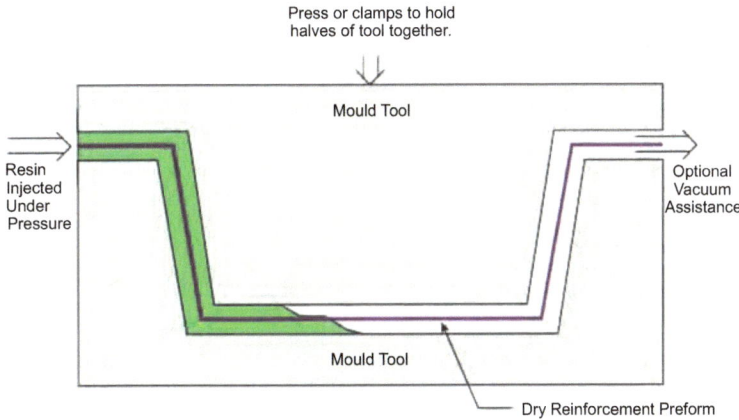

Fig. (15). Pultrusion method [88].

Autoclave (Pre-preg)

In this process, solvent or pre-catalyzed resin is impregnated prior to manufacture under pressure and heat. Aircraft and structural parts for high-performance cars are typically produced using this approach. The autoclave (Pre-preg) procedure is shown in Fig. (**16**).

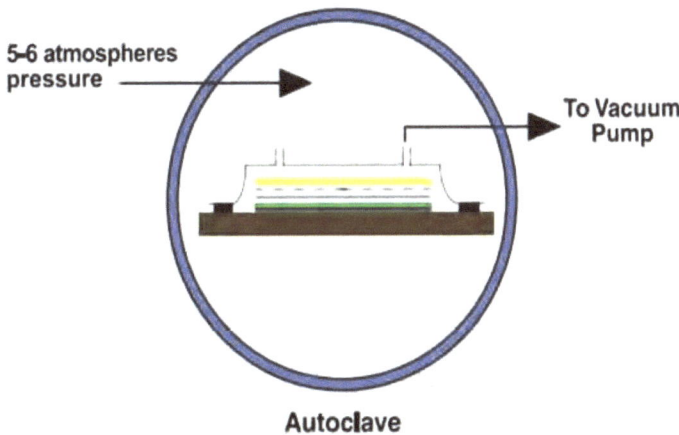

Fig. (16). Autoclave (Pre-preg) Method [90].

Other Infusion Processes (SCRIMP, RIFT, VARTM)

Vacuum-assisted resin transfer moulding (VARTM), resin infusion flexible tool (RIFT), and semen composite resin infusion manufacturing method are some examples of different infusion procedures (SCRIMP). Other infusion techniques are shown in Fig. (**17**). The hand layup process is regarded as the most appropriate and affordable way for producing hybrid GFRP out of all the ones mentioned above [41 - 55, 101 - 110].

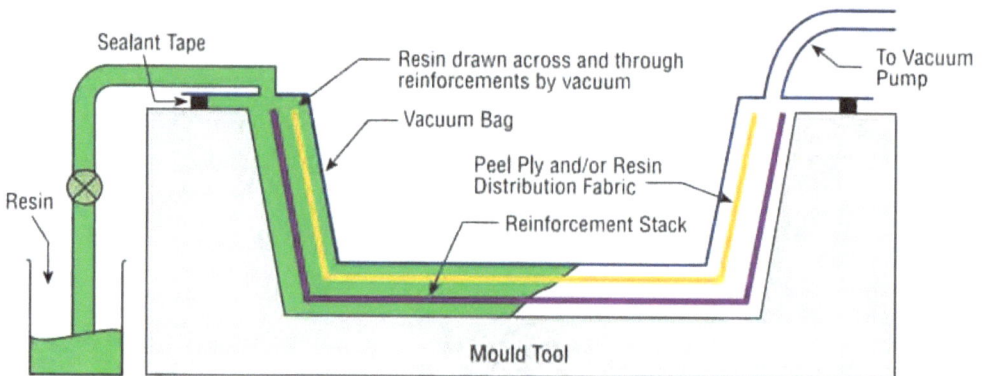

Fig. (17). Other infusion methods [90].

Applications

Several different industries, including aerospace, automotive, sports, construction, and transportation, employ composites in various ways. The utilization of composite materials in the aircraft sector is shown in Fig. (**18**).

The most often utilized applications in the aviation industry are the front nose, portions of doors, struts, *etc* [68 - 78, 123 - 132].

Sports: In the sports industry, composite materials are typically used to create items like tennis balls and rackets, bicycle components, shuttlecocks, watercraft, hockey sticks, golf balls, *etc*. Fig. (**19**) illustrates this.

Infrastructure, Building, Construction, and Industry Sectors

FRPs are utilised in construction and building because they are lightweight and need little maintenance, as shown in Fig. (**20**).

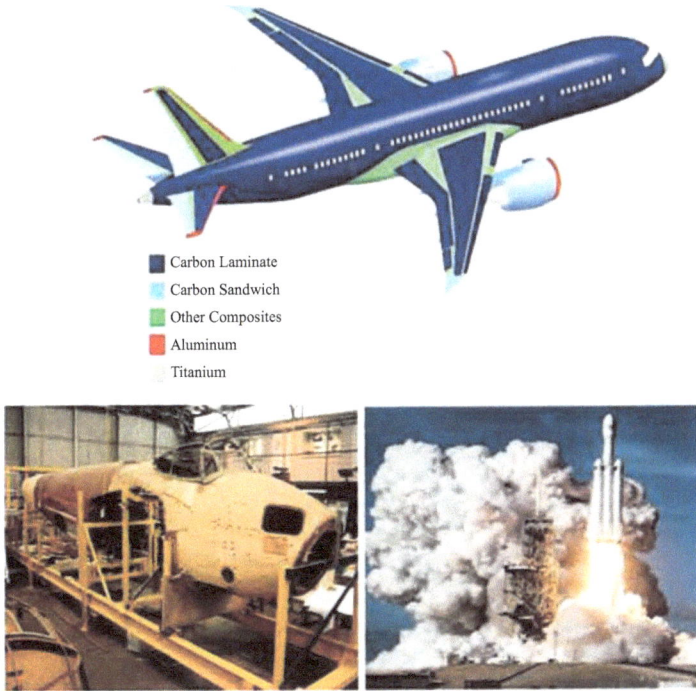

Fig. (18). Applications of Composites in Aircrafts and Aerospace [56 - 67].

Fig. (19). Applications of composites in Sports (Francesco Tornabene, 2017; Campbell FC, 2010).

Fig. (20). applicatios of composites in building and construction sector [92].

Synthetic marbles for kitchen, bathroom, and roof tiles are made by combining resins with glass fibre and fillers [79 - 89]

Marine Industry

A great example of a sector that has been entirely transformed by the usage of FRPs is the marine industry as shown in Fig. (**21**) [90-100].

Fig. (21). Applications of composites in marine sector [93].

Automobiles and Transportation

According to Fig. (**22**), FRPs are mostly employed in automobile bodywork, brake pads, shafts, fuel tanks, spoilers, *etc* [101 - 110]. These advanced materials are an obvious choice for applications in surface transportation since they are cost-effective and simple to get raw resources [111 - 125].

Moulding Applications

As shown in Fig. (**23**), FRPs are employed in the moulding of sheet metal for car bodywork. Massive, complicated objects may be created more precisely and with better strength and stiffness through moulding [126-132]. The best alternative techniques for machining difficult and difficult-to-cut materials have been considered to be non-traditional techniques [133-137].

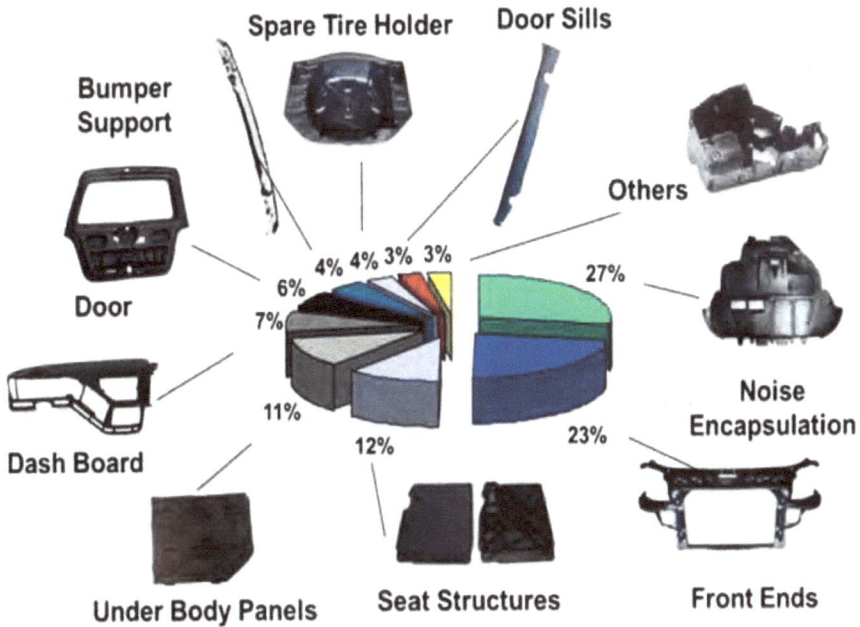

Fig. (22). Applications of composites in various automobile sector [93].

CONCLUSION

In many engineering applications, hybrid composite materials are being increasingly used because they have a variety of improved features and benefits over classic composite materials. The n (n > 2) cooperatively functioning phases make up the mechanical characteristics of hybrid composites, which are particularly significant. Because of this, it has already been mentioned that the mechanical properties of hybrid composites are described by a linear linkage of simulation models. Yet how a matrix and reinforcement react mechanically in hybrid composites also depends on the characteristics of the interface between these parts and the matrix, which must be taken into account in the numerical modelling of the mechanical properties. The effects of environmental ageing should also be taken into consideration while numerically modelling hybrid composite materials.

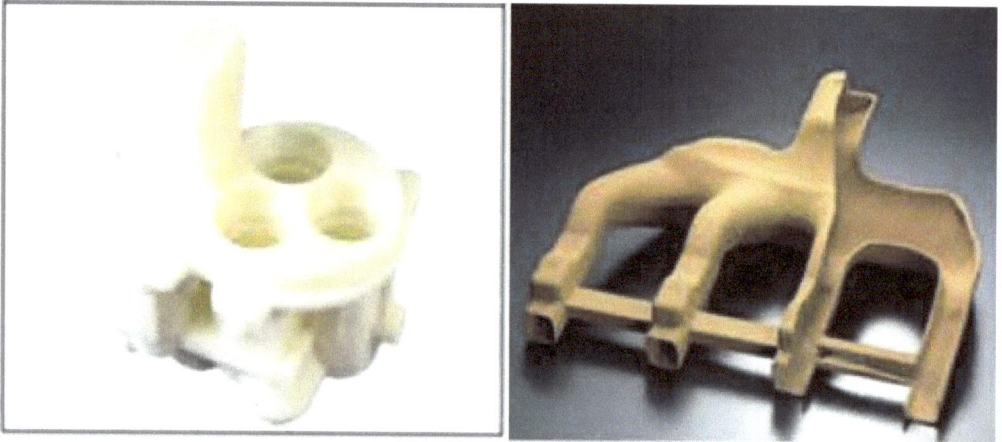

Fig. (23). Applications of composites in metals and ceramics [93].

REFERENCES

[1]   Abrate, S. *Machining of Composite Materials*; Composites Engineering Handbook: New York, **1997**, pp. 777-809.

[2]   Adalarasan, R.; Santhanakumar, M.; Rajmohan, M. Optimization of laser cutting parameters for Al6061/SiCp/Al2O3 composite using grey based response surface methodology (GRSM). *Measurement,* **2015**, *73*, 596-606.
        [http://dx.doi.org/10.1016/j.measurement.2015.06.003]

[3]   Kumar, S.; Kumar, R. Influence of processing conditions on the properties of thermal sprayed coating: A review. *Surf. Eng.,* **2021**, *37*(11), 1339-1372.
        [http://dx.doi.org/10.1080/02670844.2021.1967024]

[4]   Advani, S.; Hsiao, K.T. Manufacturing techniques for Polymer Matrix Composites (PMCs) In: *Woodhead Publishing Series in Composites Science and Engineering,* **2012**.
        [http://dx.doi.org/10.1533/9780857096258]

[5]   Singh, H.; Kumar, S.; Kumar, R.; Chohan, J.S. Impact of operating parameters on electric discharge machining of cobalt-based alloys. *Mater. Today Proc.,* **2023**.
        [http://dx.doi.org/10.1016/j.matpr.2023.01.234]

[6]   Kumar, S.; Dhingra, A.K. Multiresponse optimization of process variables of powder mixed electrical discharge machining on inconel-600 using taguchi methodology. *Int. J. Mech. Prod. Eng. Res. Development (IJMPERD),* **2015**, *60*, 6.

[7]   Kumar, S.; Dhingra, A.K. Effect of machining parameters on performance characteristics of powder

mixed EDM of Inconel-800. *Int. J. Automot. Mech.*, **2018**, *15*(2), 5221-5237.

[8]   Kumar, S.; Dhingra, A.K.; Kumar, S. Parametric optimization of powder mixed electrical discharge machining for nickel-based superalloy inconel-800 using response surface methodology. *Mech. Adv. Mater. Mod. Process.*, **2017**, *3*(1), 7.
[http://dx.doi.org/10.1186/s40759-017-0022-4]

[9]   Kumar, S.; Kumar, S. *Experimental investigation on surface characteristics of nickel-based super alloy Inconel-600 in powder mixed electric discharge machining by using response surface methodology, Advances in Engineering Materials*; Springer, **2021**, pp. 281-285.
[http://dx.doi.org/10.1007/978-981-33-6029-7_27]

[10]   Singh, Harvinder; Kumar, Manoj; Singh, Rajdeep Development of high pressure cold spray coatings of tungsten carbide composites. *Materials Science*, **2023**.
[http://dx.doi.org/10.1016/j.matpr.2022.12.210]

[11]   Singh, H. Microstructural and mechanical characterization of a cold-sprayed WC-12Co composite coating on stainless steel hydroturbine blades. *J Therm Spray Tech*, **2022**.
[http://dx.doi.org/10.1007/s11666-022-01497-8]

[12]   Singh, H.; Kumar, M.; Singh, R. An overview of various applications of cold spray coating process. *Materials Today Proceedings*, **2022**.
[http://dx.doi.org/10.1016/j.matpr.2021.10.160]

[13]   Singh, H.; Kumar, M.; Singh, R. Combating slurry and cavitation erosion of hydro turbine blades-a study. *IJAST*, **2020**, *29*, 5744-5755.

[14]   Mahajan, A.; Singh, H.; Kumar, S.; Kumar, S. Mechanical properties assessment of TIG welded SS 304 joints. *Materials Today Proceedings*, **2022**, *56*(4), 3073-3077.
[http://dx.doi.org/10.1016/j.matpr.2021.12.133]

[15]   Kumar, S.; Kumar, R.; Singh, S.; Singh, H.; Kumar, A.; Goyal, R.; Singh, S. A comprehensive study on minimum quantity lubrication. *Materials Today: Proceedings*, **2022**, *56*, 3078-3085.
[http://dx.doi.org/10.1016/j.matpr.2021.12.158]

[16]   Agrawal, R.; Wang, C.P. , **2016**. *Laser Beam Machining*;

[17]   Ahmad, I.; Baharum, A.; Abdullah, I. Effect of Extrusion Rate and Fiber Loading on Mechanical Properties of Twaron Fiber-thermoplastic Natural Rubber (TPNR) Composites. *J. Reinf. Plast. Compos.*, **2006**, *25*(9), 957-965.
[http://dx.doi.org/10.1177/0731684406065082]

[18]   Allemann, I.B.; Kaufman, J. *Laser Principles*, **2011**, *42*, 7-23.

[19]   Ashori, A.; Bahreini, Z. Evaluation of Calotropis gigantea as a Promising Raw Material for Fiber-reinforced Composite. *J. Compos. Mater.*, **2009**, *43*(11), 1297-1304.
[http://dx.doi.org/10.1177/0021998308104526]

[20]   Kumar, S.; Singh, H.; Kumar, R.; Singh Chohan, J. Parametric optimization and wear analysis of AISI D2 steel components. *Mater. Today Proc.*, **2023**.
[http://dx.doi.org/10.1016/j.matpr.2023.01.247]

[21]   Kumar, S.; Kumar, S.; Sharma, R.; Bishnoi, P.; Singh, M.; Singh, R. To evaluate the effect of boron carbide (B4C) powder mixed EDM on the machining characteristics of INCONEL-600 *Mater. Today: Proc.*, **2022**, *56*, 2794-2799.
[http://dx.doi.org/10.1016/j.matpr.2021.10.096]

[22]   Kumar, S.; Kumar, S.; Mehra, R. Parametric evaluation of PMEDM for the machining of inconel-800 using response surface methodology. *Materials Today's Proceedings*, **2023**.
[http://dx.doi.org/10.1016/j.matpr.2023.02.048]

[23]   Kumar, S.; Kumar, S.; Singh, R.; Bishnoi, P.; Chahal, V. *Analysis of PMEDM Parameters for the Machining of Inconel-800 Material Using Taguchi Methodology*; Advances in Materials and

Mechanical Engineering, **2021**, pp. 321-328.
[http://dx.doi.org/10.1007/978-981-16-0673-1_25]

[24]   Antony, *Design of experiments for engineers and scientists*; Butterworth and Heinman: USA, **2003**.

[25]   Attar, H.; Ehtemam-Haghighi, S.; Kent, D.; Dargusch, M.S.; Dargusch, M.S. Recent developments and opportunities in additive manufacturing of titanium-based matrix composites: A review. *Int. J. Mach. Tools Manuf.*, **2018**, *133*, 85-102.
[http://dx.doi.org/10.1016/j.ijmachtools.2018.06.003]

[26]   Birhan, I.; Ekici, E. Experimental investigations of damage analysis in drilling of woven glass fibre-reinforced plastic composites. *Int. J. Adv. Manuf. Technol.*, **2010**, *49*(9), 861-869.

[27]   Boldt, J.A.; Chanani, J.P. Solid-Tool Machining and Drilling *Engineered Materials handbook, ASM International. Handbook Committee*, **1987**, *1*, 667-672.

[28]   Chauhan, A.; Kumar, M.; Kumar, S. Fabrication of polymer hybrid composites for automobile leaf spring application. *Mater. Today Proc.*, **2022**, *48*(5), 1371-1377.
[http://dx.doi.org/10.1016/j.matpr.2021.09.114]

[29]   Kumar, A.; Sharma, R.; Kumar, S.; Verma, P. A review on machining performance of AISI 304 steel. *Mater. Today Proc.*, **2022**, *56*(5), 2945-2951.
[http://dx.doi.org/10.1016/j.matpr.2021.11.003]

[30]   Khanna, V.; Singh, K.; Kumar, S.; Bansal, S.A.; Channegowda, M.; Kong, I.; Khalid, M.; Chaudhary, V. Engineering electrical and thermal attributes of two-dimensional graphene reinforced copper/aluminium metal matrix composites for smart electronics. *ECS J. Solid State Sci. Technol.*, **2022**, *11*(12), 127001.
[http://dx.doi.org/10.1149/2162-8777/aca933]

[31]   Bhoopathi, R.; Ramesh, M.; Deepa, C. Fabrication and property evaluation of banana-hemp-glass fiber reinforced composites. *Procedia Eng.*, **2014**, *97*, 2032-2041.
[http://dx.doi.org/10.1016/j.proeng.2014.12.446]

[32]   Campbell, F.C. Structural Composite Materials. *ASM International*, **2010**.
[http://dx.doi.org/10.31399/asm.tb.scm.9781627083140]

[33]   Campanelli, S.L.; Casalino, G.; Contuzzi, N.; Ludovico, A.D. Taguchi optimization of the surface finish obtained by laser ablation on selective laser molten steel parts. *Procedia CIRP*, **2013**, *12*, 462-467.
[http://dx.doi.org/10.1016/j.procir.2013.09.079]

[34]   Sultan, U.; Kumar, J.; Dadra, S.; Kumar, S. Experimental investigations on the tribological behaviour of advanced aluminium metal matrix composites using grey relational analysis. *Mater. Today Proc.*, **2022**.
[http://dx.doi.org/10.1016/j.matpr.2022.12.171]

[35]   Carlos Hernández-Castañeda, J.; Kursad Sezer, H.; Li, L. The effect of moisture content in fibre laser cutting of pine wood. *Opt. Lasers Eng.*, **2011**, *49*(9-10), 1139-1152.
[http://dx.doi.org/10.1016/j.optlaseng.2011.05.008]

[36]   Cenna, A.A.; Mathew, P. Evaluation of cut quality of fibre-reinforced plastics—A review. *Int. J. Mach. Tools Manuf.*, **1997**, *37*(6), 723-736.
[http://dx.doi.org/10.1016/S0890-6955(96)00085-5]

[37]   Chan, C.L.; Mazumder, J. One-dimensional steady-state model for damage by vaporization and liquid expulsion due to laser-material interaction. *J. Appl. Phys.*, **1987**, *62*(11), 4579-4586.
[http://dx.doi.org/10.1063/1.339053]

[38]   Chandrasekharan, V.; Kapoor, S.G.; DeVor, R.E. A mechanistic approach to predicting the cutting forces in drilling: With application to fiber-reinforced composite materials. *J. Eng. Ind.*, **1995**, *117*(4), 559-570.
[http://dx.doi.org/10.1115/1.2803534]

[39]   Che, D.; Saxena, I.; Han, P.; Guo, P.; Ehmann, K.F. Machining of carbon fiber reinforced plastics/polymers: A literature review. *J. Manuf. Sci. Eng.,* **2014**, *136*(3), 034001-034022. [http://dx.doi.org/10.1115/1.4026526]

[40]   Chen, S.; Huang, J.; Xia, J.; Zhao, X.; Lin, S. Influence of processing parameters on the characteristics of stainless steel/copper laser welding. *J. Mater. Process. Technol.,* **2015**, *222*, 43-51. [http://dx.doi.org/10.1016/j.jmatprotec.2015.03.003]

[41]   Chen, Y.H.; Tam, S.C.; Chen, W.L.; Zheng, H.Y. Application of taguchi method in the optimization of laser micro-engraving of photomasks. *Int. J. Mater. Prod. Technol.,* **1996**, *11*(3-4), 333-344.

[42]   Chen, M-F.; Hsiao, W.T.; Huang, W.L.; Hu, C.W.; Chen, Y.P. Laser coding on the eggshell using pulsed-laser marking system. *J. Mater. Process. Technol.,* **2009**, *209*(2), 737-744. [http://dx.doi.org/10.1016/j.jmatprotec.2008.02.075]

[43]   Clifford, M. *Liquid Moulding Technologies*; Imperial College London, **2011**.

[44]   Dahotre, N.B.; Harimkar, S.P. *Laser fabrication and machining of materials*; Springer Science and Business Media, **2008**.

[45]   Daniel, I.M.; Ishai, O. *Engineering Mechanics of Composite Materials*; Oxford University Press, **2006**.

[46]   Davim, J.P.; Reis, P.; António, C.C. Experimental study of drilling glass fiber reinforced plastics (GFRP) manufactured by hand lay-up. *Compos. Sci. Technol.,* **2004**, *64*(2), 289-297. [http://dx.doi.org/10.1016/S0266-3538(03)00253-7]

[47]   Diaci, J.; Bračun, D.; Gorkič, A.; Možina, J. Rapid and flexible laser marking and engraving of tilted and curved surfaces. *Opt. Lasers Eng.,* **2011**, *49*(2), 195-199. [http://dx.doi.org/10.1016/j.optlaseng.2010.09.003]

[48]   Dalai, R.P.; Ray, B.C. Failure and fractography studies of FRP composites: Effects ofLoading speed and Environments. *Processing and Fabrication of Advanced Materials, ***2011**, *19*, 1-9.

[49]   Dubey, A.K.; Yadava, V. Multi-objective optimization of Nd:YAG laser cutting of nickel-based superalloy sheet using orthogonal array with principal component analysis. *Opt. Lasers Eng.,* **2008**, *46*(2), 124-132. [http://dx.doi.org/10.1016/j.optlaseng.2007.08.011]

[50]   Dumont, T.; Lippert, T.; Wokaun, A.; Leyvraz, P. Laser writing of 2D data matrices in glass. *Thin Solid Films,* **2004**, *453-454*, 42-45. [http://dx.doi.org/10.1016/j.tsf.2003.11.148]

[51]   Dutka, M.; Ditaranto, M.; Løvås, T. Application of a central composite design for the study of nox emission performance of a low NOx burner. *Energies,* **2015**, *8*(5), 3606-3627. [http://dx.doi.org/10.3390/en8053606]

[52]   Elhajjar, R.; Saponara, V.L.; Muliana, A. *Smart Composites: Mechanics and Design (Composite Materials),* 1st ed; CRC Press, **2017**.

[53]   Elkington, M.; Bloom, D.; Ward, C.; Chatzimichali, A.; Potter, K. Hand Layup: Understanding the manual process. *Advanced Manufacturing: Polymer and Composites Science,* **2015**, *1*(3), 138-151.

[54]   Enemuoh, E.U.; El-Gizawy, A.S.; Chukwujekwu Okafor, A. An approach for development of damage-free drilling of carbon fiber reinforced thermosets. *Int. J. Mach. Tools Manuf.,* **2001**, *41*(12), 1795-1814. [http://dx.doi.org/10.1016/S0890-6955(01)00035-9]

[55]   Erden, S.; Ho, K. *Fiber Technology for Fiber-Reinforced Composites*; Woodhead Publishing Series in Composites Science and Engineering, **2017**, pp. 51-79. [http://dx.doi.org/10.1016/B978-0-08-101871-2.00003-5]

[56]   Faruk, O.; Bledzki, A.K.; Fink, H.P.; Sain, M. Biocomposites reinforced with natural fibers:

2000–2010. *Prog. Polym. Sci.,* **2012**, *37*(11), 1552-1596.
[http://dx.doi.org/10.1016/j.progpolymsci.2012.04.003]

[57]   Faridnia, M.; Garbini, J.L.; Jorgensen, J.E. Machining of Graphite/epoxy materials with polycrystalline diamond tools. *J. Eng. Mater. Technol.,* **1991**, *113*(4), 430-436.

[58]   Fatimah, S.; Ishak, M.; Aqida, S.N. CO2 Laser Cutting of Glass Fibre Reinforce Polymer Composite *IOP Conference Series: Material Science Engineering,* **2012**, pp. 1-7.

[59]   Ferreira, S.L.C.; Bruns, R.E.; Ferreira, H.S.; Matos, G.D.; David, J.M.; Brandão, G.C.; da Silva, E.G.P.; Portugal, L.A.; dos Reis, P.S.; Souza, A.S.; dos Santos, W.N.L. Box-Behnken design: An alternative for the optimization of analytical methods. *Anal. Chim. Acta,* **2007**, *597*(2), 179-186.
[http://dx.doi.org/10.1016/j.aca.2007.07.011] [PMID: 17683728]

[60]   Fischer, F.; Romoli, L.; Kling, R. Laser-based repair of carbon fiber reinforced plastics. *CIRP Ann.,* **2010**, *59*(1), 203-206.
[http://dx.doi.org/10.1016/j.cirp.2010.03.075]

[61]   Fischer, F.; Kreling, S.; Dilger, K. Surface Structuring of CFRP by using Modern Excimer Laser Sources. *Phys. Procedia,* **2012**, *39*, 154-160.
[http://dx.doi.org/10.1016/j.phpro.2012.10.025]

[62]   Finger, J.; Weinand, M.; Wortmann, D. Ablation and cutting of carbon-fiber reinforced plastics using picosecond pulsed laser radiation with high average power. *J. Laser Appl.,* **2013**, *25*(4), 042007.
[http://dx.doi.org/10.2351/1.4807082]

[63]   Fisher, F.T.; Brinson, L.C. Viscoelastic interphases in polymer–matrix composites: Theoretical models and finite-element analysis. *Compos. Sci. Technol.,* **2001**, *61*(5), 731-748.
[http://dx.doi.org/10.1016/S0266-3538(01)00002-1]

[64]   Fleischer, J.; Teti, R.; Lanza, G.; Mativenga, P.; Möhring, H.C.; Caggiano, A. Composite materials parts manufacturing. *CIRP Ann.,* **2018**, *67*(2), 603-626.
[http://dx.doi.org/10.1016/j.cirp.2018.05.005]

[65]   Gaitonde, V.N.; Karnik, S.R.; Rubio, J.C.; Correia, A.E.; Abrão, A.M.; Davim, J.P. Analysis of parametric influence on delamination in high-speed drilling of carbon fiber reinforced plastic composites. *J. Mater. Process. Technol.,* **2008**, *203*(1-3), 431-438.
[http://dx.doi.org/10.1016/j.jmatprotec.2007.10.050]

[66]   Ghosal, A.; Manna, A. Response surface method based optimization of ytterbium fiber laser parameter during machining of Al/Al2O3-MMC. *Opt. Laser Technol.,* **2013**, *46*, 67-76.
[http://dx.doi.org/10.1016/j.optlastec.2012.04.030]

[67]   Gordon, S.; Hillery, M.T. A review of the cutting of composite materials. *Proc.- Inst. Mech. Eng.,* **2002**, *217*, 35-45.

[68]   Groover, M.P. *Fundamentals of Modern Manufacturing: Materials, Processes, and Systems*; Prentice Hall: New York, **1996**, pp. 543-611.

[69]   Guerrini, G.; Fortunato, A.; Melkote, S.N.; Ascari, A.; Lutey, A.H.A. Hybrid laser assisted machining: A new manufacturing technology for ceramic components. *Procedia CIRP,* **2018**, *74*, 761-764.
[http://dx.doi.org/10.1016/j.procir.2018.08.015]

[70]   Guide to Composites. Available from: http://www.gurit.com/files/documents/Gurit Guide to Composites(1).pdf (Accessed on 28th July **2019**).

[71]   Gusarov, A.V.; Grigoriev, S.N.; Volosova, M.A.; Melnik, Y.A.; Laskin, A.; Kotoban, D.V.; Okunkova, A.A. On productivity of laser additive manufacturing. *J. Mater. Process. Technol.,* **2018**, 1-59.

[72]   Hamdoun, Z.; Guillaumat, L.; Lataillade, J.L. Influence of the drilling on the fatigue behaviour of carbon epoxy laminates. *Int. J. Fatigue,* **2004**, *28*(1), 1-8.

[73]   Han, A.; Gubencu, D. Analysis of the laser marking technologies. *Non conventional. Technol. Rev.,*

2008, *4*, 18-22.

[74] Harikumar, R.; Devaraju, A. Fabrication and experimental analysis of copper wire embedded with GFRP composites. *Mater. Today Proc.,* **2018**, *5*(6), 14327-14332.
[http://dx.doi.org/10.1016/j.matpr.2018.03.015]

[75] Henning, F.; Moeller, E. *Manual lightweight construction, methods, materials and production*; HandbuchLeichtbau, Carl HanserVerlag, **2011**, p. 1289.

[76] Hirano, K.; Fabbro, R. Experimental observation of hydrodynamics of melt layer and striation generation during laser cutting of steel *Physics Procedia.,* **2011**, *12*, 555-564.
[http://dx.doi.org/10.1016/j.phpro.2011.03.070]

[77] Ho-Cheng, H.; Dharan, C.K.H. Delamination during drilling in composite laminates. *J. Eng. Ind.,* **1990**, *112*(3), 236-239.
[http://dx.doi.org/10.1115/1.2899580]

[78] El-Hofy, M.; Helmy, M.O.; Escobar-Palafox, G.; Kerrigan, K.; Scaife, R.; El-Hofy, H. Abrasive water jet machining of multidirectional CFRP laminates. *Procedia CIRP,* **2018**, *68*, 535-540.
[http://dx.doi.org/10.1016/j.procir.2017.12.109]

[79] Hu, J.; Xu, H. Pocket milling of carbon fiber-reinforced plastics using 532-nm nanosecond pulsed laser: An experimental investigation. *J. Compos. Mater.,* **2016**, *50*(20), 2861-2869.
[http://dx.doi.org/10.1177/0021998315614990]

[80] Hu, N.; Zhang, L. *Grindability of unidirectional carbon fibre-reinforced plastics*; ICCM-13: Beijing, China, **2001**.

[81] Ion, J.C. Laser processing of engineering materials. *Surf. Eng.,* **2005**, *21*(3), 347-365.

[82] Jamir, M.R.M.; Majid, M.S.A.; Khasri, A. *Natural lightweight hybrid composites for aircraft structural applications*; Woodhead Publishing Series in Composites Science and Engineering, **2018**, pp. 155-170.
[http://dx.doi.org/10.1016/B978-0-08-102131-6.00008-6]

[83] Jinyu, D.; Yu, Y.; An, X.; Shang, J.; Lei, J.; Jiang, J. 60 mm-aperture high average output power Nd: YAG composite ceramic disk laser. *Optik,* **2018**, *172*, 1-9.

[84] John, V. *Introduction to Engineering Materials*; Palgrave Macmillan, **2003**.

[85] Jones, F.R. *Glass Fibres in High-performance Fibres*; Woodhead Publishing Ltd.: Cambridge, England, **2001**, pp. 191-238.
[http://dx.doi.org/10.1533/9781855737549.191]

[86] Joshi, P.; Sharma, A. Optimization of dimensional accuracy for the Nd YAG laser cutting of aluminium alloy thin sheet using a hybrid approach. *Lasers Eng.,* **2018**, *41*(8), 263-281.

[87] Kashwani, G.A.; Al-Tamimi, A.K. Evaluation of FRP bars performance under high temperature. *Phys. Procedia,* **2014**, *55*, 296-300.
[http://dx.doi.org/10.1016/j.phpro.2014.07.043]

[88] Kasman, Ş. Impact of parameters on the process response: A Taguchi orthogonal analysis for laser engraving. *Measurement,* **2013**, *46*(8), 2577-2584.
[http://dx.doi.org/10.1016/j.measurement.2013.04.022]

[89] Keles, O.; Oner, U. A study of the laser cutting process: Influence of laser power and cutting speed on cut quality. *Lasers Eng.,* **2010**, *20*(5), 319-327.

[90] Khashaba, U.A. Notched and pin bearing strength of GFRP composite laminates. *J. Compos. Mater.,* **1996**, *30*(18), 2042-2055.
[http://dx.doi.org/10.1177/002199839603001805]

[91] Kim, G.W.; Lee, K.Y. Critical thrust force at propagation of delamination zone due to drilling of FRP/metallic strips. *Compos. Struct.,* **2005**, *69*(2), 137-141.

[http://dx.doi.org/10.1016/j.compstruct.2004.06.013]

[92]   Klocke, F.; Koenig, W.; Rummenhoeller, S.; Wuertz, C. Milling of advanced composites. In: *Machining of Ceramics and Composites*; Dekker, m., Ed.; New York, **1998**; pp. 249-266.

[93]   Klocke, F.; Wurtz, C. *Comparision of techniques for the machining of thermoplastic fibre reinforced plastics*, **1998**, *2*, 729-735.

[94]   König, W.; Wulf, C.; Graß, P.; Willerscheid, H. Machining of fibre reinforced plastics. *CIRP Ann.,* **1985**, *34*(2), 537-548.
        [http://dx.doi.org/10.1016/S0007-8506(07)60186-3]

[95]   Kornmann, X.; Rees, M.; Thomann, Y.; Necola, A.; Barbezat, M.; Thomann, R. Epoxy-layered silicate nanocomposites as matrix in glass fibre-reinforced composites. *Compos. Sci. Technol.,* **2005**, *65*(14), 2259-2268.
        [http://dx.doi.org/10.1016/j.compscitech.2005.02.006]

[96]   Koplev, A.; Lystrup, A.; Vorm, T. The cutting process, chips, and cutting forces in machining CFRP. *Composites,* **1983**, *14*(4), 371-376.
        [http://dx.doi.org/10.1016/0010-4361(83)90157-X]

[97]   Krishnamurthy, R.; Santhanakrishnan, G.; Malhotra, S.K. Machining of Polymeric Composites *Proceeding of the Machining Composite Materials Symposium,* **1992**, pp. 139-148.

[98]   Kumar, V.A.; Anil, M.P.; Rajesh, G.L.; Hiremath, V.; Auradi, V. **2018**, Tensile and compression behaviour of boron carbide reinforced 6061Al MMC's processed through conventional melt stirring. *Materials Today: Proceedings, 5*(8), 16141-16145.
        [http://dx.doi.org/10.1016/j.matpr.2018.05.100]

[99]   Kumar, R.A. Experimental investigation on mechanical behaviour and wear parameters of TiC and graphite reinforced aluminium hybrid composites. *Materials Today: Proceedings,* **2018**, *5*(6), 14244-14251.

[100]   Langella, A.; Nele, L.; Maio, A. A torque and thrust prediction model for drilling of composite materials. *Compos., Part A Appl. Sci. Manuf.,* **2005**, *36*(1), 83-93.
        [http://dx.doi.org/10.1016/S1359-835X(04)00177-0]

[101]   LATI. Thermoplastics Laser Marking. **2008**.
        Available from: https://www.lati.com/wp-content/uploads/laser_marking.pdf (Accessed on: 01/06/2019).

[102]   Lemma, E.; Chen, L.; Siores, E.; Wang, J. Study of cutting fiber-reinforced composites by using abrasive water-jet with cutting head oscillation. *Compos. Struct.,* **2002**, *57*(1-4), 297-303.
        [http://dx.doi.org/10.1016/S0263-8223(02)00097-1]

[103]   Leone, C.; Genna, S.; Tagliaferri, F.; Palumbo, B.; Dix, M. Experimental investigation on laser milling of aluminium oxide using a 30W Q-switched Yb:YAG fiber laser. *Opt. Laser Technol.,* **2016**, *76*, 127-137.
        [http://dx.doi.org/10.1016/j.optlastec.2015.08.005]

[104]   Li, L.; Lawrence, J.J.T.; Spencer, J.T. Materials Processing with a High Power Diode Laser *International Congress on Applications of Lasers and Electro-Optics (ICALEO),* **1996**, *81*(971), 38-47.
        [http://dx.doi.org/10.2351/1.5059085]

[105]   Lin, S.C.; Shen, J.M. Drilling unidirectional glass fiber-reinforced composite materials at high speed. *J. Compos. Mater.,* **1999**, *33*(9), 827-851.
        [http://dx.doi.org/10.1177/002199839903300903]

[106]   Liu, D.; Tang, Y.; Cong, W.L. A review of mechanical drilling for composite laminates. *Compos. Struct.,* **2012**, *94*(4), 1265-1279.
        [http://dx.doi.org/10.1016/j.compstruct.2011.11.024]

[107] Li, X.; Tabil, L.G.; Panigrahi, S. Chemical treatments of natural fiber for use in natural fiber-reinforced composites: A review. *J. Polym. Environ.,* **2007**, *15*(1), 25-33.
[http://dx.doi.org/10.1007/s10924-006-0042-3]

[108] Li, Z.L.; Zheng, H.Y.; Lim, G.C.; Chu, P.L.; Li, L. Study on UV laser machining quality of carbon fibre reinforced composites. *Compos., Part A Appl. Sci. Manuf.,* **2010**, *41*(10), 1403-1408.
[http://dx.doi.org/10.1016/j.compositesa.2010.05.017]

[109] Lubin, G.; Peters, S.T. *Handbook of Composites*; Springer US, **1998**.

[110] Maclean, J.O.; Hodson, J.R.; Tangkijcharoenchai, C.; Al-Ojaili, S.; Rodsavas, S.; Coomber, S.; Voisey, K.T. Laser drilling of microholes in single crystal silicon using Continuous Wave (CW) 1070 nm fibre lasers with millisecond pulse widths. *Lasers Eng.,* **2018**, *39*(1-2), 53-65.

[111] Mahamood, R.M.; Akinlabi, E.T.; Akinlabi, S. Laser power and scanning speed influence on the mechanical property of laser metal deposited titanium-alloy. *Lasers in Manufacturing and Materials Processing,* **2015**, *2*(1), 43-55.
[http://dx.doi.org/10.1007/s40516-014-0003-y]

[112] Marx, C.; Hustedt, M.; Hoja, H.; Winkelmann, T.; Rath, T. Investigations on laser marking of plants and fruits. *Biosyst. Eng.,* **2013**, *116*(4), 436-446.
[http://dx.doi.org/10.1016/j.biosystemseng.2013.10.005]

[113] Mathews, F.L.; Rawlings, R.D. *Composites materials: engineering and sciences*; Chapman &Hall, **1994**.

[114] Mathews, F.L.; Rawlings, R.D. *Composite Materials,* 1st Edition; Engineering and Science London: Woodhead Publishing, **1999**.

[115] Mazumdar, S.K. *Composites Manufacturing: Materials, Product, and Process Engineering*; CRC Press, **2002**.

[116] Meijer, J. Laser beam machining (LBM), state of the art and new opportunities. *J. Mater. Process. Technol.,* **2004**, *149*(1-3), 2-17.
[http://dx.doi.org/10.1016/j.jmatprotec.2004.02.003]

[117] Meshram, P.; Sahu, S.; Zahid Ansari, M.; Mukherjee, S. Study on mechanical properties of epoxy and nylon/epoxy composite. *Mater. Today Proc.,* **2018**, *5*(2), 5925-5932.
[http://dx.doi.org/10.1016/j.matpr.2017.12.192]

[118] Mishra, P.K. *Non-conventional machining processes,* **2005**,

[119] Mitra, B. Environment friendly composite materials: Biocomposites and green composites. *Def. Sci. J.,* **2014**, *64*(3), 244-261.
[http://dx.doi.org/10.14429/dsj.64.7323]

[120] Mohal, S.; Kumar, H. Parametric optimization of multiwalled carbon nanotube-assisted electric discharge machining of Al-10%SiC $_p$ metal matrix composite by response surface methodology. *Mater. Manuf. Process.,* **2017**, *32*(3), 263-273.
[http://dx.doi.org/10.1080/10426914.2016.1140196]

[121] Montgomery, D.C. *Design and Analysis of Experiments,* 4th ed; Wiley New York, **2002**.

[122] Montgomery, D.C. *Design and Analysis of Experiments,* 5th ed; Wiley India Private Limitid, **2010**.

[123] Müller, F.; Monaghan, J. Non-conventional machining of particle reinforced metal matrix composite. *Int. J. Mach. Tools Manuf.,* **2000**, *40*(9), 1351-1366.
[http://dx.doi.org/10.1016/S0890-6955(99)00121-2]

[124] nandaragi, S.R.; Reddy, B.; Badari Narayana, K. Fabrication, testing and evaluation of mechanical properties of woven glass fibre composite material. *Mater. Today Proc.,* **2018**, *5*(1), 2429-2434.
[http://dx.doi.org/10.1016/j.matpr.2017.11.022]

[125] Nanda, T.; Sharma, G.; Mehta, R.; Shelly, D.; Singh, K. Mechanisms for enhanced impact strength of

epoxy based nanocomposites reinforced with silicate platelets. *Mater. Res. Express,* **2019**, *6*(6), 065061.
[http://dx.doi.org/10.1088/2053-1591/ab1023]

[126] Nayak, R.K.; Dasha, A.; Ray, B.C. Effect of epoxy modifiers (Al2O3/SiO2/TiO2) on mechanical performance of epoxy/glass fibre hybrid composites. *Procedia Materials Science,* **2014**, *6*, pp. 1359-1364.

[127] Negarestani, R.; Li, L.; Sezer, H.K.; Whitehead, D.; Methven, J. Nano-second pulsed DPSS Nd:YAG laser cutting of CFRP composites with mixed reactive and inert gases. *Int. J. Adv. Manuf. Technol.,* **2010**, *49*(5-8), 553-566.
[http://dx.doi.org/10.1007/s00170-009-2431-y]

[128] Ogura, G.; Angell, J.; Wall, D. Applications test potential of laser micromachining. *Laser Focus World,* **1998**, *34*, 117-123.

[129] Okunkova, A.A.; Peretyagin, P.Y.; Podrabinnik, P.A.; Zhirnov, I.V.; Gusarov, A.V. Development of laser beam modulation assets for the process productivity improvement of selective laser melting. *Procedia IUTAM,* **2017**, *23*, 177-186.
[http://dx.doi.org/10.1016/j.piutam.2017.06.019]

[130] Parandoush, P.; Lin, D. A review on additive manufacturing of polymer-fiber composites. *Compos. Struct.,* **2017**, *182*, 36-53.
[http://dx.doi.org/10.1016/j.compstruct.2017.08.088]

[131] Deprez, P.; Melian, C.F.; Breaban, F.; Coutouly, J.F. Glass Marking with CO_2 Laser: Experimental Study of the Interaction Laser - Material. *J. Surf. Eng. Mater. Adv. Technol.,* **2012**, *2*(1), 32-39.
[http://dx.doi.org/10.4236/jsemat.2012.21006]

[132] Siddhartha, ; Patnaik, A.; Bhatt, A.D. Mechanical and dry sliding wear characterization of epoxy–TiO2 particulate filled functionally graded composites materials using Taguchi design of experiment. *Mater. Des.,* **2011**, *32*(2), 615-627.
[http://dx.doi.org/10.1016/j.matdes.2010.08.011]

[133] Dahiya, M.; Khanna, V.; Anil Bansal, S. Effect of graphene size variation on mechanical properties of aluminium graphene nanocomposites: A modeling analysis. *Mater. Today Proc.,* **2022**, (Jul)
[http://dx.doi.org/10.1016/j.matpr.2022.07.259]

[134] Gupta, P.; Ahamad, N.; Kumar, D.; Gupta, N.; Chaudhary, V.; Gupta, S.; Khanna, V.; Chaudhary, V. Synergetic effect of CeO_2 doping on structural and tribological behavior of Fe-Al_2O_3 metal matrix nanocomposites. *ECS J. Solid State Sci. Technol.,* **2022**, *11*(11), 117001.
[http://dx.doi.org/10.1149/2162-8777/ac9c92]

[135] Dahiya, M.; Khanna, V.; Anil Bansal, S. Aluminium-graphene metal matrix nanocomposites: Modelling, analysis, and simulation approach to estimate mechanical properties. *Mater. Today Proc.,* **2022**, (Nov)
[http://dx.doi.org/10.1016/j.matpr.2022.10.181]

[136] Singh, K.; Bansal, S.A.; Khanna, V.; Singh, S. Effects of performance measures of non-conventional joining processes on mechanical properties of metal matrix composites. *Metal Matrix Composites,* **2022**, (Aug), 135-165.
[http://dx.doi.org/10.1201/9781003194897-7]

[137] Khanna, V.; Kumar, V.; Bansal, S.A.; Prakash, C.; Ubaidullah, M.; Shaikh, S.F.; Pramanik, A.; Basak, A.; Shankar, S. Fabrication of efficient aluminium/graphene nanosheets (Al-GNP) composite by powder metallurgy for strength applications. *J. Mater. Res. Technol.,* **2023**, *22*, 3402-3412.
[http://dx.doi.org/10.1016/j.jmrt.2022.12.161]

CHAPTER 9

Corrosion and Wear Behaviour of Metal Matrix Composites

Rakesh Kumar[1], Harsh Kumar[2], Santosh Kumar[3,*], Mohit Kumar[1] and **Gaurav Luthra[1]**

[1] *Department of Regulatory Affair and Quality Assurance, Auxein Medical Pvt. Ltd. Sonipat, Haryana, India*

[2] *Department of Mechanical & Project Management, Kalyan Project Construction Company, Mohali, Punjab, India*

[3] *Department of Mechanical Engineering, Chandigarh Group of Colleges, Landran, Mohali, Punjab, India*

Abstract: Metal matrix composite (MMC) has several attractive characteristics (low coefficient of thermal expansion, lightweight, better abrasion, high strength-to-weight ratio, superior stiffness, thermal stability, *etc.*), when compared with monolithic materials. Due to these charming characteristics, MMC materials have received wide scope in distinct industries (marine, aerospace, defence, mineral processing industry, automotive, electronic, and recreation industries, *etc.*). But, owing to the requirement of higher ductility and brittleness in the form of reinforcement and matrix, there is a need to improve the properties of composite (MMC) that will fulfil the requirement of the engineers. In addition, MMCs are typically more prone to corrosion and wear as compared to their monolithic matrix alloys. Thus, the study of corrosion and wear behaviour of distinct composites such as Al/SiC *etc.* are highly important for better corrosion resistance for distinct applications. This chapter provides an overview of the corrosion and wear behaviour of MMCs and applications.

Keywords: Applications of MMCs, Classification, Corrosion, Historical development, Metal matrix composite, Material properties.

INTRODUCTION

Metal-matrix composites (MMCs) are metals that have had continuous or discontinuous reinforcement added to them, often in the form of fibres, mono-filaments, whiskers, short fibres, particles, *etc.* Metals (such as stainless steel, and tungsten), nonmetals (such as silicon, and carbon), or ceramics (SiC, Al_2O_3) can

* **Corresponding author Santosh Kumar:** Department of Mechanical Engineering, Chandigarh Group of Colleges, Landran, Mohali, Punjab, India; E-mail: santoshdgc@gmail.com

Virat Khanna, Prianka Sharma & Santosh Kumar (Eds.)

all be used as reinforcements. The choice of the reinforcing ingredient and matrix metal is often made based on how effectively the two work together to provide the desired qualities. MMCs are typically strong, rigid, and light, although they are sometimes designed for other characteristics. Also, reducing weight, thermal expansion, friction, and wear, reinforcements have been utilized in MMCs to enhance their properties. Even while MMCs have characteristics that are more advantageous than those of their separate components, corrosion resistance is frequently compromised. Due to the inclusion of reinforcements that change the micro-structure, electro-chemical characteristics, and corrosion morphology, the corrosion behaviour of MMCs frequently differs dramatically from that of their mono-lithic matrix alloys [1, 2].

What are MMCs

Metals are strengthened with fibres or other particles to create metal-matrix composites (MMCs), which may be used to increase or customise qualities including tribological and mechanical properties. In general, MMCs are additionally prone to corrosion than their mono-lithic matrix alloy counterparts. Various degradation issues may result from the interactions between reinforcement particles or fibres and the matrix, such as electrochemical, chemical, or physical interactions. Processing issues might potentially cause corrosion issues [3]. MMCs have improved mechanical properties over traditional materials. The materials are made up of reinforcement and metal materials, which are added to metal matrices to enhance their toughness, stiffness, strength, and wear and tear resistance. Ceramics, carbon fibres, and various metals are examples of reinforcing materials. MMCs are being employed more frequently in various sectors *i.e.* defence, sports, automotive and aerospace, for their unique qualities. They outperform traditional materials such as steel and aluminium alloys due to their higher wear resistance, strength, and remarkable thermal stability. In addition, MMCs are widely used in high-performance, lightweight, and long-lasting items such as engine parts, braking systems, drive shafts *etc.* Casting, infiltration, and powder metallurgy are all processes for producing MMCs. Some of the disadvantages of MMCs are their high cost, difficult production, and proclivity for interfacial interactions b/w the MMCs and the reinforcing material. Thus, current research focuses on developing new materials, improving existing properties such as corrosion resistance, fracture toughness, and fatigue life, and fixing their flaws. In conclusion, MMCs offer unique properties that make them appropriate for a diversity of applications [4 - 6].

HISTORICAL GROWTH OF MMCs

Early 1900s' research on the idea of strengthening metals with additional materials to enhance their mechanical properties led to the invention of MMCs. The availability of new processing methods and the stipulation for materials with enhanced mechanical qualities have, in general, driven historical growth in MMCs. To create materials with even superior qualities, researchers are at present working with new matrix material and reinforcement type combinations. The historical improvement of MMCs is represented in Fig. (1). In addition, the examples of distinct models as well as theories that have been developed to find out the characteristics as well as behaviour of MMCs are given in Table 1 [7].

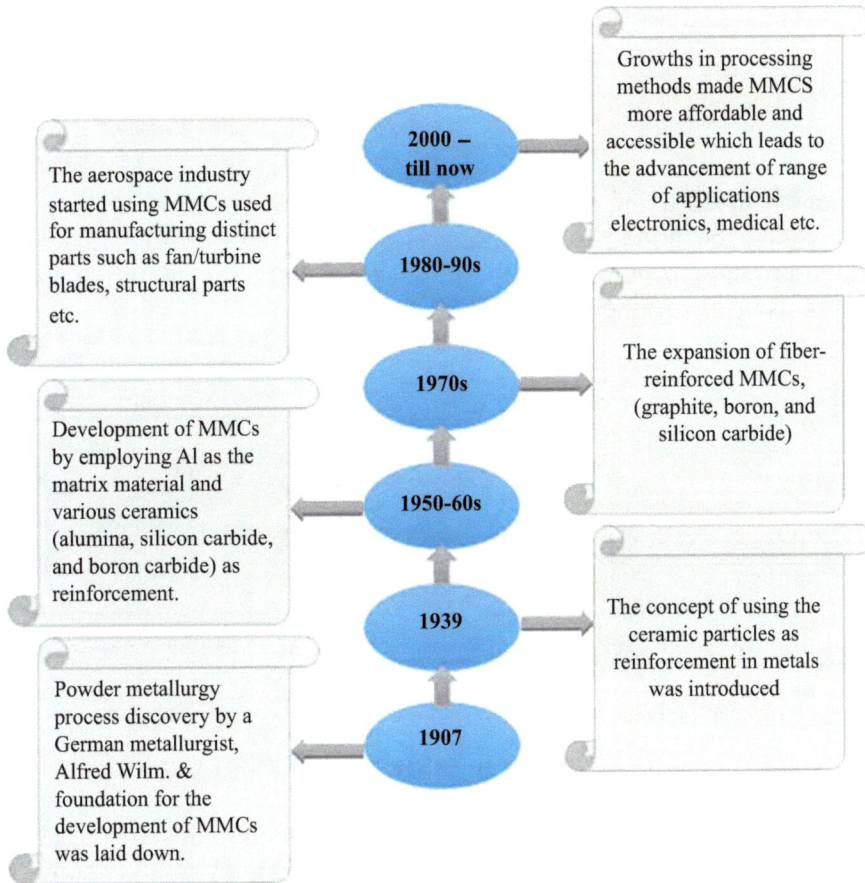

Fig. (1). Growth of MMCs [8, 9].

Table 1. Distinct models and theories employed for the forecast of MMCs behaviour [10 - 14].

Sr. No.	Authors	Year	Model Employed	Categories of Composite	Prediction	Refs.
1	Alfonso *et al.*	2016	FEM & Halpin Tsai modification model	A356-SiC(p)	Elastic-modulus	[10]
2	Hua *et al.*	2013	Mori-Tanaka	Al2080-SiC(p)	Thermo-mechanical Characteristics	[11]
3	Sanaty-Zadeh *et al.*,	2012	Modified Clyne	Particulate-reinforced composite	Evaluate the strength	[12]
4	Shoukry *et al.*	2007	FEM with the model of micro-structure	Al6061-SiC(p)	Micro-Structure	[13]
5	Hocine *et al.*	2013	Transmission line matrix	Al-SiC(p)	Temperature related characteristics	[14]

MMCs can be made using a diversity of methods and materials. The material of the matrix may be added at the process starting stage or the major process, depending on the production path. Fig. (**2**) depicts the materials processing procedures used in producing any MMC.

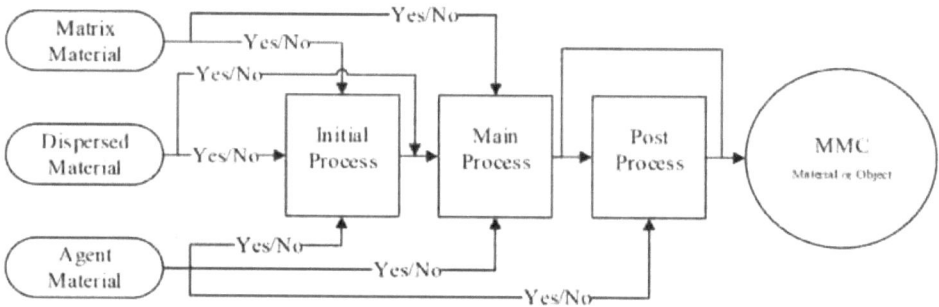

Fig. (2). Processing steps for producing MMCs [15].

CLASSIFICATION OF DISTINCT MMCs

The composite material has qualities that are better than any of the component materials since it is a mixture of two or more insoluble parts. Compared to other popular materials like steel, composite materials are stronger and lighter. To lighten the weight of a vehicle, numerous steel components are being replaced by composite materials in the automotive industry. The many matrix and reinforcing materials that may be employed to create MMCs are depicted in Fig. (**3**). In addition, some examples of reinforced material are given in Table **2** [16].

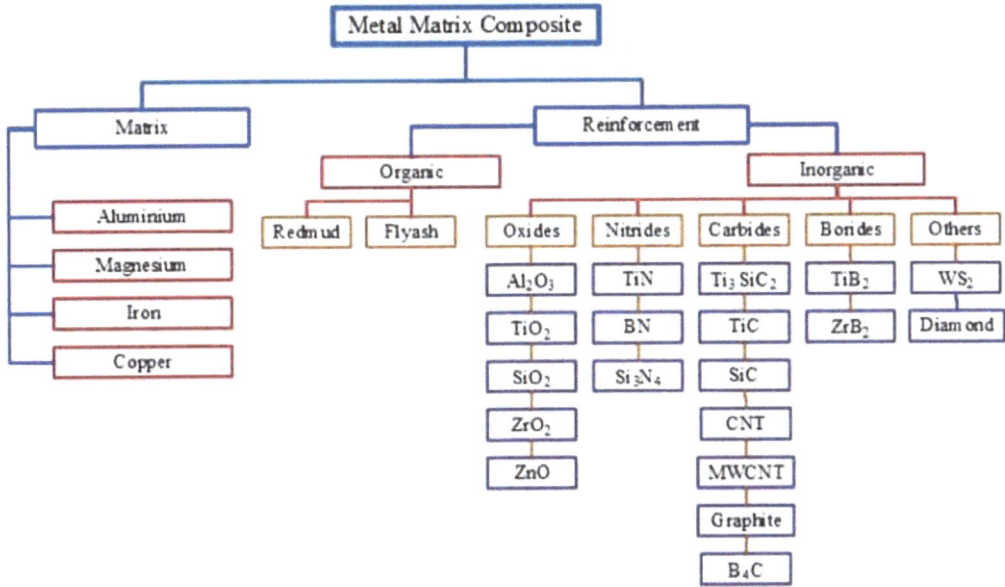

Fig. (3). Classification of MMCs [16].

Table 2. Examples of reinforced materials [16].

Aluminium MCs	Magnesium MCs	Copper MCs	Titanium MCs	Iron MCs
Silicon carbide, alumina, graphite or boron carbide	Silicon carbide, alumina, or graphite.	Silicon carbide, alumina, or graphite	Silicon carbide, alumina, or carbon fibers.	Silicon carbide, alumina, or graphite

Here, MCs: Matrix Composite; MM: Metal Matrix.

CORROSION IN THE CASE OF MMC

The manufacture of MMC might involve processing that speeds up the corrosion of the metal matrix because of the inclusion of reinforcements. However, in comparison to the corrosion of monolithic matrix alloys, specific corrosion issues become significant in MMCs. Because of their inherent qualities or those brought on by processing, MMC components may interact electrochemically, chemically, and physically to generate accelerated corrosion in MMCs. Corrosion may be sped up by galvanic contact between the interphases, the matrix, and the reinforcement. The chemical breakdown of interphases and reinforcements is not electrochemical. By promoting segregation, intermetallic formation, and dislocation development, the microstructure of MMCs can have an impact on corrosion. Processing errors might lead to deterioration in unforeseen ways. The factors that will be covered when it comes to MMC corrosion are (1) electro-chemical properties relating to the major MMC ingredients; (2) electro-chemical properties of the interphases; (3) chemical deterioration in MMCs; and (4) secondary properties resulting from the micro-structure and processing. However, the corrosion classification in the case of metal matrix compounds is represented in Fig. (**4**).

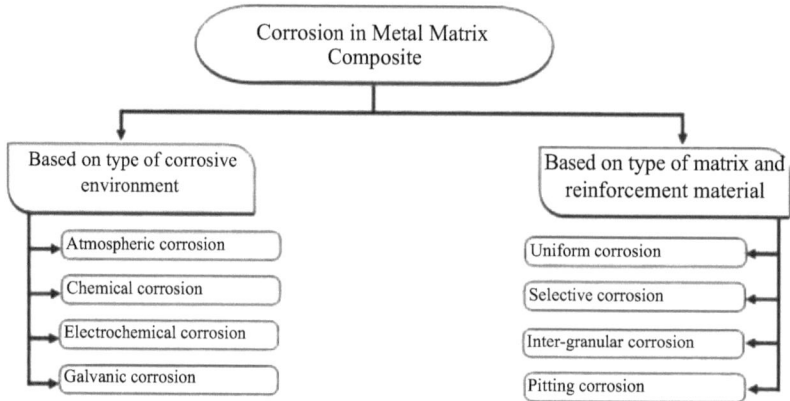

Fig. (4). Corrosion classification [17].

The following description is dependent on the type of corrosive environment, reinforcing material and matrix.

a. In atmospheric corrosion

b. Chemical corrosion

c. Electrochemical corrosion

d. Galvanic corrosion

But based on the kind of reinforcement material and matrix corrosion, it is classified as:

a. Homogeneous corrosion: where the matrix and reinforcing material are uniformly corroded.

b. Selective corrosion: This results in galvanic corrosion when only one of the materials corrodes more than the others.

c. Intergranular corrosion: In which corrosion develops, along the matrix material's grain boundaries.

d. Pitting corrosion: This type of localised corrosion results in holes or pits.

MMCs are being used in more automotive applications, which has raised questions about how well they resist corrosion. Additionally, the materials that make up MMCs, as well as how they are manufactured and processed, all affect galvanic corrosion. Only in certain circumstances can MMC corrosion behaviour be accurately predicted using the information on the electrochemical/chemical characteristics of the components of MMCs. Making MMCs with sufficient corrosion resistance can be facilitated by the judicious use of MMC components. It is more challenging to forecast the impacts of processing factors on MMC corrosion behaviour, and in certain cases, the entire effects may not be understood until after the MMC has been created and evaluated for corrosion behaviour. It should be predicted that different processing techniques would be required to create MMCs with enough corrosion resistance [18].

WEAR CLASSIFICATION IN MMCS

Metal matrix composites (MMCs) can categorise their wear depending on the wear mode, contact type and wear mechanism as depicted in Fig. (**5**).

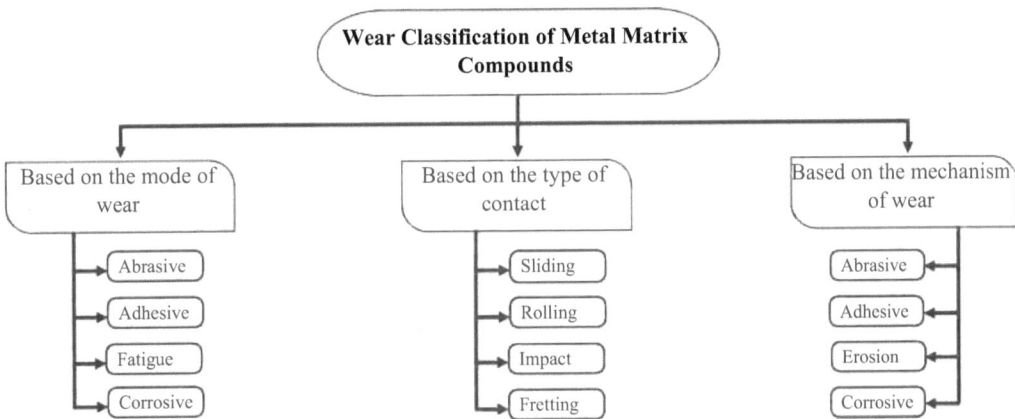

Fig. (5). Wear classification [19 - 22].

Based on the way it is worn:

a. Abrasive wear: Material is removed when there are abrasive particles.

b. Adhesive wear: When there is intense contact pressure b/w two surfaces, the material is transferred from one to the other.

c. Fatigue wear: where the material is lost as a result of continuous, cyclic loading.

d. Corrosive wear: When corrosion and wear combine forces to remove material, this form of wear results.

Based on the type of contact:

a. Sliding wear: what happens when two surfaces bump against one another?

b. Rolling wear: This type of wear is the result of two surfaces rubbing up against one another.

c. Impact wear: In this type of wear, the removal of material occurs as an outcome of the collision of particles/liquids with the surface.

d. Fretting type wear: During this type of wear, the removal of material takes place due to periodic cycle stress and small amplitude sliding.

Based on how wear takes place:

a. Abrasive wear is simply the removal of material as a result of abrasive particles' cutting activity. b. Adhesive wear: When there is an intense contact pressure b/w two surfaces, the material is transferred from one to the other.

b. Erosion wear is typically the removal of material as an outcome of the collision of particles with the substrate surface.

c. Corrosive wear: this type of wear occurs when corrosion and wear both work together to remove material.

The categorization of wear in MMCs may be utilised to improve the design of composite micro-structures and the selection of appropriate reinforcing materials, hence increasing wear resistance. Surface treatments and coatings can also improve MMC wear resistance in some applications [19 - 22].

ADVANTAGES OF MMCS

Above conventional materials, MMCs have several advantages.

The following are a few of the most important benefits of MMCs:

a. High strength and low weight: As we know MMCs are perfect for utilization where weight decline is major, like in the automobile, automotive and aerospace sectors, because they have high strength-weight ratios and are light in weight.

b. MMCs are appropriate for applications like gears where wear/tear are big problems since they outperform normal materials in terms of wear resistance.

c. Excellent thermal conductivity: MMCs have great thermal conductivity, which makes them perfect for distinct heat-dissipating devices like electronic heat sinks.

d. Increased stiffness: MMCs are more rigid than conventional materials which makes them perfect for uses when a high degree of rigidity is required, including the automotive, aerospace and defence sectors.

e. Better damping characteristics: MMCs offer better-damping characteristics, which make them perfect for use in conditions characterised by vibration as well as noise issues.

f. Improved corrosion resistance: MMCs are more resistant to corrosion than a conventional material, which makes them perfect for usage in severe settings such as in chemical processing or the marine sector. Changing the kind, size, and percentage of reinforcing material employed can alter the characteristics of MMCs, making them suitable for a variety of applications [23 - 26].

LIMITATIONS OF MMCS

It provides a lot of benefits, but they also have several uses that are restricted.

The following are some typical MMC drawbacks:

(i) Cost: MMCs may be more expensive than ordinary metals because of increased processing expenses and the cost of reinforcing materials.

(ii) Manufacturing complexity: MMCs manufacture might be a difficult technique that calls for specific tools and knowledge. Because of this, manufacturing MMCs may be more difficult and costly than manufacturing conventional metals.

(iii) Difficulty in joining: MMCs can be difficult to mix with other materials owing to differences in melting temperatures and TEC. This may prevent them from being used in situations where intricate assembly or welding is necessary.

(iv) Anisotropic properties: Anisotropic properties may be seen in MMCs, which means that they can include differing mechanical properties in distinct directions. However, the design structure of MMC machine parts may become more complex as a result, necessitating specialist testing and analysis.

(v) Limited availability of reinforcement materials: Several of the reinforcement materials utilized in MMCs, including carbon fibres, can be costly and challenging to obtain. This may restrict the use and accessibility of MMCs in particular applications.

(vi) Susceptibility to corrosion: In some situations, some MMCs, especially those with aluminium matrices, might be prone to corrosion. This might prevent them from being used in circumstances where corrosion resistance is critical.

As a result, while selecting an appropriate material for a certain application, it is critical to carefully examine the restrictions of MMCs. MMCs have a variety of benefits over traditional metals, but their disadvantages must also be recognised to make the greatest use of them [27-30].

APPLICATIONS OF MMCs

Al-MMCs are potentially popular materials in diverse industrial applications owing to their low cost, excellent strength, and low weight. In addition, Al-MMCs have favourable qualities over traditional materials [31].

Automotive Industry

MMCs are commonly used in automobile engine components such as pistons, piston rods, wrist pins, *etc.* as discussed below.

Leaf spring

One component of an automotive vehicle that absorbs stress is the leaf spring. Investigations were conducted on the leaf spring fracture failure that occurred in buses in Venezuela [32].

Gear

The many varieties of gears, which are among the most crucial parts of an automobile's power transmission system, as well as many spinning pieces of machinery, may be produced with metal matrix composite materials. Due to their low density, aluminium metal matrix material composites were selected [33].

Brake Drum

The braking drums are made of a composite material made of silicon carbide and aluminium alloy. Cast iron brake drums and composite brake drums were contrasted. To learn more about how heat treatment affects materials, the Al-SiC, MMC brake drum was examined. The primary factor used to assess the brake drum was its coefficient of friction [32].

Shaft

Finite elements are used to determine the composite driveshaft. MMCs are found to be more suitable for the manufacturing of shafts due to their high mechanical properties [33].

MMCs provide automobile manufacturers the ability to fulfil specialised and stringent design specifications to overcome the challenges [33].

Piston and Connecting Rods

The piston's operating cylinder has an extremely high temperature and pressure, so the material utilized for both needs to be most thermally conductive as well as wear-resistant.

Al-MMCs, which recommend high excellence at a reasonable price, can be utilised for this. Engines may reach operating temperatures more quickly thanks to the usage of Al-MMCs in the engine block, which also provides better wear resistance and lighter weight. Compared to the traditional piston production technique, the MMC casting process is simpler. Many benefits are obtained from using MMCs as pistons. Especially for race cars, SiC is mostly utilised as a reinforcement material. When connecting rods were made from Al-MMCs, a good weight decrease was discovered, and vibrations were minimised during the whole process. Moreover, the load on the main shaft, boosts the engine power and reduces the engine fuel consumption [34]

Break and Chassis

It is possible to significantly reduce the weight of automotive disc brakes and brake calipers. Al-MMCs are frequently used in trains of high-speed, car braking systems to cut costs and improve machinability. Brake discs and pipes are made using specialised casting techniques. Al-MMCs are increasingly being used by automakers to construct braking systems [35 - 37].

Marine and Rail

During the manufacturing process, Al-MMCs are used to lighten rail carriages. However, due to its lightweight and exceptional corrosion resistance, aluminium remains the finest material for railway carriages. Marine transportation has evolved as a result of the use of Al alloys and their composites. Because of the usage of these materials, ships and boats can now move quicker, further, and with less fuel. More maneuverability and access to low-drawn ports are frequently made possible by the use of materials based on aluminium. For better vehicle dynamics, as a result, materials that increase torsional rigidity as well as energy absorption are used. The material's hardness and strength used to make the chassis, as well as the characteristics that evaluate the occupant's safety in serious crashes, can have an impact on the performance of the vehicle [35 - 38].

Aerospace and Aircraft Applications

The demand for special properties of materials has increased due to the aerospace industry's rapid growth. Al-MMCs have numerous uses in the aerospace industry. Because of the harsh environment in space, advanced materials with required qualities are required, like high, dimensionally stable structures, superior specific stiffness, and low weight [39].

Recently, distinct MMCs are employed in the aerospace industry to make very rigid and robust pieces such as discs, landing gear, brake, and engine parts. MMCs are utilised in aerospace to reduce weight, increase fuel efficiency, and improve overall performance [40].

Aluminium-based composites have a few special qualities that allow for widespread use in the aerospace, aviation, and defence sectors. The first company to use MMC in aircraft, and military was an American one, Lockheed-Martin. Al matrix composites have been extensively utilized in the production of large aircraft wings, rudders, flaps, fuselages, and other parts. The fan outlet guiding vanes (aircraft engines) are also made from Al-MMCs. The helicopter rotor system, hydraulic pipes, landing gear, camera lens frames, and valve bodies can all be made out of these composite materials. Al-MMCs are suitable for application in satellites owing to their excellent conductivity. Al MCs are also used to make electrical and optical parts, as well as for extremely light space applications. These kinds of uses highlight the value of these composites, and a long lifespan is anticipated for them [41].

The industry of building and construction of Al-MMCs are repeatedly in the construction sector. Aluminium has two primary advantages: it is both strong and light. Aluminium and its alloys have been confirmed to be crucial materials over time since they are recyclable. In reality, recycled materials account for a large amount of the aluminium used in modern constructions. Structure capacity can be increased using advanced Al-MMCs [42].

Electronics and Electrical Industries

The most recent models of electrical machinery generate more heat than older ones. The importance of heat dissipation in electrical applications is increasing as a result. Each electrical device might fail for several reasons, including heat, so the system must eliminate or dissipate heat as much as possible. Heat sinks are often employed in electronic equipment and are essentially indispensable in processing units and computer systems [43, 44].

Military Applications

MMCs are used in big conventional armament systems on occasion. Because of their superior all-around performance, fibre-reinforced Al composites are widely used in the manufacture of firearms. However, Al-MMCs began to be utilized as significant components in previously beryllium-made missiles. In addition, MMCs provide advantages such as lower costs and the elimination of beryllium-related toxicity risks. Due to its great stiffness, MMCs are best used to make the fins of a directed gun. This helps a weapon's accuracy to rise [45].

Medical Use of MMCs

MMCs are used to make medical implants and prostheses that require high strength and biocompatibility (hip implants and dental implants). However, the usage of MMCs in medicine allows patients to live healthier lives and eliminates the need for revision surgery. Because of their unique properties, MMCs are a desirable material for a wide range of use in several sectors [46].

Thermal Power Plant Application

MMCs utilized in thermal power plants to enhance plant efficiency. The composite mainly helps heat transfer from the furnace to the turbine more efficiently, which increases the overall efficiency of the implant (corrosion, wear), *etc* [47 - 57]. Table **3** Shows the distinct reinforcement materials & their application in the development of MMCs.

Table 3. Different reinforcing materials and their use in the creation of MMCs.

Sr. No.	Authors	Year	Reinforcement Metals	Application	Refs.
1	Sajjadi *et al.*	2012	Alumina	Pistons, connecting rods & cylinder heads	[58]
2	Reddy *et al.*	2017	Silicon carbide	Propeller shaft, brake rotors, connecting rod, brake rotors, *etc.*	[59]
3	Ravichandran *et al.*	2014	TiO 2	Automobile uses	[60]
4	Bharath *et al.*	2014	ZrO	Connecting rods, pistons *etc.*	[61]
5	Gowri *et al.*	2014	SiO2	Wear-resistant applications	[62]

FABRICATION TECHNIQUES OF MMCS

MMCs can be made in a variety of ways, each with its own set of merits and demerits [63]. The fabrication methods of MMCs are categorized into three types as depicted in Fig. (**6**).

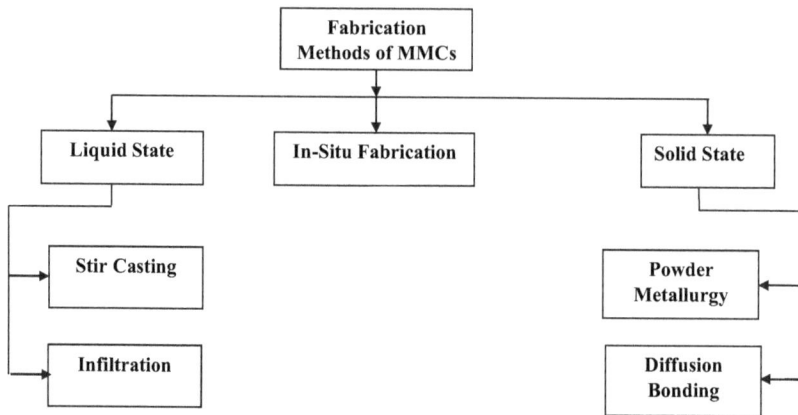

Fig. (6). Distinct methods for MMC preparation.

Liquid State Fabrication of MMCs

In this method, the reinforcement phase is incorporated into the molten metal. However, to produce improved mechanical qualities, fine interfacial bonding should be attained between reinforcements and molten metal. Wetting enhancement may be accomplished by coating the reinforcing particles (fibres) [64, 65].

Stir Casting

Stir casting consist of the stirring of molten metal with reinforcement material before casting into a mould. The reinforcement is added in fibers form. Stir casting is a easy and economical method but is limited to low volume fractions of reinforcement and relatively simple shapes [64].

Infiltration

There are three types of infiltration process.

Gas pressure infiltration

For the manufacturing of large composite pieces, this method is utilised. Because there is less time for the fibres to come into contact with molten metal, non-coated fibres are an option. Gas Pressure Infiltration causes far less fibre damage than the methods that rely on mechanical force. In this procedure, a container is used, into which the preformed material is inserted to an exacting level, and then molten metal is poured.

Squeeze Casting Infiltration

This process includes the strong penetration of MMC production in the liquid condition. However, the lower section of the fixed mould component is filled with a reinforcing phase preform (particulates, fibres).

Pressure Die Infiltration

This procedure involves the forced infiltration production of MMC. In this procedure, diecasting technology is utilised. The liquid metal is driven into a die by a movable piston (plunger). The die is filled with preformed reinforcement. Infusing molten metal with the preform results in homogenous MMCs. It is difficult to place a preform since it moves under strong pressure.

Solid state fabrication of MMCs

These procedures include bonding b/w the metal matrix and the reinforcing phase, and it is as a result of this bonding that MMCs are produced.

Diffusion Bonding

In this procedure, the metal matrix, which takes the form of foils, and the dispersion phase, which takes the form of long fibres, are stacked up in the precise sequence needed, and then pressed at a high temperature.

Powder Metallurgy

It is a popular approach for producing MMCs. It entails combining the metal powder with the reinforcing material (often fibres, particles, or whiskers) and compacting the combination into a perform material [63]. Powder metallurgy includes the below steps as shown in Fig. (7).

In-situ Fabrication of MMCs

In this procedure, the "precipitation from the melt" throughout cooling and solidification leads to the formation of the dispersed phase in the form of a matrix. An in-situ procedure may be used to manufacture different kinds of MMCs.

In addition, the particles from the reinforcement phase are distributed more uniformly in this procedure. Less costly tools and methods are employed to produce MMCs *in situ*.

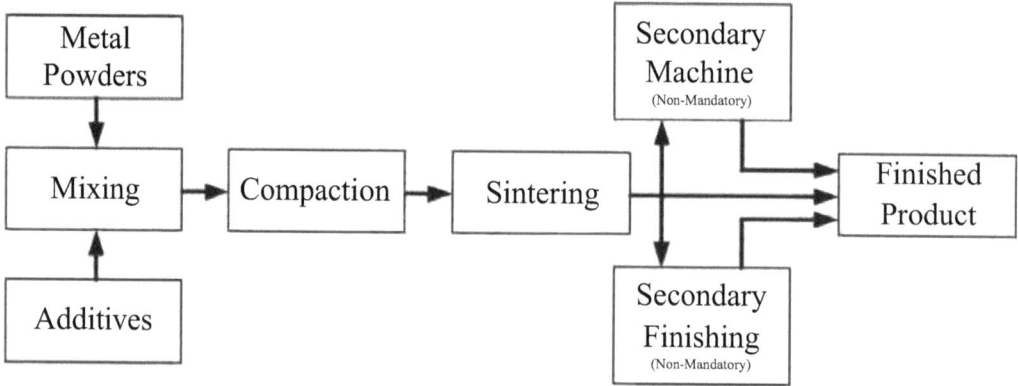

Fig. (7). Steps involved in powder metallurgy.

Other Methods

Extrusion, rolling, forging, and hot pressing are other MMC manufacturing processes. These techniques are employed to create MMCs in the shape of sheets, rods, or other forms [66, 67]. In addition, a novel method for creating composites is "Continuous Binder Powder Coating" (CBPC).

Spray Deposition

Spray deposition is the process of depositing molten metal and reinforcing particles onto a substrate to create a composite in spray form. This approach is excellent for creating coatings and thin films with large reinforcing volume percentages [66].

Metal Injection Moulding (MIM)

This technique is utilised to mass-create intricate and tiny parts. In contrast to powder metallurgy, which is only capable of producing simple form geometries, this procedure makes it straightforward to build complex shapes. Feedstock that contains very fine particles improves during the final sintering process, compaction. The price of composite materials utilised in commercial applications greatly decreased by employing this method of MMC production.

Mechanical Alloying

To create homogenous material, this method of solid-state powder processing includes fracturing, and repeatedly cold welding. This process was developed to produce MMCs for use in the aerospace sector. In summary, there are various methods to fabricate MMCs, and the technique choice mainly depends on the specific requirements of the application.

FUTURE OF MMCS

Since they provide several merits over ordinary materials in terms of distinct electrical, mechanical, physical, and thermal qualities, Metal Matrix Composites (MMCs) have a bright future. Here are some examples of prospective uses and current research projects that highlight MMCs' future potential:

Automotive/ Aerospace Industries: MMCs comprise the potential to be used in the automotive and aerospace sector due to their greater resistance against wear and strength-to-weight ratio. Research studies have demonstrated that MMCs can be used to fabricate parts (pistons, cylinder liners and connecting rods, for IC engines, which can enhance fuel efficiency and decrease emissions [68].

Electronics and Microelectronics: MMCs have major applications in the electronics as well as microelectronics industries because of their high thermal conductivity and low value of coefficient of thermal expansion. Research studies have demonstrated that MMCs can be used as packaging materials, heat sinks, and interconnects for electronic devices [69].

Biomedical Applications: MMCs have potential applications in the biomedical field owing to their biocompatibility and MMCs can be utilised to make implants and prostheses with higher mechanical qualities and a longer lifespan than standard materials, such as hip and knee replacements [70].

Renewable Energy: MMCs have potential wind turbines and solar panels are examples of uses in renewable energy systems, due to excellent high strength and corrosion resistance. Research studies have demonstrated that MMCs can be used to manufacture components for wind turbines, and gearboxes, for example, can enhance efficiency and save maintenance costs [71].

COMPOSITION OF MMCS

MMCs are composite materials made up of an MM reinforced with a 2nd phase (ceramic/carbon fibres, particles/whiskers). However, the configuration of MMCs is typically determined by the matrix type and reinforcing materials used. Here are some MMC composition examples. Composites using Al Matrix: AMCs are the most commonly investigated MMCs, with aluminium serving as the ceramic or carbon fibres or particles or matrix material serving as reinforcement. The reinforcing volume percentage in AMCs normally varies from 10% to 40% [72]. Magnesium Matrix Composites (MMCs): Magnesium matrix composites (MMCs) are a moderately recent field of research [73].

Copper Matrix Composites (CMCs): CMCs are made up of a copper matrix with ceramic fibres or particles for reinforcing. In CMCs, the volume proportion of reinforcement generally varies from 10% to 30% [74].

In summary, the composition of MMCs can vary based on the matrix and reinforcing materials utilised, and current research investigations continue to investigate novel compositions and their features. Table **4** lists the various fabrication processes and reinforcements utilised in MMCs.

Table 4. Distinct reinforcement as well as manufacturing techniques utilized for MMCs.

Sr. No	Reinforcement	Method	Refs.
1	SiC of size 40 mm	Squeeze- Casting	[75]
2	Ni-CrB-Si-based alloy is added with 10 wt% WC, Nickel chrome-based alloy & also added 10 wt. percent Cr_3C_2	HVOF and cold spray	[76]
3	B_4C	Powder metallurgy (PM)	[77]
4	SiC (7 lm)	PM	[78]
5	Al, SiC	Cold spray	[79]
6	Cu + Ti & Cu/Ti	Diffusion- bonding	[80]
7	TiB_2, Al_2O_3 by in-situ reaction of Ti & H_3BO_3 powder along with molten matrix	Stir -casting	[81]
8	SiC	Stir -casting	[82]
9	B_4C	Stir -casting	[83]
10	TiB_2 particles	Stir- casting	[84]
11	SiC & B_4C	Stir- casting	[85]

Overall, composites are especially essential materials for a wide range of applications in several engineering domains due to their superior mechanical, thermal, physical, and electrical properties [86 - 95].

CONCLUSION

MMCs, a crucial class of engineering materials, have attracted a lot of attention lately. They are distinguished by several advantageous characteristics, including lightweight, increased ductility, excellent resistance against wear, superior temperature stability, and enhanced creep resistance. Due to their improved industrial component performance without adding to the system's weight, MMCs have displaced traditional metallic materials in several industries, including the automotive as well as aerospace fields. Owing to their very low weight, good mechanical characteristics, and wear performance, MMCs have appeared as the more promising material in these industries among the different types of MMCs. However, several variables categorised as reinforcement material, operating conditions, and environmental conditions have an impact on the tribological behaviour of these MMCs.

REFERENCES

[1]   Hihara, L.H. Corrosion of metal matrix composites. *Shreir's Corrosion,* **2010**, *2*, 2250-2269.
        [http://dx.doi.org/10.1016/B978-044452787-5.00110-4]

[2]   Ravi Kumar, D.V.; Seenappa, C.R.; Prakash, R.; Bharat, V. Corrosion behavior of cenosphere reinforced Al7075 metal matrix composite —an experimental approach. *J. Miner. Mater. Charact. Eng.,* **2018**, *06*(03), 424-437.
        [http://dx.doi.org/10.4236/jmmce.2018.63030]

[3]   Hunt, W.H. Metal matrix composites. In: *Comprehensive Composite Materials*; , **2000**; pp. 57-66.
        [http://dx.doi.org/10.1016/B0-08-042993-9/00134-0]

[4]   Trethewey, K.R.; Roberge, P.R. Modelling aqueous corrosion: From individual pits to system management. *NATO Science Series E.,* **1994**, 266.

[5]   Seetharaman, S.; Gupta, M. Fundamentals of metal-matrix composites. *Reference Module in Materials Science and Materials Engineering.,* **2021**.
        [http://dx.doi.org/10.1016/B978-0-12-819724-0.00001-X]

[6]   Kelly, A.; Zweben, C. *Introduction*; Comprehensive Composite Materials, **2000**, pp. 9-14.
        [http://dx.doi.org/10.1016/B0-08-042993-9/01001-9]

[7]   Cyriac, A.J. Metal matrix composites: History, status, factors and future. *Oklahoma State University ProQuest Dissertations Publishing,* **2011**.

[8]   Kim, D.Y.; Choi, H.J. Recent developments towards commercialization of metal matrix composites. *Materials,* **2020**, *13*(12), 2828.
        [http://dx.doi.org/10.3390/ma13122828] [PMID: 32599725]

[9]   Singh, H.; Kumar, S.; Kumar, R.; Chohan, J.S. Impact of operating parameters on electric discharge machining of cobalt-based alloys. *Mater. Today Proc.,* **2023**.
        [http://dx.doi.org/10.1016/j.matpr.2023.01.234]

[10]   Alfonso, I.; Figueroa, I.A.; Rodriguez-Iglesias, V.; Patiño-Carachure, C.; Medina-Flores, A.; Bejar, L.; Pérez, L. Estimation of elastic moduli of particulate-reinforced composites using finite element and modified Halpin–Tsai models. *J. Braz. Soc. Mech. Sci. Eng.,* **2016**, *38*(4), 1317-1324.
        [http://dx.doi.org/10.1007/s40430-015-0429-y]

[11]   Hua, Y.; Gu, L. Prediction of the thermomechanical behavior of particle-reinforced metal matrix composites. *Compos., Part B Eng.,* **2013**, *45*(1), 1464-1470.
        [http://dx.doi.org/10.1016/j.compositesb.2012.09.056]

[12]   Sanaty-Zadeh, A. Comparison between current models for the strength of particulate-reinforced metal matrix nanocomposites with emphasis on consideration of Hall–Petch effect. *Mater. Sci. Eng. A,* **2012**, *531*, 112-118.
        [http://dx.doi.org/10.1016/j.msea.2011.10.043]

[13]   Shoukry, S.N.; Prucz, J.C.; Shankaranarayana, P.G.; William, G.W. Microstructure modeling of particulate reinforced metal matrix composites. *Mech. Adv. Mater. Structures,* **2007**, *14*(6), 499-510.
        [http://dx.doi.org/10.1080/15376490701410497]

[14]   Hocine, R.; Boudjemai, A.; Boukortt, A.; Belkacemi, K. 3D TLM formulation for thermal modelling of Metal matrix composite materials for space electronics systems. *2013 6th International Conference*

on Recent Advances in Space Technologies (RAST), **2013**, 47-52.
[http://dx.doi.org/10.1109/RAST.2013.6581254]

[15]   Mussatto, A.; Ahad, I.U.I.; Mousavian, R.T.; Delaure, Y.; Brabazon, D. Advanced production routes for metal matrix composites. *Eng. Rep.*, **2021**, *3*(5), e12330.
[http://dx.doi.org/10.1002/eng2.12330]

[16]   Kumar, A.; Singh, R.C.; Chaudhary, R. Recent progress in production of metal matrix composites by stir casting process: An overview. *Mater. Today Proc.*, **2019**.
[http://dx.doi.org/10.1016/j.matpr.2019.10.079]

[17]   Salvago, G.; Bestetti, M. Metal Matrix Composites: Corrosion. *Wiley Encyclopedia of Composites*, **2012**.
[http://dx.doi.org/10.1002/9781118097298.weoc140]

[18]   Hihara, L.H.; Latanision, R.M. Corrosion of metal matrix composites. *Int. Mater. Rev.*, **1994**, *39*(6), 245-264.
[http://dx.doi.org/10.1179/imr.1994.39.6.245]

[19]   Senthilkumar, M. Abrasive wear behavior of SiC reinforced aluminum metal matrix composites. *Wear*, **2009**, *267*(5-8), 1355-1363.
[http://dx.doi.org/10.1016/j.triboint.2009.12.056]

[20]   Shirvanimoghaddam, K.; Hamim, S.U.; Karbalaei Akbari, M.; Fakhrhoseini, S.M.; Khayyam, H.; Pakseresht, A.H.; Ghasali, E.; Zabet, M.; Munir, K.S.; Jia, S.; Davim, J.P.; Naebe, M. Carbon fiber reinforced metal matrix composites: Fabrication processes and properties. *Compos., Part A Appl. Sci. Manuf.*, **2017**, *92*, 70-96.
[http://dx.doi.org/10.1016/j.compositesa.2016.10.032]

[21]   Gupta, M. Erosive wear response of SiCp reinforced aluminium based metal matrix composite: Effects of test environments. *J. Mech. Eng. Sci.*, **2005**, *259*(1-6), 84-93.
[http://dx.doi.org/10.15282/jmes.11.1.2017.1.0222]

[22]   Rabiei, A.; Vendra, L.; Kishi, T. Fracture behavior of particle reinforced metal matrix composites. In: *Composites Part A: Applied Science and Manufacturing*; Elsevier, **2008**; 39, pp. (2)294-300.
[http://dx.doi.org/10.1016/j.compositesa.2007.10.018]

[23]   Milan, M.T.; Bowen, P. Tensile and fracture toughness properties of SiC$_p$ reinforced Al alloys: Effects of particle size, particle volume fraction, and matrix strength. *J. Mater. Eng. Perform.*, **2004**, *13*(6), 775-783.
[http://dx.doi.org/10.1361/10599490421358]

[24]   Ghosh, S.; Sahoo, P.; Sutradhar, G. Wear behaviour of Al-SiCp metal matrix composites and optimization using taguchi method and grey relational analysis. *J. Miner. Mater. Charact. Eng.*, **2012**, *11*(11), 1085-1094.
[http://dx.doi.org/10.4236/jmmce.2012.1111115]

[25]   Aborkin, A.V.; Elkin, A.I.; Reshetniak, V.V.; Ob'edkov, A.M.; Sytschev, A.E.; Leontiev, V.G.; Titov, D.D.; Alymov, M.I. Thermal expansion of aluminum matrix composites reinforced by carbon nanotubes with in-situ and ex-situ designed interfaces ceramics layers. *J. Alloys Compd.*, **2021**, *872*, 159593.
[http://dx.doi.org/10.1016/j.jallcom.2021.159593]

[26]   Rashad, M.; Pan, F.; Asif, M.; Chen, X. Corrosion behavior of magnesium-graphene composites in sodium chloride solutions. *J. Magnes. Alloy.*, **2017**, *5*(3), 271-276.
[http://dx.doi.org/10.1016/j.jma.2017.06.003]

[27]   Surappa, M.K. Aluminium matrix composites: Challenges and opportunities. *Sadhana*, **2003**, *28*(1-2), 319-334.
[http://dx.doi.org/10.1007/BF02717141]

[28]   Lütjering, G. *Titanium*; Springer Science & Business Media, **2007**.

[http://dx.doi.org/10.1007/978-3-540-73036-1]

[29] Kang, C.G.; Youn, S.W. Mechanical properties of particulate reinforced metal matrix composites by electromagnetic and mechanical stirring and reheating process for thixoforming. *J. Mater. Process. Technol.,* **2004**, *147*(1), 10-22.
[http://dx.doi.org/10.1016/S0924-0136(03)00606-X]

[30] Monikandan, V.V.; Rajendrakumar, P.K.; Joseph, M.A. High temperature tribological behaviors of aluminum matrix composites reinforced with solid lubricant particles. *Trans. Nonferrous Met. Soc. China,* **2020**, *30*(5), 1195-1210.
[http://dx.doi.org/10.1016/S1003-6326(20)65289-X]

[31] Sharma, A.K.; Bhandari, R.; Aherwar, A.; Rimašauskienė, R. Matrix materials used in composites: A comprehensive study. *Mater. Today Proc.,* **2020**, *21*, 1559-1562.
[http://dx.doi.org/10.1016/j.matpr.2019.11.086]

[32] Koli, D.K.; Agnihotri, G.; Purohit, R. Advanced aluminium matrix composites: The critical need of automotive and aerospace engineering fields. *Mater. Today Proc.,* **2015**, *2*(4-5), 3032-3041.
[http://dx.doi.org/10.1016/j.matpr.2015.07.290]

[33] Stojanovic, B. Application of aluminium hybrid composites in automotive industry. *Technical Gazette,* **2015**, *22*(1), 247-251.
[http://dx.doi.org/10.17559/TV-20130905094303]

[34] Chawla, K.K.; Chawla, N. Metal matrix composites: Automotive applications. In: *Encyclopedia of Automotive Engineering, John Wiley &amp*; Crolla, D.; Foster, D.E.; Kobayashi, T.; Vaughan, N., Eds.; Sons Ltd: Chichester, UK, **2014**; pp. 1-6.
[http://dx.doi.org/10.1002/9781118354179.auto279]

[35] Craciun, A.L.; Hepuţ, T.; Pinca-Bretotean, C. Aspects regarding manufacturing technologies of composite materials for brake pad application. *IOP Conf. Series Mater. Sci. Eng.,* **2018**, *294*, 012003.
[http://dx.doi.org/10.1088/1757-899X/294/1/012003]

[36] Nturanabo, F.; Masu, L.M.; Govender, G. Automotive light-weighting using aluminium metal matrix composites. *Mater. Sci. Forum,* **2015**, *828-829*, 485-491.
[http://dx.doi.org/10.4028/www.scientific.net/MSF.828-829.485]

[37] Mohan, R.; Saxena, N. V.; Kumar, S. Performance optimization and numerical analysis of boiler at husk fuel based thermal power plant. *E3S Web of Conferences,* **2023**, *405*(02010), pp. 1-12.
[http://dx.doi.org/10.1051/e3sconf/202340502010]

[38] Mouritz, A.P.; Gellert, E.; Burchill, P.; Challis, K. Review of advanced composite structures for naval ships and submarines. *Compos. Struct.,* **2001**, *53*(1), 21-42.
[http://dx.doi.org/10.1016/S0263-8223(00)00175-6]

[39] Lino Alves, F.J.; Baptista, A.M.; Marques, A.T. Metal and ceramic matrix composites in aerospace engineering.*Advanced Composite Materials for Aerospace Engineering*; Elsevier, **2016**, pp. 59-99.
[http://dx.doi.org/10.1016/B978-0-08-100037-3.00003-1]

[40] Kumar, S.; Singh, H.; Kumar, R.; Singh Chohan, J. Parametric optimization and wear analysis of AISI D2 steel components. *Mater. Today Proc.,* **2023**.
[http://dx.doi.org/10.1016/j.matpr.2023.01.247]

[41] Miracle, D.B. Aeronautical applications of metal-matrix composites. In: *Composites*; Miracle, D.B.; Donaldson, S.L., Eds.; ASM International, **2001**; pp. 1043-1049.
[http://dx.doi.org/10.31399/asm.hb.v21.a0003485]

[42] Dokšanovic, T.; Dzeba, I.; Markulak, D. Applications of aluminium alloys in civil engineering. *Technical Gazette,* **2017**, *24*, 1609-1618.
[http://dx.doi.org/10.17559/TV-20151213105944]

[43] Franck, A.G.; Majidi, A.P.; Chou, T.W. Metal matrix composites. In: *Encyclopedia of Physical Science and Technology,* 3rd ed.; Robert, A.M., Ed.; Academic Press, **2016**; pp. 485-493.

[http://dx.doi.org/10.1016/B0-12-227410-5/00424-5]

[44] Kar, K.K. *Composite Materials: Processing, Applications, Characterizations,* 1st ed.; Springer: Berlin Heidelberg, New York, **2016**.

[45] Nturanabo, F.; Masu, L.; Baptist Kirabira, J. Novel applications of aluminium metal matrix composites. In: *Aluminium Alloys and Composites*; Intechopen, **2018**.
[http://dx.doi.org/10.5772/intechopen.86225]

[46] Casati, R.; Vedani, M. Metal matrix composites reinforced by nano-particles—a review. *Metals,* **2014**, *4*(1), 65-83.
[http://dx.doi.org/10.3390/met4010065]

[47] Kumar, S.; Singh, S. Corrosion behaviour of metal, alloy and composite: An overview. In: *Metal Matrix Composites: Properties and Application*; CRC Press, **2022**.

[48] Kumar, S.; Kumar, M. Tribological and mechanical performance of coatings on piston to avoid failure—a review. *J. Fail. Anal. Prev.,* **2022**, *22*(4), 1346-1369.
[http://dx.doi.org/10.1007/s11668-022-01436-3]

[49] Kumar, S. Influence of processing conditions on the mechanical, tribological and fatigue performance of cold spray coating: A review. *Surf. Eng.,* **2022**, *38*(4), 324-365.
[http://dx.doi.org/10.1080/02670844.2022.2073424]

[50] Kumar, S.; Kumar, R. Influence of processing conditions on the properties of thermal sprayed coating: a review. *Surf. Eng.,* **2021**, *37*(11), 1339-1372.
[http://dx.doi.org/10.1080/02670844.2021.1967024]

[51] Kumar, S.; Handa, A.; Chawla, V.; Grover, N.K.; Kumar, R. Performance of thermal-sprayed coatings to combat hot corrosion of coal-fired boiler tube and effect of process parameters and post-coating heat treatment on coating performance: A review. *Surf. Eng.,* **2021**, *37*(7), 833-860.
[http://dx.doi.org/10.1080/02670844.2021.1924506]

[52] Abedini, M.; Ghasemi, H.M. Corrosion behavior of Al□brass alloy during erosion–corrosion process: Effects of jet velocity and sand concentration. *Mater. Corros.,* **2016**, *67*(5), 513-521.
[http://dx.doi.org/10.1002/maco.201508511]

[53] Kumar, S.; Kumar, M.; Handa, A. Comparative study of high temperature oxidation behavior and mechanical properties of wire arc sprayed Ni Cr and Ni Al coatings. *Eng. Fail. Anal.,* **2019**, *106*, 104173.
[http://dx.doi.org/10.1016/j.engfailanal.2019.104173]

[54] Kumar, S.; Kumar, M.; Handa, A. High temperature oxidation and erosion-corrosion behaviour of wire arc sprayed Ni-Cr coating on boiler steel. *Mater. Res. Express,* **2020**, *6*(12), 125533.
[http://dx.doi.org/10.1088/2053-1591/ab5fae]

[55] Kumar, M.; Kant, S.; Kumar, S. Corrosion behavior of wire arc sprayed Ni-based coatings in extreme environment. *Mater. Res. Express,* **2019**, *6*(10), 106427.
[http://dx.doi.org/10.1088/2053-1591/ab3bd8]

[56] Bedi, T.S.; Kumar, S.; Kumar, R. Corrosion performance of hydroxyapaite and hydroxyapaite/titania bond coating for biomedical applications. *Mater. Res. Express,* **2020**, *7*(1), 015402.
[http://dx.doi.org/10.1088/2053-1591/ab5cc5]

[57] Kumar, S.; Kumar, M.; Handa, A. Combating hot corrosion of boiler tubes – A study. *Eng. Fail. Anal.,* **2018**, *94*, 379-395.
[http://dx.doi.org/10.1016/j.engfailanal.2018.08.004]

[58] Sajjadi, S.A.; Ezatpour, H.R.; Torabi Parizi, M. Comparison of microstructure and mechanical properties of A356 aluminum alloy/Al2O3 composites fabricated by stir and compo-casting processes. *Mater. Des.,* **2012**, *34*, 106-111.
[http://dx.doi.org/10.1016/j.matdes.2011.07.037]

[59]   Reddy, P.S.; Kesavan, R.; Vijaya, R.B. Investigation of mechanical properties of aluminium 6061-silicon carbide, boron carbide metal matrix composite. *Silicon,* **2017,** *10,* 495-502.

[60]   M, R.; S, D. Synthesis of Al-TiO2 composites through Liquid powder metallurgy route. *SSRG Int. J. Mech. Eng.,* **2014,** *1*(1), 12-15.
[http://dx.doi.org/10.14445/23488360/IJME-V1I1P103]

[61]   Bharath, V.; Nagaral, M.; Auradi, V.; Kori, S.A. Preparation of 6061Al-Al2O 3 MMC's by stir casting and evaluation of mechanical and wear properties. *Procedia Materials Science,* **2014,** *6,* 1658-1667.
[http://dx.doi.org/10.1016/j.mspro.2014.07.151]

[62]   Hu, S.; Dai, Y.; Gagnoud, A.; Fautrelle, Y.; Moreau, R.; Ren, Z.; Deng, K.; Li, C.; Li, X. Effect of a magnetic field on macro segregation of the primary silicon phase in hypereutectic Al-Si alloy during directional solidification. *J. Alloys Compd.,* **2017,** *722,* 108-115.
[http://dx.doi.org/10.1016/j.jallcom.2017.06.084]

[63]   Wang, W.; Zhou, H.; Wang, Q.; Wei, B.; Xin, S.; Gao, Y. Microstructural evolution and mechanical properties of graphene-reinforced Ti-6Al-4V composites synthesized *via* spark plasma sintering. *Metals,* **2020,** *10*(6), 737.
[http://dx.doi.org/10.3390/met10060737]

[64]   Palanivendhan, M.; Chandaradass, J. Fabrication and mechanical properties of aluminium alloy/bagasse ash composite by stir casting method. *Mater. Today: Proc.,* **2021,** *45,* 6547-6552.
[http://dx.doi.org/10.1016/j.matpr.2020.11.458]

[65]   Singh, K.; Khanna, V.; Rosenkranz, A. Panorama of physico-mechanical engineering of graphene-reinforced copper composites for sustainable applications. *Materials Today Sustainability,* **2023,** *24,* 100560.
[http://dx.doi.org/10.1016/j.mtsust.2023.100560]

[66]   Hameedullah, A. Characterization of Al-based metal matrix composites produced by spray forming. *J. Mater. Sci.,* **2005,** *40*(12), 3287-3291.

[67]   Parameswaran, V. Fabrication and characterization of aluminium-based metal matrix composites by powder metallurgy and hot extrusion. *Mater. Manuf. Process.,* **2013,** *28*(6), 652-656.

[68]   Sadik, R.; Azizi, F. An overview of metal matrix composites in the aerospace industry. *J. Mater. Res. Technol.,* **2019,** *8*(2), 2296-2307.

[69]   Sajjadi, S. Recent advances in metal matrix composites for electronics applications. *J. Mater. Sci. Mater. Electron.,* **2021,** *32*(14), 18729-18752.

[70]   Bose, S.; Dasgupta, S. Metal matrix composites in biomedical applications. *Woodhead Publ. Ser. Biomater.,* **2015,** *2,* 145-164.

[71]   Moallemi, M. Metal matrix composites for renewable energy applications: a review. *Renew. Sustain. Energy Rev.,* **2021,** *145,* 111008.

[72]   Gupta, M.; Gupta, N.K. Synthesis and properties of aluminum matrix composites: a review. *J. Compos. Mater.,* **2013,** *47*(27), 3385-3402.

[73]   Ray, D. Magnesium matrix composites: A review of reinforcement materials, manufacturing technologies, and properties. *J. Compos. Mater.,* **2021,** *55*(29), 3269-3298.

[74]   Ahmad, M. Copper matrix composites: A review of synthesis, properties and applications. *J. Compos. Mater.,* **2020,** *54*(24), 3329-3348.

[75]   Gurusamy, P.; Prabu, S.B.; Paskaramoorthy, R. Influence of processingtemperatures on mechanical properties and microstructure of squeeze castaluminum alloy composites. *Mater. Manuf. Process.,* **2015,** *30*(3), 367-373.
[http://dx.doi.org/10.1080/10426914.2014.973587]

[76]   Aussavy, D.; Costil, S.; El Kedim, O.; Montavon, G.; Bonnot, A-F. Metal matrix composite coatings

manufactured by thermalspraying: influence of the powder preparation on the coating properties. *J. Therm. Spray Technol.,* **2014**, *23*(1-2), 190-196.
[http://dx.doi.org/10.1007/s11666-013-9999-3]

[77]   Zhou, Y.T.; Zan, Y.N.; Zheng, S.J.; Wang, Q.Z.; Xiao, B.L.; Ma, X.L.; Ma, Z.Y. Distribution of the microalloying element Cu in B4C-reinforced 6061Al composites. *J. Alloys Compd.,* **2017**, *728*, 112-117.
[http://dx.doi.org/10.1016/j.jallcom.2017.08.273]

[78]   Zhu, S.Z.; Ma, G.N.; Wang, D.; Xiao, B.L.; Ma, Z.Y. Suppressed negative influence of natural aging in SiCp/6092Al composites. *Mater. Sci. Eng. A,* **2019**, *767*, 138422.
[http://dx.doi.org/10.1016/j.msea.2019.138422]

[79]   Gyansah, L.; Tariq, N.H.; Tang, J.R.; Qiu, X.; Feng, B.; Huang, J.; Du, H.; Wang, J.Q.; Xiong, T.Y. Cold spraying SiC/Al metal matrix composites: Effects of SiC contents and heat treatment on microstructure, thermophysical and flexural properties. *Mater. Res. Express,* **2018**, *5*(2), 026523.
[http://dx.doi.org/10.1088/2053-1591/aaaeee]

[80]   Ren, H.S.; Ren, X.Y.; Xiong, H.P.; Li, W.W.; Pang, S.J.; Ustinov, A.I. Nano-diffusion bonding of Ti2AlNb to Ni-based superalloy. *Mater. Charact.,* **2019**, *155*, 109813.
[http://dx.doi.org/10.1016/j.matchar.2019.109813]

[81]   David Raja Selvam, J.; Dinaharan, I.; Vibin Philip, S.; Mashinini, P.M. Microstructure and mechanical characterization of *in situ* synthesized AA6061/(TiB2+Al2O3) hybrid aluminum matrix composites. *J. Alloys Compd.,* **2018**, *740*, 529-535.
[http://dx.doi.org/10.1016/j.jallcom.2018.01.016]

[82]   Sharma, V.K.; Kumar, V.; Joshi, R.S. Experimental investigation on effect of REoxides addition on tribological and mechanical properties of Al-6063 basedhybrid composites. *Mater. Res. Express,* **2019**, *6*(8), 0865d7.

[83]   Dou, Y.; Liu, Y.; Liu, Y.; Xiong, Z.; Xia, Q. Friction and wear behaviors of B4C/6061Al composite. *Mater. Des.,* **2014**, *60*, 669-677.
[http://dx.doi.org/10.1016/j.matdes.2014.04.016]

[84]   Ramesh, M.; D, J.; Ravichandran, M. Investigation on mechanical properties and wear behaviour of titanium diboride reinforced composites. *FME Transactions,* **2019**, *47*(4), 873-879.
[http://dx.doi.org/10.5937/fmet1904873R]

[85]   Das, S. Fabrication and tribological study of AA6061 hybrid metal matrixcomposites reinforced with SiC/B4C nanoparticles. *Ind. Lubr. Tribol.,* **2019**, *71*(1)

[86]   Kumar, S.; Kumar, M.; Handa, A. Erosion corrosion behaviour and mechanical properties of wire arc sprayed Ni-Cr and Ni-Al coating on boiler steels in a real boiler environment. *Mater. High Temp.,* **2020**, *37*(6), 370-384.
[http://dx.doi.org/10.1080/09603409.2020.1810922]

[87]   Sultan, U.; Kumar, J.; Kumar, S. Experimental investigations on the tribological behaviour of advanced aluminium metal matrix composites using grey relational analysis. *Mater. Today: Proc.,* **2022**, *21*(3), 1559-1562.
[http://dx.doi.org/10.1016/j.matpr.2022.12.171]

[88]   Kumar, S.; Kumar, R. Overview of 3D and 4D printing techniques and their emerging applications in medical sectors. *Current Material Science,* **2022**, *15*(2), 1-28.
[http://dx.doi.org/10.2174/2666145416666221019105748]

[89]   Kumar, R.; Kumar, M.; Singh, C.J.; Kumar, S. Effect of process parameters on surface roughness of 316L stainless steel coated 3D printed PLA parts. *Mater. Today Proc.,* **2022**, *68*(4), 734-741.
[http://dx.doi.org/10.1016/j.matpr.2022.06.004]

[90]   Kumar, R.; Kumar, M.; Chohan, J.S.; Kumar, S. Overview on metamaterial: History, types and applications. *Mater. Today Proc.,* **2022**, *56*(5), 3016-3024.

[http://dx.doi.org/10.1016/j.matpr.2021.11.423]

[91] Dahiya, M.; Khanna, V.; Anil Bansal, S. Effect of graphene size variation on mechanical properties of aluminium graphene nanocomposites: A modeling analysis. *Mater. Today Proc.,* **2022**, (Jul)
[http://dx.doi.org/10.1016/j.matpr.2022.07.259]

[92] Gupta, P.; Ahamad, N.; Kumar, D.; Gupta, N.; Chaudhary, V.; Gupta, S.; Khanna, V.; Chaudhary, V. Synergetic effect of CeO_2 doping on structural and tribological behavior of $Fe-Al_2O_3$ metal matrix nanocomposites. *ECS J. Solid State Sci. Technol.,* **2022**, *11*(11), 117001.
[http://dx.doi.org/10.1149/2162-8777/ac9c92]

[93] Dahiya, M.; Khanna, V.; Anil Bansal, S. Aluminium-graphene metal matrix nanocomposites: Modelling, analysis, and simulation approach to estimate mechanical properties. *Mater. Today Proc.,* **2022**, (Nov)
[http://dx.doi.org/10.1016/j.matpr.2022.10.181]

[94] Singh, K.; Bansal, S.A.; Khanna, V.; Singh, S. Effects of performance measures of non-conventional joining processes on mechanical properties of metal matrix composites. *Metal Matrix Composites,* **2022**, (Aug), 135-165.
[http://dx.doi.org/10.1201/9781003194897-7]

[95] Khanna, V.; Kumar, V.; Bansal, S.A.; Prakash, C.; Ubaidullah, M.; Shaikh, S.F.; Pramanik, A.; Basak, A.; Shankar, S. Fabrication of efficient aluminium/graphene nanosheets (Al-GNP) composite by powder metallurgy for strength applications. *J. Mater. Res. Technol.,* **2023**, *22*, 3402-3412.
[http://dx.doi.org/10.1016/j.jmrt.2022.12.161]

Metal Matrix Composites, 2024, 249-282

CHAPTER 10

An Experimental Investigation of Process Optimization of EDM for Newly Developed Aluminium Metal Matrix Composites

Jatinder Kumar[1,*]**, Gurpreet Singh**[2] **and Santosh Kumar**[3]

[1] *Department of Mechanical Engineering, Modern Group of Colleges, Mukerian, Punjab, India*

[2] *Department of Mechanical Engineering, St. Soldier Institute of Engineering and Technology, Jalandhar, Punjab, India*

[3] *Department of Mechanical Engineering, Chandigarh Group of Colleges, Landran, Mohali, Punjab, India*

Abstract: The aim of this investigation is to investigate the contribution of controllable input parameters (*viz.* pulse on times, peak currents) on the performance of two newly developed MMCs (Al-8.5%SiC-1.5%Mo and Al-7%SiC-3%Mo). Both the metal matrix composites were fabricated using the stir-casting method. Thereafter, various tests such as microhardness test, tensile test, and porosity analysis of the newly developed composite were performed. To carry out the machining trials, an L_{18} orthogonal array (OA) was chosen. Optimization of the machining process was performed according to Taguchi analysis followed by grey relational analysis (GRA). The results showed that with increasing weight fraction of the molybdenum particulates, microhardness and density of the composites increase with a small reduction in the tensile strength. In addition, pulse on time is the most contributing parameter among others to obtain optimal process performance. The optimum setting of input variables suggested by GRA to obtain optimal responses is a molybdenum composition of 3%, Pulse on time of 70 μs, and a peak current of 9A. Based on the interaction plot, it is evident that process performance measures of EDM depend on controllable input parameters.

Keywords: AMMC, EDM, GRA, Molybdenum, Optimization, Silicon Carbide.

INTRODUCTION

Today's materialistic world has become very advanced in terms of research and developing almost new kinds of infrastructural elements and manufacturing them

* **Corresponding author Jatinder Kumar:**Department of Mechanical Engineering, Modern Group of Colleges, Mukerian, Punjab, India; E-Mail ID: jatinderbahal@gmail.com

Virat Khanna, Prianka Sharma & Santosh Kumar (Eds.)

based on unique scientific requirements. To fulfill the need and specific requirements, scientists and engineers are upfront to find out new materials for specific use [1, 2]. This phenomenon encouraged them to invent and make new materials by combining two or more materials. During combination, both materials exchange their chemical, physical, and mechanical properties with each other thus possessing new materials with new properties. Recently, composite materials are becoming too popular in the industrialized world to satisfy and fulfill the desired needs as per requested requirements.

Composite materials are fabricated with two or more distinguishable materials defining totally dissimilar mechanical, physical and chemical properties which are combined together in an orderly manner to create an almost new material. This newly developed material is specialized for specific use. Based on matrix materials, these are classified as a polymer matrix, metal matrix, ceramic matrix and natural composites such as wood, which is composed of cellulose (a polymer – C6H10O5) and lignin [3]. However, in the manufacturing methodology of composite material, which material's property is to be changed is taken as a matrix material and the material which is added into matrix material to change its properties is called the reinforcement material [4]. When the matrix material is polymer, then the fabricated material will be PMCs; if the base material is metal, then the manufactured one will be MMCs and CMCs, if ceramic is taken as the matrix material. Matrix material is considered a monolithic material in which the fiber system of other materials is to be embedded. Both matrix materials and reinforcement materials are identical at macroscopic and microscopic levels.

When the base or matrix material is an alloy or a metal, then the composite material is known as MMCs. When two or more reinforcement materials are added to the matrix material, then the resulting material is called "Hybrid MMCs (HMMC's). These advanced HMMCs are developed for lightweight and heavy-duty applications [5]. The changes in the properties are relied upon by both materials-matrix and reinforcement. Constituent materials can be added into the matrix material in the following forms – continuous fibre and discontinuous fibre form, particulate form and monocrystalline whiskers form.

Among the family of metal matrix composites, AMMCs are one of them. This type of composite material is well-renowned for its low weight and heavy load work abilities. By combining the strength of reinforcement with the ductility of matrix material, composite defines the properties of both materials – the matrix and the reinforcement. For compositions, usually silicon carbide (SiC) or Aluminium Oxide (Al_2O_3) is taken as a reinforcement material. However, the reinforcements are still distinguishable at macroscopic or microscopic levels after composing. AMMCs have significant importance in the manufacturing of long-lasting applications where higher specific stiffness and strength are required. Usually, high-speed and continuous-running components of machines and robots are manufactured from this composite material. After being composed of silicon carbide, it can withstand higher elevated temperatures with excellent thermal conductivity thus highly useful for making automotive engines.

Aluminium 6061 alloy has low specific gravity and possesses a good strength-to-weight ratio. It possesses significant engineering applications in the field of aerospace and automotive industries. The combination of good strength at low weight and its high weldability properties, allows the use of Al 6061 alloy for making the fuselage structure of aeroplanes, helicopter rotor skins, structural towers, bridges, frame making of vehicles, and rail coaches. It is highly weldable by two different fusion welding methods (TIG and MIG welding). Throughout the time of welding, there is a loss of strength magnitudes to 40% of the main strength of the material that occurs at joints made by welding. Thanks to its heat treatment ability by which, 30% of its strength loss can be recovered by 2-phase T6 heat treatment. The Al 6061 grade composite material has higher thermal conductivity than 7075 Al and possesses low resistivity in the electric current flow thus having higher electrical conductivity. Therefore, it has indicative importance in the manufacturing of conductive materials. The Al 6061 alloy has predominant engineering importance in the manufacturing processes of extrusion, forging, and casting.

Fig. (1). SEM Images of Silicon Carbide (SiC).

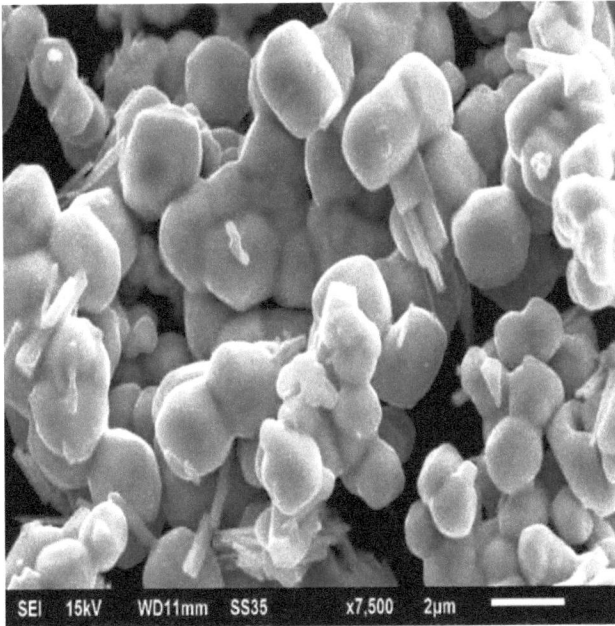

Fig. (2). Molybdenum (Mo) particulates, respectively.

"Liquid stir casting" is one of the liquid phase fabrication techniques in which matrix material is melted inside the furnace resting in a crucible. A rotating stirrer impeller is used to form a vortex of molten metal or alloy under certain temperatures and mixing parameters. A motor is used to rotate the impeller at the required speed. The speed of the impeller can be changed by the regulatory device mounted on the driving controls. The reinforcement materials are added into the matrix material in the form of fibres, particulates, and whiskers through the feed mechanism [6]. A stirrer material is normally chosen as graphite because it defines good sustainability at elevated temperatures. Mostly the vertical position of the stirrer is used to produce a vortex in matrix material. Before adding, reinforcing materials are heated at 150 - 1100°C temperature for a specific time, (temperature range varies from material to material) to remove moisture content from them. It also helps in the uniform mixing of materials during fabrication. Then these materials are fed into matrix material at the desired feed rate. Therefore, preheating of reinforcements removes casting defects such as blow holes (usually occur due to the presence of moisture in reinforcement materials). The constituents can be added at a continuous feed rate or an alternative feed rate. After mixing for a certain time, the molten composite material is poured into the mould *via* gates; thus casting is produced in the cavity of the mould. This fabrication method is very simple and every component of the setup is easily accessible and controllable. It is economic and simple. Every constituent mixes up in an orderly manner and possesses high portability in the whole casting process [7, 8].

In the EDM process, thermo-electric sparks are generated b/w the electrode or wire and substrate, thus resulting in the erosion of the workpiece to machine the workpiece. Electric discharge machining is also called spark eroding. In other words, the material is removed regardless of the contact between the material and the substrate. This machine is highly renowned for its incredible low-tolerance machining. On an average of 0.00254 mm, tolerance can be obtained in EDM. As low as 0.001 mm and as high as 0.005 mm are the ranges of tolerance of EDM that eliminate the post-machining of the workpiece. Sometimes, EDM is also referred to as zero-tolerance machining. It is used to cut any conductive material despite its hardness. While machining with EDM, the tool and workpieces are completely immersed into a dielectric medium. Superalloys, composites and advanced ceramics can be machined with great precision and surface finish. EDM has a significant role in manufacturing of the cutting tools and the die-making industrial field. The tool and dies can be machined with very high precision and accuracy as compared to traditional drilling and milling machine operations. Several researchers have reported their valuable work related to the fabrication and EDM of aluminium 6061 alloys-based MMCs [9, 10]. Khalid and Kuppan [11] examined the recast layer of the finished workpiece by mixing Al in powder form in the dielectric fluid. In their study, electrolytic copper is chosen as a tool electrode and W300 die steel as a work material. The authors observed that a lower white layer of 17.14 μm was formed at 4g/l powder mixture with a low 6A peak current. Velmurugan *et al.* [12] analyzed an HMMCs site fabricated with stir casting consisting of Al6061 reinforced with 4% graphite and 10% SiC. Also, Minitab R14 was used to develop a mathematical model. An electron beam scanning approach such as an SEM instrument is employed to analyze the surface characteristics of the fabricated surface. Shen *et al.* [13] performed experimentation to evaluate the efficiency of a dry EDM considering standard air as a dielectric medium. In this investigation, Ti-6Al-4V alloy was machined to determine MRR and SR at higher machining rates. Kumar *et al.* [14] in their study took an Al-Mg alloy containing 0.6% Si, and 0.35% of Fe as the base matrix. A 15% Rice Husk Ash (RHA) and 5% Copper were introduced to the base matrix to form a hybrid MMC. ANOVA and TOPSIS were used after machining the work piece on WEDM to find the optimized surface roughness, MRR and radial overcut. Input parameters such as current (A), pulse duration in μs units, pulse interval in μs units, and wire feed rate at mm/min were taken as 150A, 125μs, 50μs and 8mm/min respectively. The study confirmed that current (A) was the most significant factor for output parameters contributing 24.09% alone which is overtaken by pulse duration (16.38%) and pulse interval with 15.18% for optimization. Dar *et al* [15] prepared AMMCs and explored the role of SiC reinforcement on the base matrix. Experimentation attempted on LM25 alloy that is reinforced with 7% of SiC and 3% Gr in the particulates form. Variables such as pulse-on time, pulse-off time, voltage and current are painstaking as input variables for consideration. TWR, OC (over-cut) and MRR were considered as response measures. For fabrication, the stir casting was adopted to fabricate the composite and the process was optimized using Taguchi-based ANOVA. It was found that peak current has a significant impact on both TWR and MRR at 88.84% and 81.77% respectively. In an investigation conducted by Sultan *et al.* [16], Al6061 was reinforced with 10%SiC and (2, 4, 6 and 8)% Gr. The purpose of this study was to explore the impact of Gr particulates on the wear behaviour of the base Al-SiC composite. Stir casting was used to develop AMCs. The Friction coefficient and wear rate index were evaluated using a Pin-on-Disc instrument. It was determined that the content of Gr is the most significant factor for wear loss having the highest contribution of 59.82% and for co-efficient of friction, normal load was the most significant factor contributing about 66.34%. Grey Relational Analysis approach was implemented and experiment results revealed that disc speed (RPM), Gr content (%age) and normal load were the most impacting factors and contributed to 46.87%, 36.63% and 14.09% respectively. Shen *et al.* [17] studied electric discharge machining of Cr-embedded TiNi SMA (shape memory alloy) for biological purposes using dry EDM process. Authors observed that SMA has more resistance to wear and corrosion. Boopathi *et al.* [18] investigated Al6061 with 3% SiC using pure oxygen as the dielectric medium and pure Mo as electrode during wire electro-discharge machining process. The parameters MRR and SR were examined with the Taguchi method *via* OVAT (one variable at one time) approach and he found that each of them is affected by open circuit voltage. Dhakar and Devedi [19] performed dry EDM on Inconel 718 and observed that the increase in pulse-on time up to some extent results in higher MRR, but up to some extent and after that MRR decreases when pulse-on time further increases. Teimouri and Besari [20] performed EDM experiments on cold working steel with Cu and brass tool considering air as a dielectric medium. Evaluated as a brass tool, it has better MRR but lower SR and EWR than the Cu electrode. Singh and Bhatia [21] reviewed the optimization of discharge current and pulse time process parameters of electric discharge machining. However, the pulse interval has not been examined independently. Jeykrishnan [22] performed a deep analysis to optimize MRR and TWR while taking EN24 tool steel for evaluation. Pulse on time, pulse off time and current are investigated and analyzed by Minitab and Taguchi approaches. Morankar and Shelke [23] presented a review article on the effect of input parameters on MRR and evaluated peak current as the most significant factor for MRR and TWR. Ohdar and Jena [24] both evaluated higher MRR and less significant TWR with the copper material electrode, Peak current at 12 amperes, pulse on time 15 μsec., pulse of time 3μsec and flushing pressure of

$0.3kg/cm^2$ using Taguchi method. Reddy and Krishna [25] studied the electric discharge machining of 304 stainless steel using the Taguchi method. The authors evaluated that MRR, TWR and SR are directly proportional to peak current and pulse on time. On the contrary, MRR and SR both are decreased with an increase in the value of "pulse off time". Kale and Khedekar [26] illustrate the relation between several input parameters along with their influence on output parameters. Experimentation was performed on Inconel 718 and MRR and TWR were evaluated and both increased with elevated current and pulse on time. Babu and Soni [27] machined Inconel 625 to optimize process parameters related to surface roughness (SR) with the help of Taguchi and ANOVA techniques. They determined that peak current is not a prominent parameter for SR because it enhances with the rise in pulse on time and decreases with pulse off time. The next research was performed by Kumar *et al.* [28] who developed AMMCs that were reinforced with SiC and Gr. The SiC remained constant during fabrication while Gr (graphite) varied as 1%, 2% and 3%, respectively, during composition *via* stir casting methodology. Taguchi L9 orthogonal array (OA) and GRA approaches were employed to find out the role of stirring speed, stirring time and graphite content responsible for the resulting hardness of AMC. Authors observed that Gr content is a predominant factor responsible for the resultant hardness of AMC with having 63.83% role weightage followed by stirring speed at 17.27 wt. role % and stirring time at 11.1%.

Fig. (3). Schematic diagram of stir casting setup.

Chandramouli and Eswaraiah [29] performed their investigation on 17-4 PH steel with Cu tungsten (W) electrode to optimize EDM process parameters. In this case study, pulse on time and discharged current play a vital role in MRR and SR, whereas pulse off time has shown the least influence on MRR and SRR. Yanzhen *et al.* [30] analyzed the generation of porosity on the recast layer by water/oil, kerosene and deionized dielectric water. The result evaluated that the generation of porosity of both internal and external holes on the recast layer was maximum when machining was performed in a water/oil emulsion. Roth and Hartmi [31] performed experimentation with copper and cemented carbide tools to study the breakdown behaviour of DEDM by using oxygen as a dielectric medium. The authors reported that the cemented carbide tool has more MRR than Cu tool. Mahesh *et al.* [32] conducted an experiment to examine the surface quality of MMCs (Al-SiC-B4C). Stir casting was employed for reinforcement to form cylindrical rods with specific dimensions. Taguchi design of the experiment and L9 orthogonal array were used before the S/N ratio technique. It was concluded that the feed rate parameter was the most prominent factor for the surface finish. Rajarshi *et al.* [33] carried out research with several optimization algorithms and methodologies on two WEDM processes simultaneously with single and multi-objective styles. Norfadzlan *et al.* [34] studied the previously used process optimization techniques to demonstrate optimal optimization methodology to customize traditional and non-traditional parameters considered between the years of 2007 and 2011. Researchers evaluated genetic algorithms as the most widely used optimization technique for effective machining.

Kumar *et al.* [35] conducted a wear test on AISI D2 steel at different parameters. The three independent parameters employed during the study were sliding speed, slide duration and normal load. The results showed that the wear rate reduced at optimum parameters. Singh *et al.* [36] also studied the effect of operating variables on EDM on Co-based alloys. Results showed that the discharge current has a maximum share (67.8%) of overall performance. Kumar *et al.* [37] examined the influence of process variables on the surface roughness of 3D-printed PLA parts. The results revealed that at optimum parameters, high strength, and low roughness were achieved. Further Kumar *et al.* [38, 39] provide a comprehensive review of the role of MQL on the quality of material. However, the recent advancement in machining is summarized by various researchers [40 - 45]. However, in the future, metamaterials may be used in place of alloys owing to their excellent mechanical and thermal characteristics for diverse applications [46]. Kumar *et al.* [47] performed a study of Al-Si alloy-based MMCs, which were reinforced with SiC and Cr particulates through the liquid metal vortex casting methodology. Reinforced AMMCs were examined for standability against forces and wear by various tests such as micro-hardness test, porosity analysis, sliding wear test, scratch test and tensile test. Chromium was added from 0-3 weight percentage in the steps of 1.5 percentage and silicon carbide of 10% was kept unvaried during reinforcements. The resulting hybrid composite exhibited a significant increase in hardness, strength, abrasive and wear resistance whereas the ductility was decreased with a slight increase in friction coefficient. The porosity level confirmed that there were no defects in the final casting. In another investigation performed by Kumar *et al*. [48], Al/SiC/Mo hybrid metal matrix composites were formulated with varying Mo wt.% along with constant wt.% of SiC. For fabrication, the stir-casting setup was used. Newly developed composites were examined using advanced testing techniques *viz*. SEM, EDS, XRD, *etc*. Hardness test, tensile test, wear test, and porosity test were also conducted to explore their stability against wear and load. The machinability of the composites was explored using the turning process. Microstructure was obtained by SEM, XRD (X-ray diffraction), and EDS (energy dispersive spectroscopy) approaches. It was obtained that tool wear and machined surface roughness were impacted by all inputs and speed-feed, respectively.

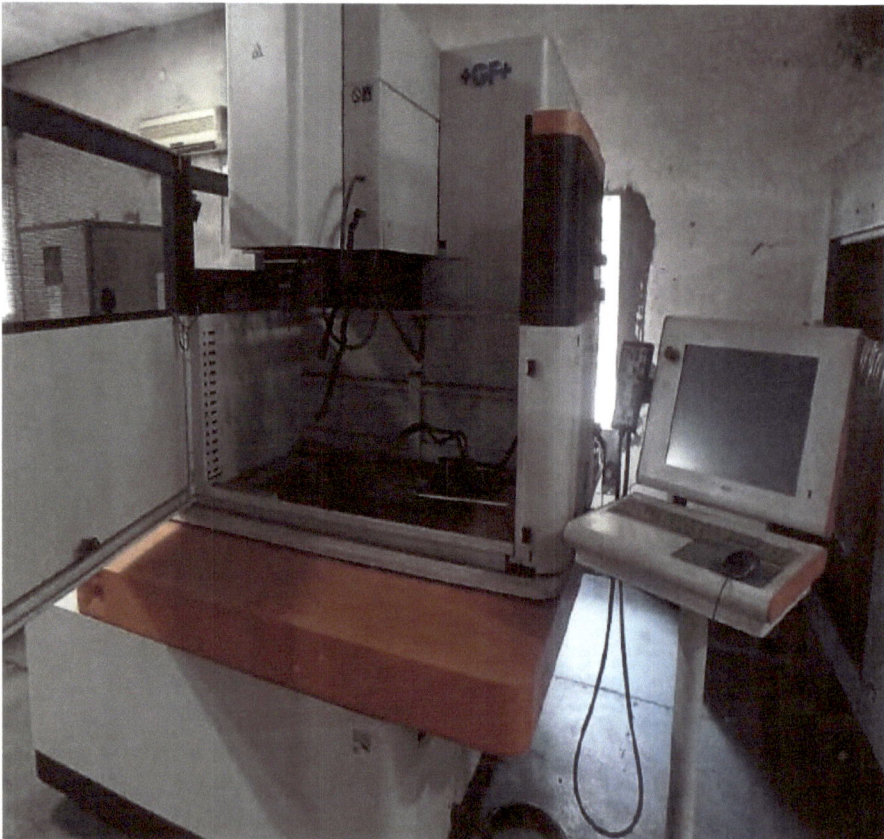

Fig. (4). EDM used during machining of AMMCs.

Machine learning is a type of Artificial intelligence (AI) that enables software applications to forecast outcomes more accurately without having that capability explicitly coded into them. To forecast new output values, machine learning algorithms use historical data as input. The various machine learning techniques can be classified as:

Supervised Learning

It is the most typical kind of machine learning. In supervised learning, a set of labelled data with known outputs for each data point is provided to the algorithm. The algorithm then develops a mapping between the input and output data. Classification, regression, and prediction tasks are examples of tasks that use this form of learning.

Unsupervised Learning

When there is no labelled data available, it is used. In unsupervised learning, the algorithm learns to identify patterns in a set of unlabeled data. Tasks like grouping, dimensionality reduction, and anomaly detection require this form of learning.

Semi-supervised Learning

It combines supervised learning and unsupervised learning. The algorithm is given a set of labelled data and a set of unlabeled data when learning semi-supervised. The algorithm then learns to enhance its performance by combining the knowledge from the labeled and unlabeled data.

Reinforcement Learning

It is a kind of machine learning in which the algorithm picks up new information through error. In reinforcement learning, the algorithm learns to select the actions that maximize its reward from a list of possible actions. For tasks like playing video games and building robots, this kind of learning is employed.

Apart from these, the other most popular machine-learning techniques are Linear regression, Logistic regression, Decision trees Support vector machines, K-means clustering, Principal component analysis, Ensemble methods, Neural networks, Deep learning, *etc.* These are just a few of the many machine-learning techniques that are available. The best technique for a particular task will depend on the data that is available and the desired outcome.

A statistical technique called Grey Relational Analysis (GRA) is used to examine the connections between various sequences. It is a non-parametric approach;

therefore, it doesn't assume anything about how the data are distributed. The grey system theory, which Deng Julong created in 1982, is the foundation of GRA. A mathematical framework called the grey system theory was developed to deal with systems that have missing or ambiguous information. Constructing a grey relational coefficient for each sequence is the fundamental tenet of GRA. The degree to which a sequence resembles a reference sequence is indicated by the grey relationship coefficient. The sequences that are closest comparable to the reference sequence have the highest grey relational coefficients.

From literature review, it has been found that the applied peak current and pulse on time are the most important input variables during the EDMed of composite materials. Also, the type and composition of reinforcements (used in composite) exhibited an impact on the machining of these composites and can be considered an important factor during the machining of AMMCs using the EDM process. The main motive of this investigation is to fabricate Al-SiC$_p$-Mo$_p$ hybrid composites through the stir casting method and optimization of the electrical discharge machining process is carried out using the MADM method.

MATERIALS AND METHODOLOGIES

Materials

In this experimental approach, stir casting is employed to fabricate MMCs using an aluminium Al6061 alloy as a base matrix. After that, electric discharge machining was applied to the cast composites. Al 6061 alloy is one of the most frequently used aluminium alloys due to its weldability and formability. This alloy possesses many applications such as welded constructions, frameworks for ships, frameworks for trucks and aircraft, chemical apparatus, electronic components, *etc*. Keeping this in view, Al 6061 aluminium alloy having a density of 7.2 g/mm^3 was chosen as the matrix material for the current study based on a review of the literature, and its composition is mentioned in Table **1**.

Table 1. Al 6061 Composition.

Elements	Cu	Mn	Fe	Mg	Cr	Ti	Zn	Si	Al
Wt. %	0.21	0.04	0.25	0.89	0.25	0.1	0.11	0.6	Remaining

The reinforcement is a material's discontinuous phase, which is usually stronger and heavier than the material's continuous phase. For Al6061 alloy reinforcement, silicon carbide powder with a density of 3.21 g/mm^3 and a particle size of 60 μm (Fig. **1**) and molybdenum with a density of 8.9 g/mm^3 and a particle size of 80 μm (Fig. **2**) were selected.

Fabrication

Two hybrid composites *viz.* Al-8.5%SiC-1.5%Mo and Al-7%SiC-3%Mo MMCs were fabricated. Fig. **(3)** shows the stir casting setup that was utilized to develop AMCs that contain SiC (particle size 60 µm) and molybdenum particulates (particle size 80 µm). The procedure was adopted in accordance with ref [49]. The desired quantity of Al6061 alloy was put in a crucible made up of ceramic and heated in an electrical furnace. Silicon carbide and molybdenum powders were also heated to a temperature of around 400°C before incorporating into the base matrix. Al6061 was totally melted in a crucible by heating it to 730°C for 15 minutes. Then, SiC and Mo powders were preheated and gradually added to Al6061in three steps. With the aid of an electric motor-integrated graphite stirrer, the mixture was stirred for approximately 12 minutes at a speed of 500 rpm to ensure that the slurry was properly mixed. The prepared mixture was poured into the preheated steel mould. The casting was removed from the mold and finally samples were cut-off for pilot study followed by final experimentation by using wire-EDM. Fig. **(3)** illustrates the setup for the casting process as shown above.

Properties

After fabrication, composites were tested for hardness, tensile strength, and density analysis as per ASTM standards using advanced testing techniques. Microhardness is evaluated in accordance with ASTM 384 standards using a Vicker hardness tester. The tensile test was performed in conformity with ASTM E8 standard and conducted on a computer integrated double column universal testing machine. Actual density of the newly developed composites is calculated on the basis of Archimedes principle by using Mettler Toledo apparatus Table **2**.

Table 2. Properties of the newly developed composites.

Composite	Microhardness (HV)	UTS (N/mm²)	Actual Density (g/cm³)
Al-8.5%SiC-1.5%Mo	110	174	2.852
Al-7%SiC-3%Mo	116	131	2.957

Machining

The machining is performed at the Central Institute of Hand Tools, Jalandhar City, Punjab, India. An Agie Charmilles FORM 300 (die-sinking type) Electric Discharge Machine is used and the setup is shown in Fig. **(4)**. As a dielectric fluid, commercial grade oil for the EDM process having a specific gravity of 0.763 and freezing temperature of 94°C is employed. Newly developed AMMCs are machined with a copper tool as an electrode having a positive polarity.

The material's MRR is defined in terms of the ratio of the distinct b/w the weight of the workpiece before machining and after machining to the machining time and density of the material as expressed in Equation 1.

$$\text{MRR (Material Removal Rate)} = \frac{W_{jb} - W_{ja}}{t * \rho} \tag{1}$$

Whereas

W_{jb} = This term refers to the weight of workpiece before machining.

W_{ja} = This term refers to weight of workpiece after machining.

t = This term refers to machining time =Five minutes.

ρ = This term refers to the Density of a particular composite material Table **2**.

Acronym TWR refers to the Tool wear rate and is defined as the ratio of the weight difference between the tool before machining (W_{eb}) and after machining ((W_{ea}) to the amount of machining time (t). Terms are explained in Equation 2.

$$EWR = \frac{W_{eb} - W_{ea}}{t} \tag{2}$$

Whereas

W_{eb} = This term refers to the weight of the tool measured before machining.

W_{ea} = This term refers to the weight of the tool taken after machining.

t = Machining time (In this regard machining time is five minutes).

Radial overcut (ROC) is defined as half of the difference in the diameter of the hole/cavity produced in the workpiece after machining to the electrode diameter as shown in Equation 3.

$$Radial\ Overcut = \frac{D_h - D_e}{2} \tag{3}$$

Where

D_h = Is the dia. of hole produced in the workpiece

D_e = Is the dia. of tool

The Taguchi DOE (Design of Experiment) is an approach which helps to design an experiment based on different levels. Levels can be two, three, four, five, and as many as possible along with mixed levels. This is also called the orthogonal array for experimental procedures. In this experiment, there are three levels taken along with two input parameters. The Mo composition is the third input parameter which has two levels of percentage in the steps of 1.5% to 3% maximum. The first composing value of Mo was taken as 1.5% in this methodology. Pulse on time and peak current both have three levels of each. In DOE, there are eighteen experiments that are evaluated and the term is defined as OA L_{18} setup. This led to the less required number of observations for final results and this L_{18} setup has affected the later stage of calculations. This level of DOE is expressed in Table 3.

Table 3. Input Variables and their Levels.

Input Factors	Symbol	Levels		
		L-I	L-II	L-III
Molybdenum Composition (wt.%)	Mo%	1.5	3	-
Pulse on Time (µs)	T_{on}	70	130	200
Peak Current (A)	I_p	3	6	9

There are three factors tackled and a total of 18 experiments are conducted on the die-sinking EDM as per Taguchi L_{18} OA. The electronic weight balance machine is used for the calculation of TWR and MRR. The capacity of the machine is 300 grams maximum and accuracy is up to 0.001grams. With the advantage of toolmaker microscope, the overcut can be measured with an accuracy of 0.0001mm.

Table **4** shows the matrix designed and observations taken table of this study.

Table 4. Matrix designed and Observations taken Table.

Exp. No	Mo% (Wt.%)	P_{on} (µs)	I_p (A)	Wt. of workpiece (g)		Wt. of electrode (g)		Dia of Cavity mm
				W_{jb}	W_{ja}	W_{eb}	W_{ea}	
1	1.5	70	3	22.535	19.239	6.732	6.661	9.6429
2	1.5	70	6	23.069	22.321	6.734	6.722	9.3590
3	1.5	70	9	22.789	22.675	6.738	6.731	8.7180
4	1.5	130	3	22.80	12.981	6.807	6.574	9.8361

(Table 4) cont.....

Exp. No	Mo% (Wt.%)	P_{on} (µs)	I_p (A)	Wt. of workpiece (g)		Wt. of electrode (g)		Dia of Cavity mm
5	1.5	130	6	23.193	20.328	6.801	6.739	9.7005
6	1.5	130	9	23.197	20.671	6.812	6.758	9.4098
7	1.5	200	3	23.201	13.701	6.764	6.597	9.8137
8	1.5	200	6	23.20	17.566	6.760	6.639	9.7151
9	1.5	200	9	23.063	20.528	6.759	6.747	9.8576
10	3	70	3	23.073	21.056	6.748	6.705	8.2800
11	3	70	6	23.069	22.100	6.479	6.462	8.1746
12	3	70	9	22.789	22.559	6.751	6.747	8.2361
13	3	130	3	22.798	12.941	6.741	6.533	9.1298
14	3	130	6	22.794	19.140	6.746	6.671	9.1024
15	3	130	9	22.789	19.590	6.753	6.605	9.1161
16	3	200	3	22.798	13.368	6.758	6.554	9.1512
17	3	200	6	23.069	15.713	6.752	6.598	9.1083
18	3	200	9	22.789	18.518	6.768	6.724	8.9990

Newly developed composites were EDMed using cylindrical copper electrode (diameter 8mm) under commercial grade dielectric fluid (specific gravity= 0.763, freezing point= 94°C). Throughout the experiment, the voltage and duty cycle are kept unchanged from 50 v and 8, respectively. Machining time for all tests is kept at 5 minutes.

RESULTS AND DISCUSSION

Taguchi Analysis

For Taguchi analysis, S/N ratios were calculated as per criteria (maximization or minimization type) according to Equation 4 and Equation 5.

$$S/_N \, ratio = -10 \log \frac{1}{k} \sum_{i=1}^{k} \frac{1}{y_i^2} \tag{4}$$

$$S/_N \, ratio = -10 \log \frac{1}{k} \sum_{i=1}^{k} y_i^2 \tag{5}$$

Where 'k' defines the total no. of experiments performed and 'y' illustrates the response value of the respective performed experiment.

According to the Taguchi method, the experiment results in the form of MRR, TWR or EWR (electrode wear rate) and ROC are transformed into the signal-to-noise noise ratios (S/N ratios) and analyzed using Minitab software with an aim to explore the impact of input parameters on them. ANOVA is employed to evaluate three types of plots such as main effects plots, interaction plots and residual plots which are explained in Table **5** as under:

Table 5. Layout of Experiments and their corresponding Responses.

Exp No.	Mo%	P_{on} (µs)	I_p (A)	MRR (g/min)	SNRA1	EWR (g/min)	SNRA2	ROC (mm)	SNRA3
1	1.5	70	3	0.23111	-12.7236	0.01417	36.97465	0.821463	1.7082
2	1.5	70	6	0.05244	-25.606	0.00250	52.0412	0.679512	3.3561
3	1.5	70	9	0.00800	-41.9382	0.00133	57.50123	0.359024	8.8975
4	1.5	130	3	0.68844	-3.24262	0.04667	26.61986	0.918049	0.7427
5	1.5	130	6	0.20089	-13.9409	0.01250	38.0618	0.850244	1.4091
6	1.5	130	9	0.17711	-15.0351	0.01083	39.30476	0.704878	3.0377
7	1.5	200	3	0.66622	-3.52762	0.03333	29.54243	0.906829	0.8495
8	1.5	200	6	0.39489	-8.0705	0.02417	32.33566	0.857561	1.3347
9	1.5	200	9	0.17778	-15.0025	0.00250	52.0412	0.928780	0.6417
10	3	70	3	0.13644	-17.3009	0.00858	41.32688	0.140000	17.0774
11	3	70	6	0.06556	-23.6678	0.00333	49.54243	0.087317	21.1780
12	3	70	9	0.01556	-36.1623	0.00083	61.58362	0.118049	18.5588
13	3	130	3	0.66667	-3.52183	0.04167	27.60422	0.564878	4.9609
14	3	130	6	0.24711	-12.1422	0.01500	36.47817	0.551220	5.1735
15	3	130	9	0.21640	-13.2949	0.02967	30.55462	0.558049	5.0666
16	3	200	3	0.63778	-3.90661	0.04083	27.7797	0.575610	4.7974
17	3	200	6	0.49756	-6.06317	0.03083	30.21959	0.554146	5.1275
18	3	200	9	0.28889	-10.7854	0.00875	41.15984	0.499512	6.0291

ANOVA for Responses

The signal-to-noise ratios for MRR, EWR and ROC are evaluated and analyzed by using Taguchi method considering MRR as "larger is better" type criteria, while EWR and ROC were chosen as "smaller is best" type criteria and the results are shown in Table **6**. For MRR, pulse-on-time, peak-current along with interaction between them are the most significant parameters and contribute to 58.34%, 31.44% and 8.13% to obtain maximum MRR, respectively. Moreover, other factors are not significant.

Table 6. ANOVA Responses.

Source	DF	Seq SS	Adj SS	Adj MS	F	P	% Cont.
MRR							
Mo%	1	8.33	8.33	8.326	3.55	0.1330	0.41%
Pulse on Time (µs)	2	1197.73	1197.73	598.863	255.06	0.0000	58.34%
Peak Current (A)	2	645.44	645.44	322.720	137.45	0.0000	31.44%
Mo%*Pulse on Time (µs)	2	0.78	0.78	0.390	0.17	0.8530	0.04%
Mo%*Peak Current (A)	2	24.69	24.69	12.343	5.26	0.0760	1.20%
Pulse on Time (µs)*Peak Current (A)	4	166.84	166.84	41.711	17.77	0.0080	8.13%
Residual Error	4	9.39	9.39	2.348			0.46%
Total	17	2053.19					
EWR							
Mo%	1	18.35	18.35	18.349	2	0.23	0.91%
Pulse on Time (µs)	2	980.87	980.87	490.437	53.44	0.001	48.74%
Peak Current (A)	2	710.7	710.7	355.351	38.72	0.002	35.32%
Mo%*Pulse on Time (µs)	2	38.4	38.4	19.201	2.09	0.239	1.91%
Mo%*Peak Current (A)	2	30.48	30.48	15.239	1.66	0.299	1.51%
Pulse on Time (µs)*Peak Current (A)	4	196.73	196.73	49.182	5.36	0.066	9.78%
Residual Error	4	36.71	36.71	9.178			1.82%
Total	17	2012.25					
ROC							
Mo%	1	241.941	241.941	241.941	72.73	0.001*	35.39%
Pulse on Time (µs)	2	291.383	291.383	145.691	43.8	0.002*	42.62%
Peak Current (A)	2	12.407	12.407	6.204	1.86	0.268	1.81%
Mo%*Pulse on Time (µs)	2	109.543	109.543	54.772	16.47	0.012*	16.02%
Mo%*Peak Current (A)	2	6.335	6.335	3.167	0.95	0.459	0.93%
Pulse on Time (µs)*Peak Current (A)	4	8.821	8.821	2.205	0.66	0.65	1.29%
Residual Error	4	13.306	13.306	3.327			1.95%
Total	17	683.736					

Similarly, for EWR, the pulse-on-time, peak-current and interaction between them showed the most significant parametric importance for EWR, while other parameters are insignificant. Pulse on time contributes to 48.74% to achieve minimum EWR followed by peak current (35.32%) and interaction between current and pulse duration (9.78%). However, the weight fraction of molybdenum is not important for influencing EWR. From ANOVA table for ROC, it is evident that the weight fraction of molybdenum and pulse-on-time and the interaction between them have significant parametric importance to obtain dimensional accuracy. Pulse on time is the highly contributing parameter (42.62%) followed by weight fraction of molybdenum which contributes 35.39% to achieve minimum radial overcut. Other parameters are insignificant for ROC.

RESPONSE TABLE FOR RESPONSES

From the delta values in the response table, it is vividly clear that pulse-on-time (P_{on}) is the most important parametric factor for MRR and EWR followed by peak current (I_p) and weight fraction of the molybdenum (Mo%). However, RoC is significantly influenced by P_{on}, which is followed by Mo% and I_p. For indicative purposes, the responses for MRR are illustrated below in Table 7:

Table 7. Response Table for Responses.

	MRR			EWR			RoC		
Level	Mo%	P_{on} (µs)	I_p (A)	Mo%	P_{on} (µs)	I_p (A)	Mo%	P_{on} (µs)	I_p (A)
1	-15.454	-26.233	-7.371	40.49	49.83	31.64	2.442	11.796	5.023
2	-14.094	-10.196	-14.915	38.47	33.1	39.78	9.774	3.398	6.263
3		-7.893	-22.036	-	35.51	47.02		3.13	7.039
Delta	1.36	18.341	14.666	2.02	16.72	15.38	7.332	8.666	2.016
Rank	3	1	2	3	1	2	2	1	3

MAIN EFFECTS AND INTERACTION PLOTS FOR RESPONSES

From the main effects plots shown in Fig. (**5**) (a_1-c_1), it is evident that controllable variables possess significant effects on the responses during the electrical discharge machining of newly developed composites. MRR followed a proportional trend w.r.t. the pulse on time (P_{on}) in the range of 70 to 130 µs and beyond this value, it increased slightly. This value is exhibited due to the increase in pulse-on-time that results in the production of a long spark. This phenomenon further increases the higher temperature that causes the material to melt and erode more from the substrate. However, MRR reduces back to a rise in peak-current. In addition, it is evident that no other factor influenced as much higher than peak current and pulse-on-time. For EWR, it has been found that increasing the composition of Mo resulting in a reduced S/N ratio means an increase in electrode wear rate. On the other hand, increasing pulse on time from 70 to 130 µs caused an increase in EWR and beyond this limit, EWR starts to decrease. However, EWR varies proportionally with respect to peak current because it increases the pulse energy into the tool material thus more heat energy is developed in the tool substrate which further leads to enhanced the melting behavior and evaporation of the electrode. Similar results were also reported by Dhar and Purohit [50 - 55].

It has also been observed that ROC is affected by all input parameters. When the composition of molybdenum varied from 1.5 to 3 wt.%, the S/N ratio increased which reduced the ROC value. However, the increase in pulse on time, overcut increases rapidly from 70 to 130 µs, and beyond this limit, this rate is reduced. This is due to the long pulse time which causes rapid heat and evaporation of the material at the interface and reduces the dimensional accuracy. Moreover, an increase in peak current reduces the ROC with an approximate proportionality trend. In interaction plots shown in Fig. (**5**) (a$_2$-c$_2$), each one's interaction between three different controllable input parameters is shown."This confirms that the effect of one factor is being impacted by another factor. It can be further evaluated by the ANOVA table.

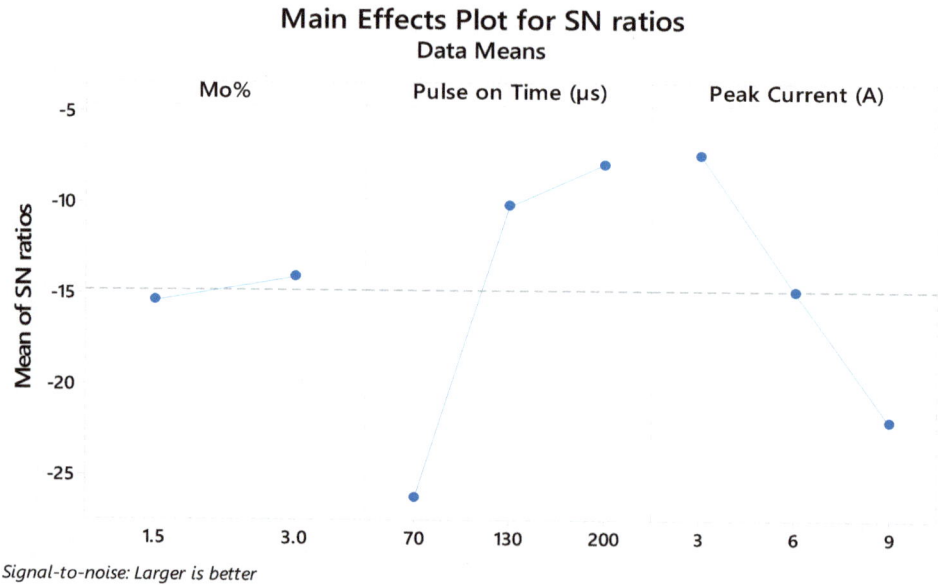

Fig. (5). -a$_1$. Main effects plot for MRR.

Interaction Plot for SN ratios
Data Means

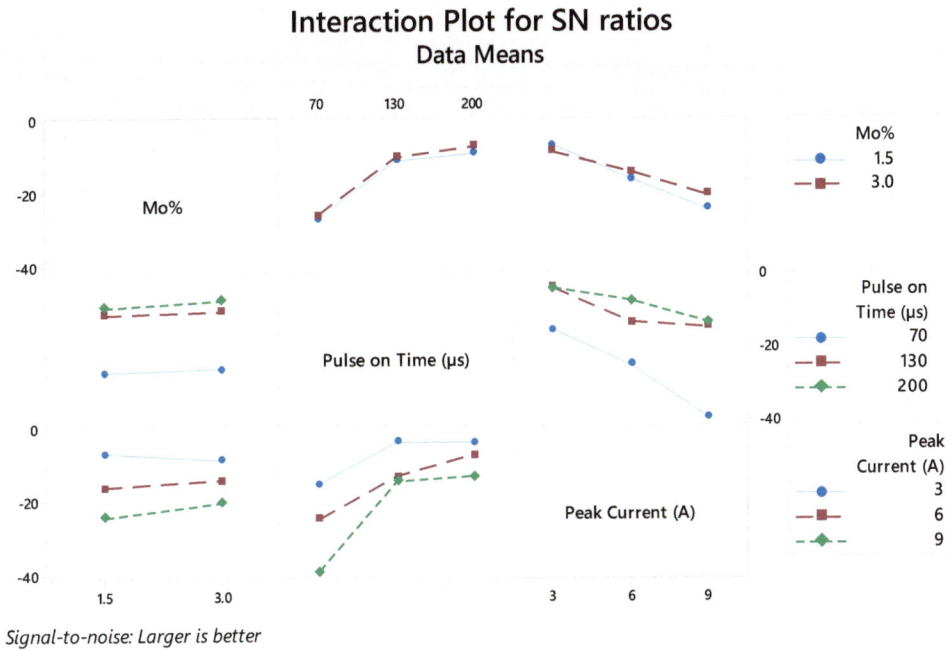

Signal-to-noise: Larger is better

Fig. (5). -a₂ Interactions plot for MRR.

Main Effects Plot for SN ratios
Data Means

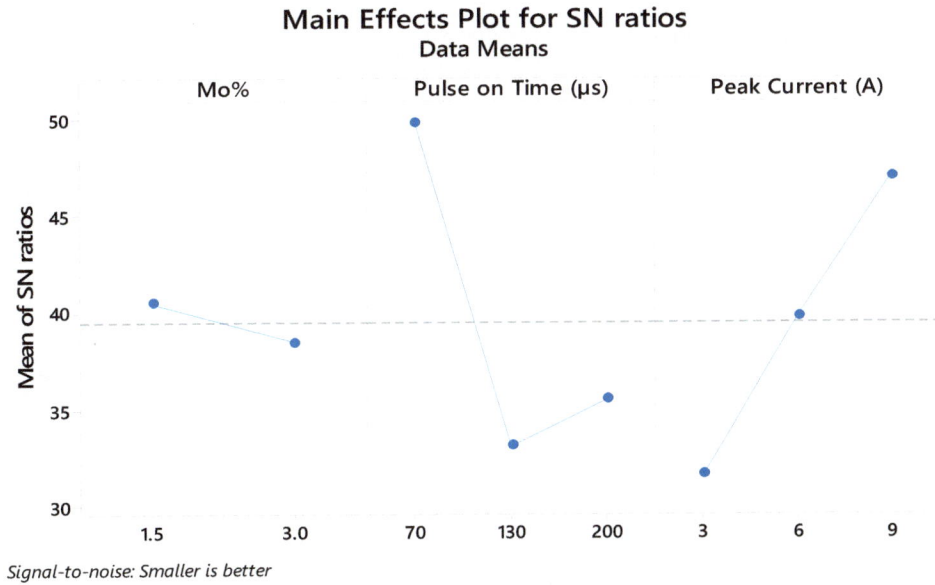

Signal-to-noise: Smaller is better

Fig. (5). -b₁ Main effects plot for EWR.

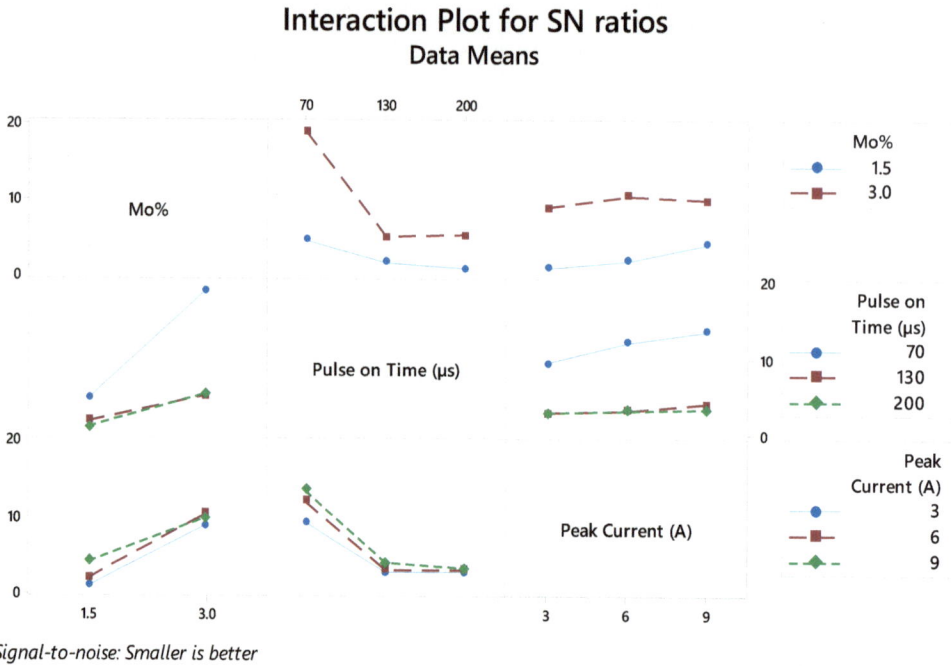

Fig. (5). -b₂ Interactions plot for EWR.

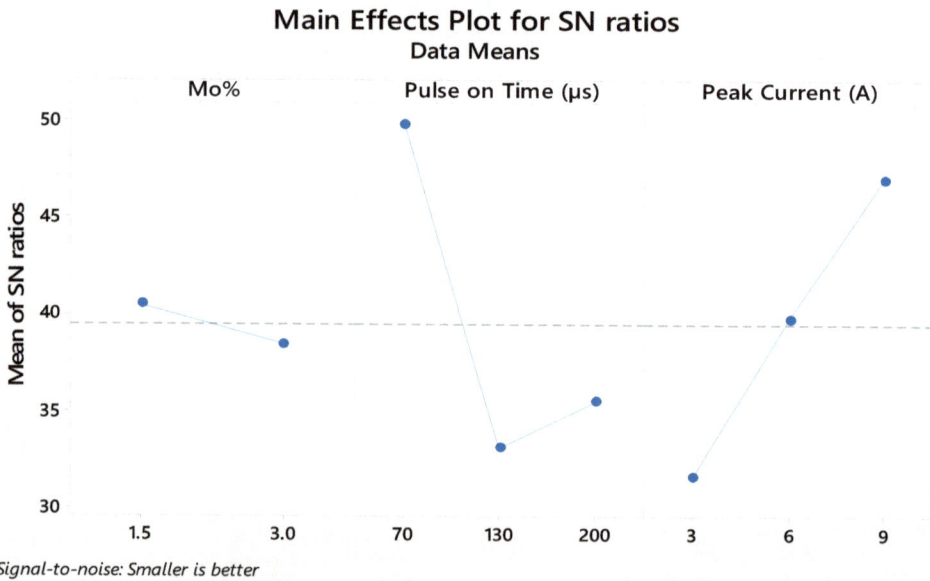

Fig. (5). -c₁ Main effects plot for ROC.

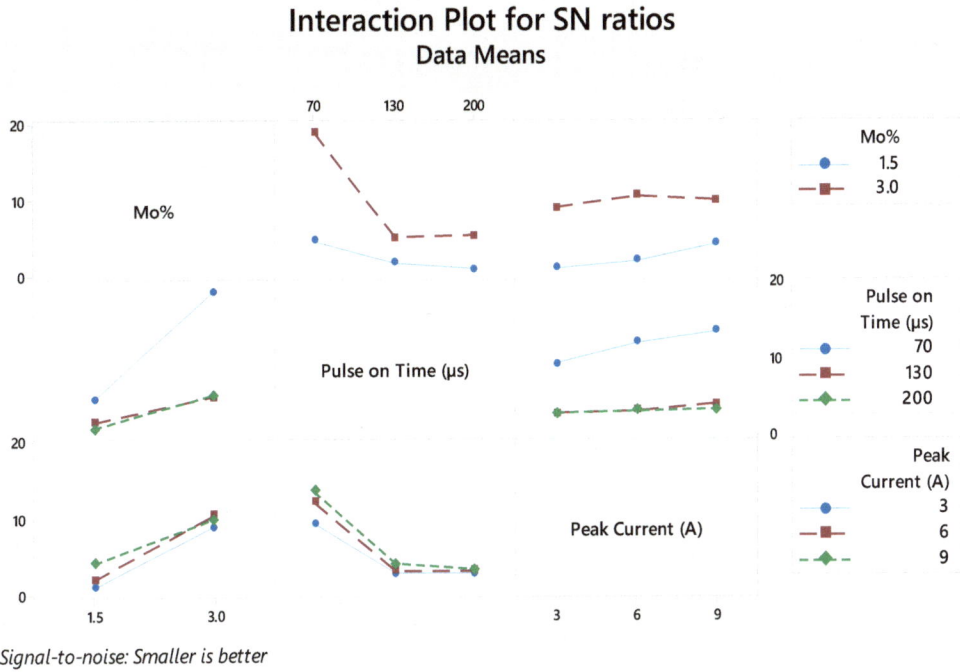

Fig. (5). -c₂. Interactions plot for ROC.

The Residual Plot of Responses

The residual plots for S/N ratios of several responses can be seen in Fig. (6) for MRR, EWR, and ROC. This layout of responses is made to determine the analytic behavior to meet the assumptions. Both residual plots in graphs and the interpretation of each individual residual plot are shown in Figs. (6) (a-c).

Residual Plots for SN ratios

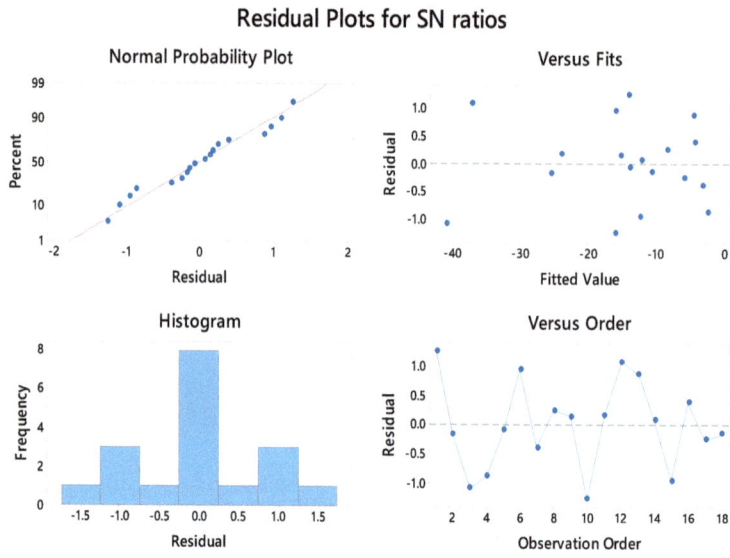

Fig. (6A). Residual Plots for MRR.

Residual Plots for SN ratios

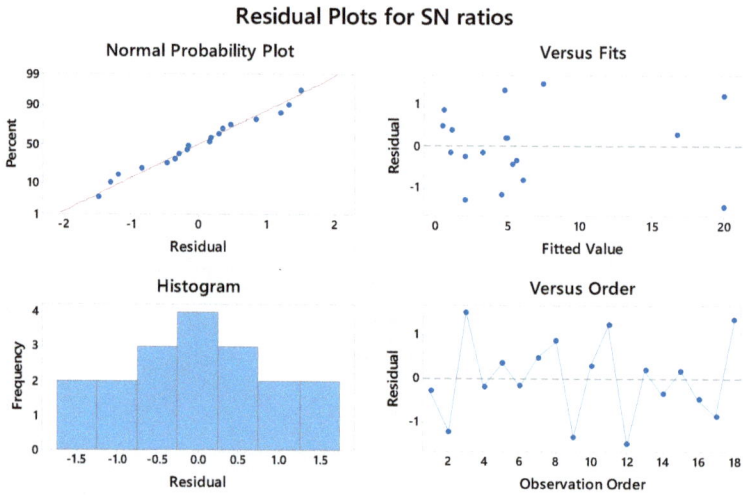

Fig. (6B). Residual Plots for EWR.

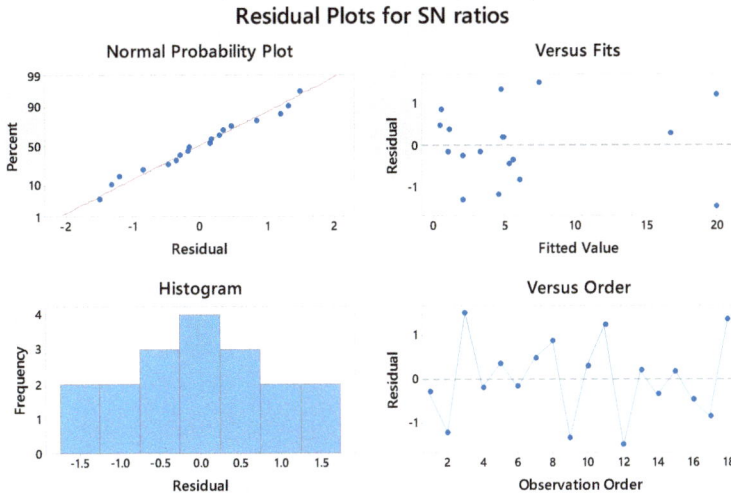

Fig. (6C). Residual Plots for ROC.

First of all, the normal probability plot illustrates the output data distribution and the influence of variables on responses. The outliers are not recorded in the data due to the range of residues that are taken as -2 to 2. The second dotted plot of residuals *versus* fitted values indicates that variation is kept constant. This results in a non-linear relationship without the existence of outliers in the prescribed data. Furthermore, the histogram graph proves the data illustrated here is accurate and not skewed and no outliers exist.

Grey Relational Analysis

The Taguchi method is limited to single-variable optimization for every individual parametric response. To fix this issue, an advanced method of optimization is used. The GRA approach is studied for the uniqueness in the capability for optimization of multiple variables. As a result, an output set of all common settings can be determined after calculations on the basis of eligibility criteria. An observation chart is formed and a set of multiple variables for optimization is determined. Aluminium matrix composite machining optimization is firstly performed by the GRA approach and later all calculated data is transferred for the optimization of GRG (Grey Relational Grades) only. The ranking is done in the descending order value from maximum to minimum value. MRR was considered as maximization type criteria while EWR and ROC were chosen as minimization type. Based on the criteria, normalization was performed according to Equation 6 and Equation 7 for minimization type and maximization type criteria, respectively.

$$X_i(f) = \frac{\max x_i(f) - x_i(f)}{\max x_i(f) - \min x_i(f)} \tag{6}$$

$$X_i(f) = \frac{x_i(f) - \min x_i(f)}{\max x_i(f) - \min x_i(f)} \tag{7}$$

The normalization is examined by a deviation sequence according to Equation 8.

$$\Delta_{i(f)} = \|x^*_o(f) - x^*_i(f)\| \tag{8}$$

Later, Grey relational coefficients are calculated based on Equation 9.

$$\xi_{i(f)} = \frac{\min \Delta_{i(f)} - \Psi \max \Delta_{i(f)}}{\Delta_{i(f)} - \Psi \max \Delta_{i(f)}} \tag{9}$$

Where Ψ (psi) lies between 0 and 1, $0 \leq \Psi \leq 1$. For this study, $\Psi = 0.5$ is taken.

Eventually, the GRG was determined according to GRC by using equation 10. Finally, distinct experiments were ranked from 1 to 18.

$$\gamma = \frac{1}{k} \sum_{i=1}^{k} \xi_{i_f}(f) \tag{10}$$

A confirmatory experiment was also performed. The predicted theoretical value of the grey rational grade was determined by using Equation 11.

$$\gamma_{Predicted} = \gamma_n + \sum_{i=1}^{m} (\overline{\gamma} - \gamma_n) \tag{11}$$

In the above prescribed equation, 'γ_n' depicts the mean value of GR grades, '$\tilde{a}^-$' indicates the mean of GR grades at the optimum level, and 'm' represents the no. of input variables for GRGs.

Based on Table **8**, the optimal combination of MRR, EWR, and ROC is obtained at the 12th number of experimental runs for maximum GRG's value which indicates the best suitable combined objective.

ANOVA for S/N ratios for GRG is given in Table **9**. It is evident that the weight fraction of molybdenum has the maximum influence that is followed by pulse on time. Also, the interaction of pulse on time with peak current and the weight fraction of molybdenum are the major influencing parameters affecting the performance measures. Pulse on time is the most significant input parameter among other parameters and is followed by others.

Table 8. Results of Grey Relational Analysis.

	Performance Parameters			Normalized Values			Grey Relational Coefficients			GRG	Rank
	MRR in (g/min) A	EWR in (g/min) B	ROC (mm) C	A	B	C	A	B	C		
1	0.23111	0.01417	0.821463	0.328	0.709	0.128	0.427	0.632	0.364	0.474	16
2	0.05244	0.0025	0.679512	0.065	0.964	0.296	0.349	0.932	0.415	0.565	8
3	0.008	0.00133	0.359024	0.000	0.989	0.677	0.333	0.979	0.608	0.640	4
4	0.68844	0.04667	0.918049	1.000	0.000	0.013	1.000	0.333	0.336	0.557	10
5	0.20089	0.0125	0.850244	0.283	0.745	0.093	0.411	0.663	0.355	0.476	15
6	0.17711	0.01083	0.704878	0.249	0.782	0.266	0.400	0.696	0.405	0.500	14
7	0.66622	0.03333	0.906829	0.967	0.291	0.026	0.939	0.414	0.339	0.564	9
8	0.39489	0.02417	0.857561	0.569	0.491	0.085	0.537	0.495	0.353	0.462	17
9	0.17778	0.0025	0.92878	0.250	0.964	0.000	0.400	0.932	0.333	0.555	11
10	0.13644	0.00858	0.14	0.189	0.831	0.937	0.381	0.747	0.889	0.672	3
11	0.06556	0.00333	0.087317	0.085	0.945	1.000	0.353	0.902	1.000	0.752	2
12	0.01556	0.00083	0.118049	0.011	1.000	0.963	0.336	1.000	0.932	0.756	1
13	0.66667	0.04167	0.564878	0.968	0.109	0.432	0.940	0.359	0.468	0.589	5
14	0.24711	0.015	0.55122	0.351	0.691	0.449	0.435	0.618	0.476	0.510	13
15	0.2164	0.02967	0.558049	0.306	0.371	0.441	0.419	0.443	0.472	0.445	18
16	0.63778	0.04083	0.57561	0.926	0.127	0.420	0.870	0.364	0.463	0.566	7
17	0.49756	0.03083	0.554146	0.719	0.346	0.445	0.641	0.433	0.474	0.516	12
18	0.28889	0.00875	0.499512	0.413	0.827	0.510	0.460	0.743	0.505	0.569	6

Table 9. ANOVA evaluation of %age contribution for signal-to-noise ratios to GRG.

Contributing Source	DF	Seq. SS	Adj. SS	Adj.MS	F	P	% Cont.
Mo%	1	3.7564	3.7564	3.7564	25.48	0.007*	12.34%
Pulse on Time (μs)	2	11.7951	11.7951	5.8975	40	0.002*	38.75%
Peak Current (A)	2	0.8348	0.8348	0.4174	2.83	0.171	2.74%

(Table 9) cont.....

Mo%*Pulse on Time (µs)	2	4.5482	4.5482	2.2741	15.42	0.013*	14.94%
Mo%*Peak Current (A)	2	1.1192	1.1192	0.5596	3.8	0.119	3.68%
Pulse on Time (µs)*Peak Current (A)	4	7.7945	7.7945	1.9486	13.22	0.014*	25.61%
Residual Error	4	0.5897	0.5897	0.1474			1.94%
Total	17	30.4379					

Fig. (**7a**) provides the main effect plot for GRG and shows that for the sake of getting higher GRG, the process parameters should be at a high level for the weight fraction of molybdenum along with the low level of pulse-on time and a high level of peak current yields for the optimum responses during EDM of AMMCs. Further, it is also observed that all input parameters have a strong impact on the process performance measures. The optimum setting of input parameters is necessary to obtain optimal responses to molybdenum composition of 3%, pulse on time of 70 µs, and peak current of 9A. Based on the interaction plot, it is evident that process performance measures of EDM depend on controllable input parameters as shown in Fig. (**7b**).

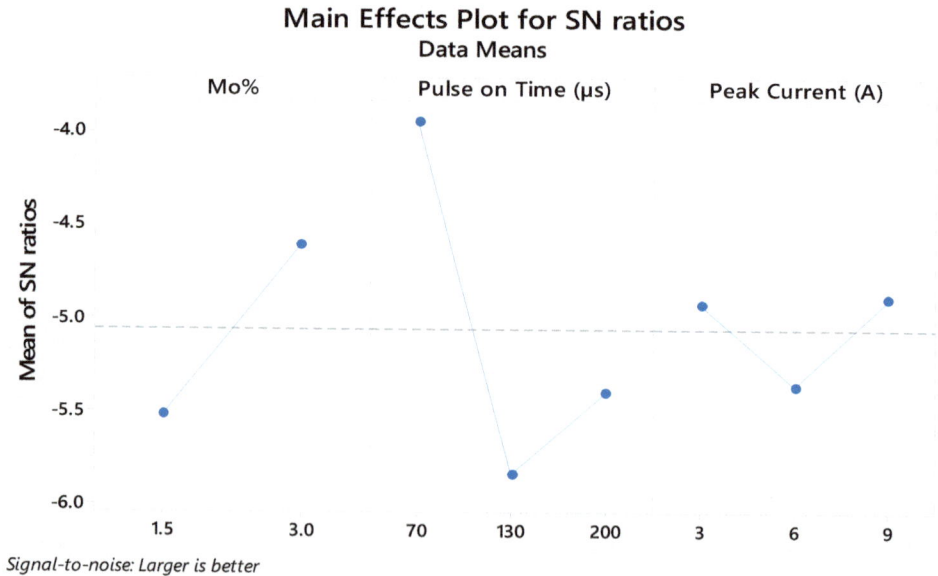

Fig. (7a). Main effects plot for GRG.

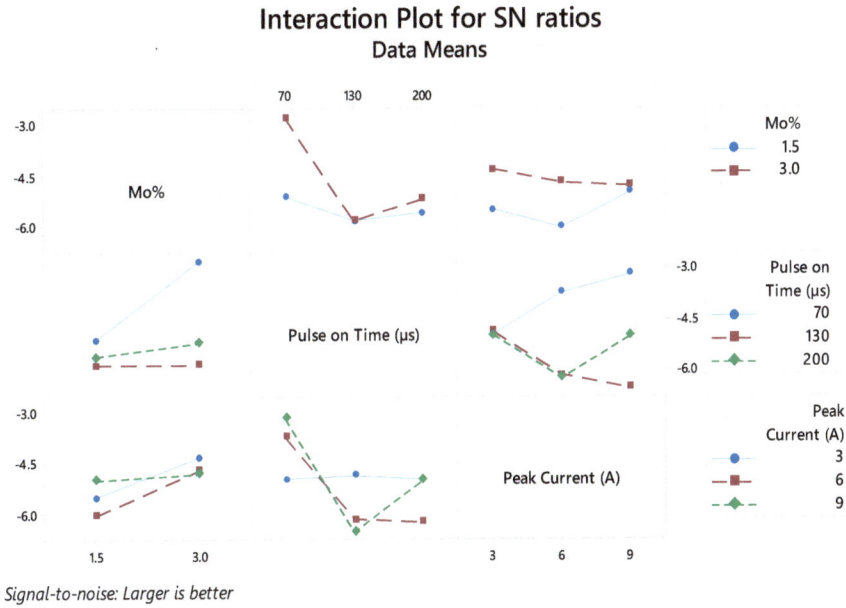

Fig. (7b). Interactions plot for GRG.

The relationship of GRG (means response measures *viz.* Mo%, P_{on} and I_p) with other process parameters is determined by using multiple regression analysis. This analysis is commonly used in industries for the sake of getting an overall understanding of how the typical parametric value is being affected by other variables' responses that have been predicted. Furthermore, it should be verified that the regression model must be able to fit into the set of data that is investigated. Later, the correlations between the processing parameters and the output responses are obtained by several linear regressions by using Minitab 18 version software. Equation 5.1 is the regression equation that is fitted to the response parameters with an amount of variation (R^2) value of 98.06%, adjusted amount of variation ($AdjR^2$) equal to 91.77% and standard deviation of error (S) equal to 0.3840.

GRG $= 0.5639 + 0.0430$ Mo% $- 0.000772$ P_{on} (μs) $+ 0.00119$ I_p (A) 5.1

Where,

P_{on}–Pulse on time (70, 130, 200 μs)

I_p- Peak Current (3, 6, 9 A)

Mo% - Weight percent of Molybdenum (1.5, 3 wt.%)

The result of GRG and their delta values Table **10** are used to rank process parameters to investigate their consequences on the combined objective function. The descending order of effecting parametric factors is identified as the pulse on time, the composition of molybdenum and the peak current from GRG. It is clear that the pulse on time has the most significant influence on the combined objective function among the other variables.

Experiments were conducted according to Taguchi's L_{18} orthogonal array using the electric discharge machining set-up and the cylindrical copper electrodes. Finding the result of MRR and EWR, pulse on time is evaluated as the most significant input parameter that is followed by peak current. However, the weight fraction of molybdenum and pulse on time affect the dimensional accuracy (ROC) significantly. From GRA, it is evident that the weight fraction of molybdenum, pulse-on time, the interaction of pulse-on time with the peak current (I_p), and the weight fraction of Mo are the most significant parameters that affect the performance characteristics. The pulse-on time is evaluated as the most effective contributing parameter among all other parameters.

CONCLUSION

A recent effort has been made on the phenomenon of machining responses to three output variables such as MRR, EWR, and ROC of the newly developed AMMC by the use of copper electrodes on EDM. The experiments were conducted and observations were made under some parametric values of pulse On-Time (P_{on}), peak Current (I_p), and weight fraction of Molybdenum. L_{18} OA DOE is performed as per Taguchi design and with the aid of Minitab software that was employed for the analysis of responses. This analysis is partially validated by an experimental approach. The Taguchi optimization was followed by Grey Relational Analysis to obtain optimal values of process measures. The following main findings can be derived from the present work:

Table 10. Response Table for GRG.

-	Mo%	Pulse on Time (µs)	Peak Current (A)
Level-1	0.5326	0.64325*	0.57036
Level-2	0.59717*	0.51277	0.54678
Level-3	-	0.53865	0.57752*
Diff	0.0646	0.1305	0.0236
Rank	2	1	3

* Optimum level

1. Hybrid Al-SiC-Mo AMMC was fabricated successfully by using stir casting - a type of liquid fabrication methodology.
2. Finding the result of MRR, the pulse on time, peak current, and interaction between them are the significant parameters and contribute 58.34%, 31.44%, and 8.13% to obtain maximum MRR, respectively.
3. Peak current, pulse on time, and interaction between them are significant variables for EWR, while other parameters are insignificant. Pulse on time contributes 48.74% to achieve minimum EWR followed by peak current (35.32%) and interaction between current and pulse duration (9.78%).
4. In the case of radial overcut, the weight fraction of molybdenum, the pulse-on time and the interaction between them are significant parameters to obtain dimensional accuracy. Pulse on time is the highly contributing parameter (42.62%) followed by the weight fraction of molybdenum which contributes 35.39% to achieve the minimum radial overcut.
5. Findings of the grey relational analysis revealed that the weight fraction of molybdenum, pulse-on time, the interaction of peak current with a pulse-on time and weight fraction of Molybdenum are significant parameters affecting performance measures. Pulse on time is the most significant contributing parameter among all others. The optimal setting of input parameters suggested by GRA to obtain optimal responses is a molybdenum composition of 3%, pulse on time of 70 μs, and peak current of 9A. Based on the interaction plot, it is evident that process performance measures of EDM depend on controllable input parameters.

REFERENCES

[1]   Das, D.K.; Mishra, P.C.; Singh, S.; Pattanaik, S. Fabrication and heat treatment of ceramic-reinforced aluminium matrix composites - a review. *Int. J. Mech. Mater. Eng.,* **2014**, *9*(1), 6.
        [http://dx.doi.org/10.1186/s40712-014-0006-7]

[2]   Dhavamani, C.; Alwarsamy, T. Optimization of machining parameters for aluminium and silicon carbide composites using Genetic Algorithm. *Procedia Eng.,* **2012**, *38*, 1994-2004.
        [http://dx.doi.org/10.1016/j.proeng.2012.06.241]

[3]   Surappa, M.K. Aluminium matrix composites: Challenges and opportunities. *Sadhana,* **2003**, *28*(1-2), 319-334.
        [http://dx.doi.org/10.1007/BF02717141]

[4]   Tjong, S.C. *Processing and deformation characteristics of metals reinforced with ceramic nanoparticles. Nanocrystalline materials,* 2nd ed; Elsevier: Oxford, **2014**, pp. 269-304. Internet

[5]   Dunia, A.S. Aluminium silicon carbide and aluminium graphite particulate composites. *J. Eng. Appl. Sci.,* **2011**, *6*, 41-46.

[6]   TonyThomas, A.; Parameshwaran, R.; Muthukrishnan, A.; Arvindkumaran, M. Development of feeding and stirring mechanism for stir casting of aluminium matrix composites. *Proced. Mater. Sci.,* **2014**, *5*, 1182-1191.
        [http://dx.doi.org/10.1016/j.mspro.2014.07.415]

[7]   Kumar, J.; Singh, D.; Kalsi, N.S. Tribological, physical and microstructural characterization of silicon carbide reinforced aluminium matrix composites: A review. *Mater. Today Proc.,* **2019**, *18*, 3218-3232.

[http://dx.doi.org/10.1016/j.matpr.2019.07.198]

[8] Khanna, V.; Singh, K.; Kumar, S.; Bansal, S.A.; Channegowda, M.; Kong, I.; Khalid, M.; Chaudhary, V. Engineering electrical and thermal attributes of two-dimensional graphene reinforced copper/aluminium metal matrix composites for smart electronics. *ECS J. Solid State Sci. Technol.,* **2022**, *11*(12), 127001.
[http://dx.doi.org/10.1149/2162-8777/aca933]

[9] Singh, D.P.; Mishra, S.; Yadav, S.K.S.; Porwal, R.K.; Singh, V. Comparative analysis and optimization of thermoelectric machining of alumina and silicon carbide-reinforced aluminum metal matrix composites using different electrodes. *J. Adv. Manuf. Syst.,* **2022**, *22*(02), 373-401.

[10] Khajuria, A.; Akhtar, M.; Pandey, M.K.; Singh, M.P.; Raina, A.; Bedi, R.; Singh, B. Influence of ceramic Al_2O_3 particulates on performance measures and surface characteristics during sinker EDM of stir cast AMMCs. *World J. Eng.,* **2019**, *16*(4), 526-538.
[http://dx.doi.org/10.1108/WJE-01-2019-0015]

[11] Syed, K.H.; Kuppan, P. Studies on recast-layer in EDM using aluminium powder mixed distilled water dielectric fluid. *IACSIT Int. J. Eng. Technol.,* **2013**, *5*(2), 1775-1780.

[12] Velmurugan, C.; Subramanian, R.; Thirugnanam, S.; Ananadavel, B. Experimental investigations on machining characteristics of Al 6061 hybrid metal matrix composites processed by electrical discharge machining. *Int. J. Eng. Sci. Technol.,* **1970**, *3*(8), 87-101.
[http://dx.doi.org/10.4314/ijest.v3i8.7]

[13] Shen, Y.; Liu, Y.; Sun, W.; Dong, H.; Zhang, Y.; Wang, X.; Zheng, C.; Ji, R. High-speed dry compound machining of Ti6Al4V. *J. Mater. Process. Technol.,* **2015**, *224*, 200-207.
[http://dx.doi.org/10.1016/j.jmatprotec.2015.05.012]

[14] Kumar, J.; Sharma, S.; Singh, J.; Singh, S.; Singh, G. Optimization of Wire-EDM process parameters for Al-Mg-0.6Si-0.35Fe/15%RHA/5%Cu hybrid metal matrix composite using TOPSIS: Processing and characterizations. *J. Manuf. Mater. Process.,* **2022**, *6*(6), 150.
[http://dx.doi.org/10.3390/jmmp6060150]

[15] Dar, S.A.; Kumar, J.; Sharma, S.; Singh, G.; Singh, J.; Aggarwal, V.; Chohan, J.; Kumar, R.; Sharma, A.; Mishra, M.; Obaid, A.J. Investigations on the effect of electrical discharge machining process parameters on the machining behavior of aluminium matrix composites. *Mater. Today Proc.,* **2022**, *48*, 1048-1054.
[http://dx.doi.org/10.1016/j.matpr.2021.07.126]

[16] Sultan, U.; Kumar, J.; Dadra, S.; Kumar, S. Experimental investigations on the tribological behaviour of advanced aluminium metal matrix composites using grey relational analysis. *Mater. Today Proc.,* **2022**.
[http://dx.doi.org/10.1016/j.matpr.2022.12.171]

[17] K., Singh; V., Khanna; A., Rosenkranz; V., Chaudhary, Sonu; G., Singh; S., Rustagi Panorama of physico-mechanical engineering of graphene-reinforced copper composites for sustainable applications, *Materials Today Sustainability,* **2023**, *24*, 100560.
[http://dx.doi.org/10.1016/j.mtsust.2023.100560]

[18] Boopathi, S.; Sivakumar, K. Experimental investigation and parameter optimization of near-dry wire-cut electrical discharge machining using multi-objective evolutionary algorithm. *Int. J. Adv. Manuf. Technol.,* **2013**, *67*(9-12), 2639-2655.
[http://dx.doi.org/10.1007/s00170-012-4680-4]

[19] Dhakar, K.; Dvivedi, A.; Dhiman, A. Experimental investigation on effects of dielectric mediums in near-dry electric discharge machining. *J. Mech. Sci. Technol.,* **2016**, *30*(5), 2179-2185.
[http://dx.doi.org/10.1007/s12206-016-0425-x]

[20] Teimouri, R.; Baseri, H. Experimental study of rotary magnetic field-assisted dry EDM with ultrasonic vibration of workpiece. *Int. J. Adv. Manuf. Technol.,* **2013**, *67*(5-8), 1371-1384.
[http://dx.doi.org/10.1007/s00170-012-4573-6]

[21] Singh, N.; Bhatia, O.S. Optimization of process parameters in die sinking EDM-A review. *IJSTE,* **2016**, *2*(11), 2349-2784.

[22] Jeykrishnan, J.; Vijaya, R.B.; Akilesh, S.; Pradeep, K.R.P. Optimization of process parameters on EN24 Tool steel using Taguchi technique in Electro-Discharge Machining (EDM). *IOP Conf. Series Mater. Sci. Eng.,* **2016**, *149*(1), 012022.
[http://dx.doi.org/10.1088/1757-899X/149/1/012022]

[23] Morankar, K.S.; Shelke, R.D. Influence of process parameters in EDM process-A Review. *Int. J. Innovations in Eng. and Tech.,* **2017**, *8*, 147-143.

[24] Ohdar, N.K.; Jena, B.K.; Sethi, S.K. Optimization of EDM process parameters using Taguchi Method with Copper Electrode. *Int. Res. J. Eng. Technol,* **2017**, *4*, 2428-2431.

[25] Vikram Reddy, V.; Vamshi Krishna, P.; Jawahar, M.; Shiva kumar, B. Optimization of process parameters during EDM of SS304 using taguchi-grey relational analysis. *Mater. Today Proc.,* **2018**, *5*(13), 27065-27071.
[http://dx.doi.org/10.1016/j.matpr.2018.09.011]

[26] Kale, M.S.M.; Khedekar, M.D. Optimization of Process parameters in EDM for machining of inconel 718 using response surface methodology. *IJIET,* **2016**, *7*(3), 2319-1058.

[27] Babu, T.V.; Soni, J.S. Optimization of process parameters for surface roughness of Inconel 625 in Wire EDM by using Taguchi and ANOVA method. *Int J Recent Technol Eng,* **2017**, *7*(3), 1127-1131.

[28] Kumar, J.; Kumar, V.; Sharma, S.; Chohan, J.; Kumar, R.; Singh, S.; Obaid, A.J.; Akram, S.V. Optimizations of reinforcing particulates and processing parameters for stir casting of aluminium metal matrix composites for sustainable properties. *Mater. Today Proc.,* **2022**, *68*, 1172-1179.
[http://dx.doi.org/10.1016/j.matpr.2022.10.109]

[29] Chandramouli, S.; Eswaraiah, K. Optimization of EDM process parameters in machining of 17-4 PH steel using Taguchi method. *Mater. Today Proc.,* **2017**, *4*(2), 2040-2047.
[http://dx.doi.org/10.1016/j.matpr.2017.02.049]

[30] Zhang, Y.; Liu, Y.; Ji, R.; Cai, B.; Li, H. Influence of dielectric type on porosity formation on electrical discharge machined surfaces. *Metall. Mater. Trans., B, Process Metall. Mater. Proc. Sci.,* **2012**, *43*(4), 946-953.
[http://dx.doi.org/10.1007/s11663-012-9653-3]

[31] Roth, R.; Balzer, H.; Kuster, F.; Wegener, K. Influence of the anode material on the breakdown behavior in dry electrical discharge machining. *Procedia CIRP,* **2012**, *1*, 639-644.
[http://dx.doi.org/10.1016/j.procir.2012.05.013]

[32] Babu, T.S.M.; Sugin, M.S.A.; Muthukrishnan, N. Investigation on the characteristics of surface quality on machining of hybrid metal matrix composite (Al-SiC-B4C). *Procedia Eng.,* **2012**, *38*, 2617-2624.
[http://dx.doi.org/10.1016/j.proeng.2012.06.308]

[33] Mukherjee, R.; Chakraborty, S.; Samanta, S. Selection of wire electrical discharge machining process parameters using non-traditional optimization algorithms. *Appl. Soft Comput.,* **2012**, *12*(8), 2506-2516.
[http://dx.doi.org/10.1016/j.asoc.2012.03.053]

[34] Yusup, N.; Zain, A.M.; Hashim, S.Z.M. Evolutionary techniques in optimizing machining parameters: Review and recent applications (2007–2011). *Expert Syst. Appl.,* **2012**, *39*(10), 9909-9927.
[http://dx.doi.org/10.1016/j.eswa.2012.02.109]

[35] Kumar, S.; Singh, H.; Kumar, R.; Singh Chohan, J. Parametric optimization and wear analysis of AISI D2 steel components. *Mater. Today Proc.,* **2023**.
[http://dx.doi.org/10.1016/j.matpr.2023.01.247]

[36] Singh, H.; Kumar, S.; Kumar, R.; Chohan, J.S. Impact of operating parameters on electric discharge machining of cobalt-based alloys. *Mater. Today Proc.,* **2023**, 1-10.
[http://dx.doi.org/10.1016/j.matpr.2023.01.234]

[37] Kumar, R.; Kumar, M.; Singh Chohan, J.; Kumar, S. Effect of process parameters on surface roughness of 316L stainless steel coated 3D printed PLA parts. *Mater. Today Proc.,* **2022**, *68*(4), 734-741.
[http://dx.doi.org/10.1016/j.matpr.2022.06.004]

[38] Kumar, S.; Kumar, R.; Singh, S.; Singh, H.; Kumar, A.; Goyal, R.; Singh, S. A comprehensive study on minimum quantity lubrication. *Mater. Today Proc.,* **2022**, *56*(5), 3078-3085.
[http://dx.doi.org/10.1016/j.matpr.2021.12.158]

[39] Singh, S.; Kumar, H.; Kumar, S.; Chaitanya, S. A systematic review on recent advancements in Abrasive Flow Machining (AFM). *Mater. Today Proc.,* **2022**, *56*(5), 3108-3116.
[http://dx.doi.org/10.1016/j.matpr.2021.12.273]

[40] Kumar, A.; Sharma, R.; Kumar, S.; Verma, P. A review on machining performance of AISI 304 steel. *Mater. Today Proc.,* **2022**, *56*(5), 2945-2951.
[http://dx.doi.org/10.1016/j.matpr.2021.11.003]

[41] Kumar, R.; Kumar, H.; Kumar, S.; Chohan, J.S. Effects of tool pin profile on the formation of friction stir processing zone in AA1100 aluminium alloy. *Mater. Today Proc.,* **2022**, *48*(5), 1594-1603.
[http://dx.doi.org/10.1016/j.matpr.2021.09.491]

[42] Chauhan, A.; Kumar, M.; Kumar, S. Fabrication of polymer hybrid composites for automobile leaf spring application. *Mater. Today Proc.,* **2022**, *48*(5), 1371-1377.
[http://dx.doi.org/10.1016/j.matpr.2021.09.114]

[43] Singh, S.; Kumar, S.; Singh, S.; Kumar, R.; Sidhu, H.S. Advance technologies in fine finishing and polishing processes: A study. *J. Xidian Univ.,* **2020**, *14*(4), 1387-1399.
[http://dx.doi.org/10.37896/jxu14.4/161]

[44] Mehra, R.; Kumar, S. A review on lasers assisted machining methods – types, mode of operations, comparison and applications. *CGC Int. J. Contemp. Technol.,* **2022**, *4*(2), 307-315.
[http://dx.doi.org/10.46860/cgcijctr.2022.07.31.307]

[45] Birdi, A.; Procha, A.; Brar, A.S.; Kumar, R.; Kumar, S. Effect of tool pin profile on mechanical characteristics of friction stir welded al alloys: A critical review. *IRJNST,* **2019**, *1*(5), 15-21.

[46] Kumar, R.; Kumar, M.; Chohan, J.S.; Kumar, S. Overview on metamaterial: History, types and applications. *Mater. Today Proc.,* **2022**, *56*(5), 3016-3024.
[http://dx.doi.org/10.1016/j.matpr.2021.11.423]

[47] Kumar, J.; Singh, D.; Kalsi, N.S.; Sharma, S.; Pruncu, C.I.; Pimenov, D.Y.; Rao, K.V.; Kapłonek, W. Comparative study on the mechanical, tribological, morphological and structural properties of vortex casting processed, Al–SiC–Cr hybrid metal matrix composites for high strength wear-resistant applications: Fabrication and characterizations. *J. Mater. Res. Technol.,* **2020**, *9*(6), 13607-13615.
[http://dx.doi.org/10.1016/j.jmrt.2020.10.001]

[48] Kumar, J.; Singh, D.; Kalsi, N.S.; Sharma, S.; Mia, M.; Singh, J.; Rahman, M.A.; Khan, A.M.; Rao, K.V. Investigation on the mechanical, tribological, morphological and machinability behavior of stir-casted Al/SiC/Mo reinforced MMCs. *J. Mater. Res. Technol.,* **2021**, *12*, 930-946.
[http://dx.doi.org/10.1016/j.jmrt.2021.03.034]

[49] Kumar, J.; Singh, D.; Kalsi, N.S.; Sharma, S.; Pruncu, C.I.; Pimenov, D.Y.; Rao, K.V.; Kapłonek, W. Comparative study on the mechanical, tribological, morphological and structural properties of vortex casting processed, Al–SiC–Cr hybrid metal matrix composites for high strength wear-resistant applications: Fabrication and characterizations. *J. Mater. Res. Technol.,* **2020**, *9*(6), 13607-13615.
[http://dx.doi.org/10.1016/j.jmrt.2020.10.001]

[50] Dhar, S.; Purohit, R.; Saini, N.; Sharma, A.; Kumar, G.H. Mathematical modeling of electric discharge machining of cast Al–4Cu–6Si alloy–10wt.% SiCP composites. *J. Mater. Process. Technol.,* **2007**, *194*(1-3), 24-29.
[http://dx.doi.org/10.1016/j.jmatprotec.2007.03.121]

[51] Dahiya, M.; Khanna, V.; Anil, B.S. Effect of graphene size variation on mechanical properties of aluminium graphene nanocomposites: A modeling analysis. *Mater. Today Proc.*, **2022**, (Jul)
[http://dx.doi.org/10.1016/j.matpr.2022.07.259]

[52] Gupta, P.; Ahamad, N.; Kumar, D.; Gupta, N.; Chaudhary, V.; Gupta, S.; Khanna, V.; Chaudhary, V. Synergetic effect of CeO_2 doping on structural and tribological behavior of $Fe-Al_2O_3$ metal matrix nanocomposites. *ECS J. Solid State Sci. Technol.*, **2022**, *11*(11), 117001.
[http://dx.doi.org/10.1149/2162-8777/ac9c92]

[53] Dahiya, M.; Khanna, V.; Anil Bansal, S. Aluminium-graphene metal matrix nanocomposites: Modelling, analysis, and simulation approach to estimate mechanical properties. *Mater. Today Proc.*, **2022**, (Nov)
[http://dx.doi.org/10.1016/j.matpr.2022.10.181]

[54] Singh, K.; Bansal, S.A.; Khanna, V.; Singh, S. Effects of performance measures of non-conventional joining processes on mechanical properties of metal matrix composites. *Metal Matrix Composites,* **2022**, (Aug), 135-165.
[http://dx.doi.org/10.1201/9781003194897-7]

[55] Khanna, V.; Kumar, V.; Bansal, S.A.; Prakash, C.; Ubaidullah, M.; Shaikh, S.F.; Pramanik, A.; Basak, A.; Shankar, S. Fabrication of efficient aluminium/graphene nanosheets (Al-GNP) composite by powder metallurgy for strength applications. *J. Mater. Res. Technol.*, **2023**, *22*, 3402-3412.
[http://dx.doi.org/10.1016/j.jmrt.2022.12.161]

SUBJECT INDEX

www.ingramcontent.com/pod-product-compliance
Lightning Source LLC
Chambersburg PA
CBHW050812220326
41598CB00006B/189